Pitman Research Notes in Mathematics Series

Submission of proposals for consideration

Suggestions for publication, in the form of outlines and representative samples, are invited by the Editorial Board for assessment. Intending authors should approach one of the main editors or another member of the Editorial Board, citing the relevant AMS subject classifications. Alternatively, outlines may be sent directly to the publisher's offices. Refereeing is by members of the board and other mathematical authorities in the topic concerned, throughout the world.

Preparation of accepted manuscripts

On acceptance of a proposal, the publisher will supply full instructions for the preparation of manuscripts in a form suitable for direct photo-lithographic reproduction. Specially printed grid sheets can be provided and a contribution is offered by the publisher towards the cost of typing. Word processor output, subject to the publisher's approval, is also acceptable.

Illustrations should be prepared by the authors, ready for direct reproduction without further improvement. The use of hand-drawn symbols should be avoided wherever possible, in order to maintain maximum clarity of the text.

The publisher will be pleased to give any guidance necessary during the preparation of a typescript, and will be happy to answer any queries.

Important note

In order to avoid later retyping, intending authors are strongly urged not to begin final preparation of a typescript before receiving the publisher's guidelines. In this way it is hoped to preserve the uniform appearance of the series.

Addison Wesley Longman Ltd
Edinburgh Gate
Harlow, Essex, CM20 2JE
UK
Telephone (0) 1279 623623)

Titles in this series. A full list is available from the publisher on request.

François Ledrappier

Ecole Polytechnique, Palaiseau, France

Jorge Lewowicz

I.M.E.R.L., Montevideo, Uruguay

and

Sheldon Newhouse

Michigan State University, USA

(Editors)

International conference on dynamical systems

Montevideo 1995 – a tribute to Ricardo Mañé

 LONGMAN

Addison Wesley Longman Limited
Edinburgh Gate, Harlow
Essex CM20 2JE, England
and Associated Companies throughout the world.

Published in the United States of America
by Addison Wesley Longman Inc.

© Addison Wesley Longman Limited 1996

First published 1996

AMS Subject Classifications   (Main) 34-06
                                (Subsidiary) 34C, 58F

ISSN 0269-3674

ISBN 0 582 30296 X

British Library Cataloguing in Publication Data

  A catalogue record for this book is
  available from the British Library

Library of Congress Cataloging-in-Publication data

A catalog record for this book is available

Printed and bound in Great Britain
by Biddles Ltd, Guildford and King's Lynn

Contents

Ricardo Mañé

Preface

These are the proceedings of the first international scientific meeting organized by the Dynamical Systems Group at Montevideo, Uruguay. Nearly 150 specialists from all around the world participated in this meeting.

In our view, this congress was academically successful due to the breadth and depth of the two different types of lectures: those presenting new developments and those of survey nature. We believe that these proceedings reflect quite accurately these virtues.

Unfortunately, not all the participants were able to present their results in plenary sessions and we were not able to offer mini-courses. Both facts were due to our inexperience and economic constraints, which forced us to organize a shorter meeting. We would like to note that in spite of these difficulties the organization of the meeting was possible thanks to the work and efforts of all the faculty and administrative staff. The organization was carried out in a fraternal atmosphere that allowed us on various ocassions to surmount what seemed to be unavoidable obstacles.

The congress was dedicated to the memory of Ricardo Mañé who died a few days before the meeting began. At this time, we shall say no words about his contributions to mathematics or his inspiring influence in forming the Dynamical Systems Group at Montevideo, Uruguay. We all know Mañé's sense of humour, his fine irony, his opinions and devotions through his words and gestures. We would like to relate an episode that describes him more characteristically. One afternoon, a few days before he died, Mañé was being visited by some colleagues, students and relatives. Everybody in the room knew about his very serious condition. Mañé was feeling very ill, but he searched and found in his memory the subjects and topics of most interest to his visitors and he talked about them, thus provoking the enthusiasm of every person in the room. As always, he did not want anybody to spend a bad time with him.

Jorge Lewowicz

Acknowledgements

Acknowledgements : the Congress received financial support from the following Institutions.

Proyecto Sistemas Dinamicos (CONICYT/BID)

Programa Recursos Humanos (CONICYT/BID)

Universidad de la Republica :

Dpto de Cooperacion Internacional
Comision Sectorial de Investigacion Cientifico
Facultad de Ingenieria
Facultad de Ciencias

National Science Foundation

International Science Foundation

Coopération Régional Française

UNESCO (Oficina Regional para America Latina y el Caribe)

Ambassade de France, United Kingdom Embassy

Instituto Italiano de Cultura

International Mathematical Union

International Centre for Theoretical Physics and the following companies

BID

UTE

ANTEL

ANCAP

BULL, COASIN, INTERFASE

SEMM

La Oficina (MITA)

RODRIGO BAMÓN[*]
Singular cycles of vector fields

1 Introduction.

Vector fields with generic singular cycles may lie in regular parts of the boundary of the Morse-Smale systems. One-parameter families of vector fields crossing transversally this boundary describes, varying the parameter, different mechanisms through which a vector field may evolve from Morse-Smale systems into others with non-trivial forms of recurrences. See for example [BLMP], [L], [PR], [Ri], [Ro], [S].

In this note we present a characterization and classification of all those singular cycles. We provide the proofs in a forthcoming paper. Our vector fields are C^r−differentiable, $r \geq 3$, and the ambient space n-dimensional.

2 Definitions and a Lemma.

A **cycle** Γ of a vector field X is a chain recurrent set formed by a finite collection of critical elements (singularities and periodic orbits) and a finite collection of regular orbits. More precisely, we can write $\Gamma = \left(\bigcup_{i=0}^{k-1} \sigma_i \right) \bigcup \left(\bigcup_{i=0}^{k-1} \gamma_i \right)$ where $\{\sigma_i\}_{i=0}^{k-1}$ are critical elements and $\{\gamma_i\}_{i=0}^{k-1}$ are regular orbits with $\alpha(\gamma_i) = \sigma_i$ and $\omega(\gamma_i) = \sigma_{i+1}$ (mod k). By $\alpha(\gamma)$ and $\omega(\gamma)$ we mean the α and the ω limits sets of γ. A cycle Γ is **singular** if at least one of its critical element is a singularity. A cycle is **hyperbolic** if all its critical elements are hyperbolic, otherwise it is called **non-hyperbolic**. A singular cycle Γ is **simple** if either of the following hold:

a) It is hyperbolic. It contains a unique singularity and every regular orbit γ_i in Γ, (with the exception of exactly one (say γ_j)) is a transveral intersection between $W^s(\sigma_{i+1})$ and $W^u(\sigma_i)$. Furthermore, $\dim(W^s(\sigma_{j+1}))+ \dim (W^u(\sigma_j)) = n$ and $W^s(\sigma_{j+1}), W^u(\sigma_j)$ are in general position (that is, $T_p W^s(\sigma_{j+1}) + T_p W^u(\sigma_j)$ is a hyperspace in $T_p M$ $\forall p \in \gamma_j$).

b) Only one of the critical elements of Γ is non-hyperbolic and of saddle-node type. There is a unique singularity in Γ and every regular orbit σ_i in Γ is a transversal intersection between $W^s(\sigma_{i+1})$ and $W^u(\sigma_i)$.

[*]Proyecto Fondecyt 1930863

Lemma 1 *a) Vector fields with ssc (simple singular cycles) lie in codimension one submanifolds in the space of differentiable vector fields. They involve only one degeneracy. b) Vector fields with non simple singular cycles lie in higher codimensional submanifolds. They have more than one degeneracy.*

Our problem is to decide which codimension one submanifolds defined by a ssc are in fact a regular part of the boundary of the class of Morse-Smale systems. This depends both on some generic conditions and on some local properties to be satisfied by the ssc. Before stating our results we describe these generic and local properties.

3 Generic and local properties.

Generically, the real part of the eigenvalues of a linear vector field are either simple (with no multiplicity) or of multiplicity two (two complex conjugate eigenvalues). A similar generic property holds for the modulus of the eigenvalues of a linear automorphism.

The first generic property for a ssc is:

(G1) The eigenvalues of the linear part of the vector field at the singularity, and of the linear part of the return map at the periodic orbits, verifies the generic property stated above.

Given a hyperbolic singularity σ veryfying (G1), the weakest contracting (expanding) eigenvalue is real or complex. The weakest contracting (expanding) eigenvalue has naturally associated a weakest contracting (expanding) direction which is one or two demensional. Let u (s) be the dimension of $W^u(\sigma)$ ($W^s(\sigma)$) and let c be the dimension of the weakest contracting (expanding) direction (c = 1 or 2). Then any u+c (s+c)- dimensional invariant submanifold containing $W^u(\sigma)$ ($W^s(\sigma)$) which is tangent at σ to the weakest contracting (expanding) direction will be denoted by $W^{cu}(\sigma)$ ($W^{cs}(\sigma)$) and called the **center-unstable (center-stable)** manifold of σ. Similar notions are defined for hyperbolic periodic orbits.

The next two generic conditions are:

(G2) At every hyperbolic critical element σ_i of a ssc Γ, the regular orbit γ_{i-1} (resp. γ_i) reaches (resp. leaves) σ_i tangentially to the weakest contracting (expanding) direction of σ_i. At a saddle-node type critical element σ_j (if it exists) the regular orbit γ_{j-1}(resp. σ_j) reaches (resp. leaves) σ_j tangentially to the 1-dimensional neutral direction.

(G3) If γ_j is the non-transversal intersection between $W^u(\sigma_j)$ and $W^s(\sigma_{j+1})$ in a hyperbolic ssc, then $W^{cu}(\sigma_j) \pitchfork W^s(\sigma_{j+1})$ and $W^u(\sigma_j) \pitchfork W^{cs}(\sigma_{j+1})$.

At the hyperbolic critical element σ_j (resp. σ_{j+1}), $W^{cu}(\sigma_j)$ (resp. $W^{cs}(\sigma_{j+1})$) can be chosen so that $\gamma_{j-1} \subseteq W^{cs}(\sigma_j)$ (resp. $\gamma_{j+1} \subseteq W^{cu}(\sigma_{j+1})$).

Our last generic condition is not essential and is introduced for technical reasons only.

(G4) There is a neighborhood $\mathcal{U}$ of X in the C^3-topology such that the analytic continuations of the hyperbolic critical elements of X are C^2-linearizable for every $Y \in \mathcal{U}$.

A ssc veryfing these generic conditions will be called a **generic ssc** (gssc).

Now we define some concepts which are necessary for the statement of our results. Consider a hyperbolic singularity veryfing condition (G1). Let $a < 0$ (resp. $b > 0$) be the real part of the weakest contracting (resp. expanding) eigenvalue of the corresponding linear part. We say that σ is **centrally contracting (centrally expanding)** if $a + b < 0$ (resp. $a + b > 0$).

Now consider a hyperbolic gssc Γ and assume that $\gamma_j \subseteq W^u(\sigma_j) \cap W^s(\sigma_{j+1})$ is the non-transversal intersection of Γ. Assume further that the weakest contracting eigenvalue of σ_j and the weakest expanding eigenvalue of σ_{j+1} are real. In addition, suppose also that whenever σ_j (σ_{j+1}) is a periodic orbit then this eigenvalue is positive.

Then $W^u(\sigma_j)(W^s(\sigma_{j+1}))$ separates $W^{cu}(\sigma_j)(W^{cs}(\sigma_{j+1}))$ into two invariant submanifolds, one of which contains $\gamma_{j-1}(\gamma_{j+1})$. Call this manifold $W^{cu}_+(\sigma_j)$ ($W^{cs}_+(\sigma_{j+1})$). At each point $p \in \gamma_j$ choose vectors v_p and w_p with $v_p(w_p)$ tangent to $W^{cu}(\sigma_j)$ ($W^{cs}(\sigma_{j+1})$) and pointing towards $W^{cu}_+(\sigma_j)$ ($W^{cs}_+(\sigma_{j+1})$).

By (G3) neither v_p nor w_p are contained in the hyperspace $T_p W^u(\sigma_j) + T_p W^s(\sigma_{j+1})$. We say that γ_j has **positive (negative) signature** if v_p and w_p points towards the same (resp. different) side of such hyperplane.

For a gssc Γ, we say that a cycle Γ_0 is **subordinated** to Γ if the critical elements of Γ_0 are contained in those of Γ and if furthermore Γ_0 contains either the non-transversal orbit of Γ, in case Γ is hyperbolic, or the non-hyperbolic critical elements of Γ if Γ is not hyperbolic. A subordinated cycle which contains the singularity is itself a gssc. A gssc Γ is **isolated** with **isolating block** U, if U is a neighborhood of Γ such that $\cap_{t \in \mathbb{R}} \varphi^t(U)$ is the collection of subordinated cycles of Γ contained in U. As usual, φ^t denotes the flow of the underlying vector field.

3

We say that a vector field X with a gssc Γ is in the **boundary of local Morse-Smale systems** if there is an isolating block U of Γ and a small neighborhood $\mathcal{U}$ of X such that for every Y in one of the components of $\mathcal{U} \setminus \mathcal{N}$ the non-wandering set $\Omega(Y/U)$ is just the analytic continuation of the critical elements of Γ. Here $\mathcal{N} = \{Y \in \mathcal{U}/Y$ has a gssc Γ_Y, Γ_Y the analytic continuation of $\Gamma\}$.

The analytic continuation of a hyperbolic critical element is a well known concept. The analytic continuation of a gssc is clear.

4 Statements of results.

Theorem 1 *Let X be a vector field with a hyperbolic homoclinic gssc Γ (that is, with exactly one critical element σ, namely a singularity). Then:*
a) If X is in the boundary of the class of Morse-Smale systems, either σ is centrally contracting with a real weakest expansion or σ is centrally expanding with a real weakest contraction.
b) If one of the situations in a) holds then the cycle Γ is in the boundary of local Morse-Smale systems.

Theorem 2 *Let X be a vector field with a hyperbolic heteroclinic gssc Γ (that is, with more than one critical element). Let γ_j be the non transversal orbit of Γ, $\gamma_j \subseteq W^u(\sigma_j) \cap W^s(\sigma_{j+1})$. Then:*
a) If X is in the boundary of the Morse-Smale systems, the following two properties holds:

a.1) The weakest contracting (resp. expanding) eigenvalue of σ_j (resp. σ_{j+1}) is real, and positive if σ_j (resp. σ_{j+1}) is a periodic orbit.

a.2) The non-transversal orbit γ_j has negative signature
b) If a.1) and a.2) hold, then X is in the boundary of local Morse-Smale systems.

Theorem 3 *Every vector field with a non-hyperbolic gssc is in the boundary of local Morse-Smale systems.*

Theorem 1 is mentioned in [A, pag.117], and the proof of Theorem 3 is quite easy. Theorem 2 is our main result.

5   Classification of ssc.

In this section we give the main elements that classify the simple singular cycles. Recall that a cycle Γ of a vector field X is a set $\Gamma = \left(\bigcup_{i=0}^{k-1} \sigma_i \right) \cup \left(\bigcup_{i=0}^{k-1} \gamma_i \right)$ where $\{\sigma_i\}_{i=0}^{k-1}$ are critical elements and $\{\gamma_i\}_{i=0}^{k-1}$ are regular orbits of X with $\alpha(\gamma_i) = \sigma_i$ and $\omega(\gamma_i) = \sigma_{i+1}$ (mod k). If Γ is a ssc we assume that σ_0 is the singularity and that $s_i = \dim W^s(\sigma_i)$ and $u_i = \dim W^u(\sigma_i)$.

5.1   Hyperbolic ssc.

The hyperbolic ssc are classified by a four-tuple (n, k, j, s_0) where n is the dimension of the ambient space, k is the number of critical elements in Γ, j is the subindex of the non-transversal intersection in Γ and $s_0 = \dim W^s(\sigma_0)$.

Their range of variation is: $n \in \mathbb{N}, n \geq 2$; $k \in \mathbb{N}, k \geq 1$; $j \in \mathbb{N}, 0 \leq j \leq k - 1$; $s_0 \in \mathbb{N}, 1 \leq s_0 \leq n - 1$. If $s_0 = 1$ then $j = k - 1$ and if $s_0 = n - 1$ then $j = 1$.

Given a four-tuple (n, k, j, s_0) then $s_0 + u_0 = n$ and
$s_i = s_0 + 1, u_i = u_0 \; \forall \, i = 1, \ldots, j$ and
$s_i = s_0, u_i = u_0 + 1 \; \forall \, i = j + 1, \ldots, k - 1$.

Hyperbolic gssc of vector fields at the boundary of the class of Morse-Samle systems are the above ones together with the conditions in Theorems 1 and 2.

5.2   Non-hyperbolic ssc.

The non-hyperbolic ssc are also classified by a four-tuple (n, k, j, s_0) where n, k and s_0 are as before, and j is now the index of the non-hyperbolic critical element. Their range of variation is: $n \in \mathbb{N}, n \geq 2$; $k \in \mathbb{N}, k \geq 1$; $j \in \mathbb{N}, 0 \leq j \leq k - 1$; $s_0 \in \mathbb{N}$. If $j = 0$ then $1 \leq s_0 \leq n$ and if $s_0 = 1$ or $s_0 = n$ then $k = 1$. If $j \neq 0$ then $2 \leq s_0 \leq n - 2$.

Given a four-tuple (n, k, j, s_0) we have:
If $j = 0$ then $s_0 + u_0 = n + 1$ and $s_i = s_0, u_i = u_0 \; \forall \, i = 0, \ldots, k - 1$.
If $j \neq 0$ then $s_0 + u_0 = n$ and
$\quad s_i = s_0 + 1, \, u_i = u_0 \, \forall \, i = 1, \ldots, j - 1$,
$\quad s_j = s_0 + 1, \, u_j = u_0 + 1, \text{ and}$
$\quad s_i = s_0, \, u_i = u_0 + 1, \, \forall \, i = j + 1, \ldots, k - 1$.

Acknowledgement: I am grateful to J. Palis, L. Diaz and B. San Martin for helpful conversation. I am also grateful to Brazil's IMPA for its hospitality while this work was being carried out.

References

[A] V.I.Arnold (Ed.), Dynamical Systems V, Encyclopaedia of Mathematical Sciences Vol. 5, Springer Verlag, 1994.

[BLMP] R.Bamón, R.Labarca, R.Mañe, M.J.Pacífico, The explosion of singular cycles, Publ. Math. I.H.E.S. 78 (1993), 207 - 232.

[L] R.Labarca, Bifurcation of contracting singular cycles, Ann. Scient. Éc. Norm. Sup. 4º Série, t. 28 (1995), 705 - 745.

[PR] M.J.Pacífico, A.Rovella, Unfolding contracting singular cycles, Ann. Scient. Éc. Norm. Sup. 4º Série, t. 26 (1993), 691 - 700.

[Ri] M.R.Richlik, Lorenz attactor through Sil'nikov - type bifurcation I, Erg. Th. and Dyn. Sys. 10, 4 (1990), 793 - 822.

[Ro] A.Rovella, The dynamics of perturbations of the contracting Lorenz attractor, Bol. Soc. Bras. Mat. Vol. 24, 2 (1993), 233 - 259.

[S] L.P.Sil'nikov, Generation of periodic motions froma trajectory exiting from an equilibrium of saddle - node and re - entering it, Sov. Math. Dokl. 7 (1966), 1155 - 1158.

Departamento de Matemática
Facultad de Ciencias
Universidad de Chile
Casilla 653 Santiago - CHILE
e-mail: rbamon@abello.dic.uchile.cl

KEITH BURNS[*] AND GABRIEL P PATERNAIN[†]

On the growth of the number of geodesics joining two points

To Ricardo Mañé, in Memoriam

1   *Introduction*

Let M be a closed connected C^∞ manifold endowed with a C^∞ Riemannian metric g. Given p and q in M and $T > 0$, define $n_T(p, q)$ as the number of geodesic segments joining p and q with length $\leq T$. If $(p, q) \in M \times M$ is a pair of non-conjugate points, $n_T(p, q)$ is finite and counts the number of non-degenerate critical points of the length functional on the path space $\Omega^T(p, q)$ given by all the curves joining p and q with length $\leq T$. By Morse theory, the sum of the Betti numbers of $\Omega^T(p, q)$ bounds $n_T(p, q)$ from below. Using these tools, Serre [9] showed that if M is complete and non-contractible (in particular if M is compact), then

$$\lim_{T \to \infty} n_T(p, q) = \infty.$$

In [2] Berger and Bott observed that for each $p \in M$ and $T > 0$, the function $q \to n_T(p, q)$ is integrable in M and, if B_T is the ball about the origin in $T_p M$ with radius T, then

$$\int_M n_T(p, q)\, dq = \int_{B_T} |\det d_v \exp_p|\, dv. \tag{1}$$

It follows immediately from (1) that

$$\limsup_{T \to \infty} \frac{1}{T} \log \int_M n_T(p, q)\, dq < \infty.$$

Recent work [5, 6, 7, 8], has uncovered intimate connections between the growth rate of $n_T(p, q)$ and the topological entropy h_{top} of the geodesic flow. G.P. Paternain [6] showed that

$$\limsup_{T \to \infty} \frac{1}{T} \log \int_M n_T(p, q)\, dq \leq h_{top} \quad \text{for all } p \in M. \tag{2}$$

[*]Partially supported by NSF grant DMS-9206923
[†]Partially supported by grants from CSIC and CONICYT #301

In [5] Mañé obtained the remarkable formula:

$$\lim_{T\to\infty} \frac{1}{T} \log \int_{M\times M} n_T(p,q)\, dpdq = h_{top}.$$

Mañé also observes in [5] that a simple argument using the Borel-Cantelli Lemma shows that for every $p \in M$ one has

$$\limsup_{T\to\infty} \frac{1}{T} \log n_T(p,q) \le \limsup_{T\to\infty} \frac{1}{T} \log \int_M n_T(p,q)\, dq \quad \text{for a.e. } q \in M. \qquad (3)$$

It is immediate from (2) and (3) that for all $p \in M$ one has

$$\limsup_{T\to\infty} \frac{1}{T} \log n_T(p,q) \le h_{top} \quad \text{for a.e. } q \in M. \qquad (4)$$

This inequality and the above results naturally raise the following questions:

Question I *Can inequality (4) be improved to equality, at least for typical points p and q?*

Question II *Is it true that*

$$\limsup_{T\to\infty} \frac{1}{T} \log n_T(p,q) \le h_{top}$$

whenever p and q are not conjugate?

Question I was asked by Mañé in [5]. Its answer is negative. In [3] we constructed an open set of C^∞ metrics on S^2 for which there exists a positive measure set $U \subset M \times M$, such that for all $(p,q) \in U$,

$$\limsup_{T\to\infty} \frac{1}{T} \log n_T(p,q) < h_{top}.$$

The present paper answers Question II in the negative. We can construct a C^∞ metric on S^2 with a point p not conjugate to itself for which $n_T(p,p)$ grows **as fast as we wish**. In particular, it is possible for $n_T(p,p)$ to have superexponential growth. More precisely, suppose we are given a sequence $n_1, n_2, \ldots$ of positive integers. Then we can find a C^∞ Riemannian metric g on S^2 with a point p such that p is not conjugate to itself along any geodesic and there are n_k distinct closed geodesics of minimum period $2^{k-1} \cdot 2\pi$ that pass through p for all integers $k \ge 1$. This ensures that

$$\liminf_{T\to\infty} \frac{1}{T} \log n_T(p,p) \ge \liminf_{k\to\infty} \frac{1}{2^{k-1} \cdot 2\pi} \log n_k.$$

Since the sequence n_k was arbitrary, the right hand side of this inequality can be made as large as we wish, even infinite.

Our construction uses an iterative procedure that begins with the standard round metric g_0 on S^2. For this metric $n_T(p, p) = \infty$ for all $T \geq 2\pi$. We construct a small perturbation g_1 of g_0 for which n_1 closed geodesics of length 2π are preserved. We also arrange the perturbation such that g_1 possesses a closed geodesic γ_1 of minimal period 4π having p as a simple point and there exists a neighborhood of γ_1 of constant curvature $+1$. Moreover, by a careful application of the transversality theorem we can choose g_1 so that p is not conjugate to itself along any geodesic of length $< 2\pi + 1$.

Next using the same type of perturbation as before, we construct a small perturbation g_2 of g_1 such that n_2 closed geodesics of minimal period 4π are preserved, but p is not conjugate to itself along any geodesic of length $< 4\pi + 1$. We also arrange the perturbation so that there exists a closed geodesic γ_2 of minimal period 8π having p as a simple point and there exists a neighborhood of γ_2 of constant curvature $+1$.

Continuing this procedure we obtain a sequence of metrics $g_1, g_2, \ldots$ such that the metric g_k has the properties:

- there are n_k closed geodesics $\gamma_{k,1}, \ldots, \gamma_{k,n_k}$ of minimal period $2^{k-1} \cdot 2\pi$ having p as a simple point;

- all of the g_i-geodesics $\gamma_{i,j}$ for $i < k$ are also geodesics for the metric g_k and the Jacobi equation along each $\gamma_{i,j}$ is the same in the metric g_k as it was in the metric g_i;

- there exists a closed geodesic γ_k of minimal period $2^k \cdot 2\pi$ having p as a simple point and there exists a neighborhood of γ_k of constant curvature $+1$;

- p is not conjugate to itself along any geodesic of length $< 2^{k-1} \cdot 2\pi + 1$.

By making sufficiently small perturbations at each step we can show that the sequence $g_1, g_2, \ldots$ converges to a smooth metric g_∞. The construction procedure ensures that p is not conjugate to itself along any geodesic and that for each $k = 1, 2, \ldots$ there are at least n_k closed geodesics with minimal period $2^{k-1} \cdot 2\pi$ passing through p.

This construction is performed in section 3 of the paper. Section 2 develops some of the geometry that is needed for the construction.

Acknowledgements: The second author is grateful to Northwestern University for hospitality while this work began.

2   *Spherical strips*

It is convenient to say that a closed geodesic γ is the centre of a *spherical strip* if the length of γ is an integral multiple of 2π and γ has a neighbourhood in which the curvature is everywhere 1. If γ is the centre of a spherical strip, there are $\delta > 0$, an integer $m \geq 1$ and a local diffeomorphism $\Phi : (-\delta, \delta) \times [0, 2m\pi] \to S^2$ such that $t \mapsto \Phi(0, t)$ parametrizes γ by arclength, the maps $s \mapsto \Phi(s, t)$ parametrize the geodesics orthogonal to γ by arclength, and the image of Φ has constant positive curvature 1. Thus γ has a neighbourhood that is an m-fold covering of a great circle on the unit sphere. Notice that the geodesic γ that is the centre of the strip may have self intersections. Observe also that since the length of the spherical strip Σ is an integral multiple of 2π and Σ has curvature 1, all geodesics that cross γ at a small enough angle will be closed geodesics that lie in Σ and have the same least period as γ, namely $2m\pi$.

A crucial ingredient in the construction that we will make in section 3 is

Lemma 2.1 *Let h be a Riemannian metric on S^2, p a point on S^2 and m and n positive integers. Suppose that (S^2, h) has a closed geodesic γ of least period $2m\pi$ having p as a simple point and is the centre of a spherical strip Σ. Arbitrarily close to h in the C^∞ topology, we can find a metric h' such that*

- *h' agrees with h outside Σ;*

- *(Σ, h') contains n closed geodesics of least period $2m\pi$ that pass through p and have the property that p is not conjugate to itself along any of them;*

- *(Σ, h') contains a closed geodesic of least period $4m\pi$ that is the centre of a spherical strip and has p as a simple point.*

Proof: As we saw above, all geodesics that pass through p at a small enough angle to γ will be closed geodesics that lie in Σ and have least period $2m\pi$. Let $\gamma_1, \ldots, \gamma_n$ be n of these closed geodesics such that $\gamma_i \neq \gamma$, $1 \leq i \leq n$. Pick points $x_1, \ldots, x_n$, such that x_i is a simple point of the closed geodesic γ_i and there is $\varepsilon > 0$ such that the geodesics γ and γ_j, $j \neq i$, do not pass through the ε-neighborhood of x_i. We may also assume that the ε-neighborhood of x_i lies in Σ. Lemma 2.2 below tells us that we can make arbitrarily small perturbations of the metric h inside the ε-neighborhoods of x_i so that p is not conjugate to itself along any of the geodesics γ_i.

Choose a point x on γ that is not conjugate to p. Pick a geodesic β in Σ that is close to γ and passes through x but does not pass through p. Then β is closed and has length $2m\pi$. We may assume that the ε chosen above is small enough so that the closed ε ball B about x lies in Σ and does not intersect the ε-neighborhood of any of the closed geodesics γ_i. We shall modify the metric in B as shown in Figure 1 and explained below.

Suppose for convenience that β and γ are parametrized so that they have unit speed, $\beta(0) = x = \gamma(0)$ and $\dot{\beta}(0)$ is close to $\dot{\gamma}(0)$. We choose disjoint smooth curves $\alpha_1, \alpha_2 : (-2\varepsilon, 2\varepsilon) \to S^2$ that map $(-\varepsilon, \varepsilon)$ into the interior of B and join up the points where β and γ cross the boundary of B in the manner shown in Figure 1. We choose α_1 so that $\alpha_1(t) = \beta(t)$ for $-2\varepsilon < t \leq -\varepsilon$ and $\alpha_1(t) = \gamma(t)$ for $\varepsilon \leq t < 2\varepsilon$; and we choose α_2 so that $\alpha_2(t) = \gamma(t)$ for $-2\varepsilon < t \leq -\varepsilon$ and $\alpha_2(t) = \beta(t)$ for $\varepsilon \leq t < 2\varepsilon$. Now choose a small $\delta > 0$ and diffeomorphisms $A_i : (-2\varepsilon, 2\varepsilon) \times (-\delta, \delta) \to S^2$, $i = 1, 2$, with the following properties, illustrated in Figure 1:

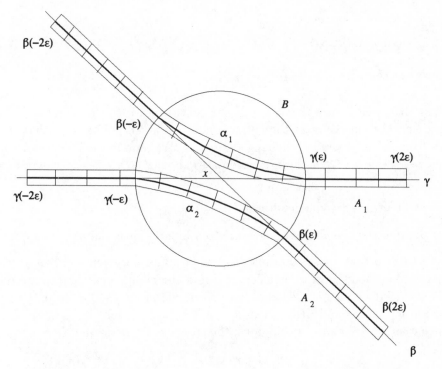

Figure 1: Modifying the metric to create the new spherical strip.

- the images of A_1 and A_2 are disjoint;

- $A_i(t,0) = \alpha_i(t)$ for $-2\varepsilon < t < 2\varepsilon$;

- when $|t| \geq \varepsilon$, the curve $A_i(t,\cdot)$ is the geodesic orthogonal to β or γ at $\alpha_i(t)$, so the restrictions of A_i to $(-2\varepsilon, -\varepsilon] \times (-\delta, \delta)$ and $[\varepsilon, 2\varepsilon) \times (-\delta, \delta)$ agree with the normal exponential maps for β and γ.

Now we change the metric in the images of the A_i so that the new metric agrees with the old metric at $A_i(t,s)$ whenever $|t| \geq \varepsilon$ or $|s| \geq 2\delta/3$, the new metrics have curvature 1 at $A_i(t,s)$ whenever $|s| \leq \delta/3$, each α_i is a geodesic for the new metric, and inverses of the maps A_i define Fermi coordinates on the $\delta/3$ neighbourhoods of the α_i. The new metric contains a closed geodesic of minimim period $4m\pi$ formed by α_1 and α_2 inside the ball B and β and γ outside B. The curvature of the new metric is 1 everywhere in the $\delta/3$ neighbourhood of this geodesic.

◇

Lemma 2.2 *Let (M,h) be a surface and $\gamma : \mathbf{R} \to M$ a closed geodesic. Let p and x be distinct simple points of γ and let U be a neighborhood of x such that $p \notin U$. Arbitrarily close to h in the C^∞ topology, we can find a metric h' such that*

- h and h' agree outside U;

- γ is a geodesic for h';

- p is not conjugate to itself along γ.

Proof: It is well known that one can change the metric so that γ remains a closed geodesic and the curvature of the surface increases at all points on γ that are close enough to x and is otherwise unchanged along γ. We remind the reader how this can be achieved. Let σ and τ be Fermi coordinates about γ in a neighbourhood of x such that:

- τ parametrizes γ by arclength and $\tau = 0$ at x;

- σ parametrizes geodesics orthogonal to γ by arclength and $\sigma = 0$ on γ.

Choose $\delta > 0$ so that these coordinates define a diffeomorphism $\alpha : [-\delta, \delta] \times [-\delta, \delta]$ onto a neighbourhood V of x that lies in U and the points where $\sigma = 0$ are the only points in V that lie on γ. Also choose a C^∞ function $\psi : [-\delta, \delta] \to [0, 1]$ such that $\psi(r) = 0$ if $|r| \geq 2\delta/3$ and $\psi(r) = 1$ if $|r| \leq \delta/3$.

Since σ and τ are Fermi coordinates, the components of the metric h are given by

$$h_{\sigma,\sigma}(\sigma, \tau) = 1, \qquad h_{\sigma,\tau}(\sigma, \tau) = 0, \qquad h_{\tau,\tau}(\sigma, \tau) = \|Y_\tau(\sigma)\|_h^2,$$

where $Y_\tau(\sigma) = \partial\alpha/\partial\tau(\sigma, \tau)$ is a Jacobi field along the geodesic $\alpha_\tau(\sigma) = \alpha(\sigma, \tau)$ that is orthogonal to α_τ and satisfies the initial conditions $\|Y_\tau(0)\|_h = 1$ and $Y_\tau'(0) = 0$. It follows from the Jacobi equation that the curvature of the metric h at the point $\alpha(0, \tau)$, which lies on γ, is given by

$$K_h(0, \tau) = -\left.\frac{d^2}{d\sigma^2}\right|_{\sigma=0} \|Y_\tau(\sigma)\|_h.$$

Define a new metric h^ε by requiring that $h^\varepsilon_{\sigma,\sigma} = h_{\sigma,\sigma} = 1$, $h^\varepsilon_{\sigma,\tau} = h_{\sigma,\tau} = 0$ and

$$h^\varepsilon_{\tau,\tau}(\sigma, \tau) = \left(\|Y_\tau(\sigma)\|_h - \frac{\varepsilon}{2}\psi(\sigma)\psi(\tau)\sigma^2\right)^2.$$

Then γ and α_τ are geodesics for h^ε and σ and τ are still Fermi coordinates for γ; in particular $Y_\tau(\sigma)$ is still a Jacobi field along α_τ that is orthogonal to α_τ and satisfies the initial conditions $\|Y_\tau(0)\|_h = 1$ and $Y_\tau'(0) = 0$. It follows from the Jacobi equation that the curvature of h^ε at $\alpha(0, \tau)$, which is on γ, is

$$
\begin{aligned}
K_h(0, \tau) &= -\left.\frac{d^2}{d\sigma^2}\right|_{\sigma=0} \|Y_\tau(\sigma)\|_{h^\varepsilon} \\
&= -\left.\frac{d^2}{d\sigma^2}\right|_{\sigma=0} \left(\|Y_\tau(\sigma)\|_h - \frac{\varepsilon}{2}\psi(\sigma)\psi(\tau)\sigma^2\right) \\
&= K_h(0, \tau) + \varepsilon\psi(\tau).
\end{aligned}
\tag{5}
$$

We now show that we can find an arbitrarily small $\varepsilon > 0$ such that the point p is not conjugate to itself along the closed geodesic γ in the geometry of the metric h^ε. For each ε, consider γ parametrized by arclength so that $\gamma(0) = p$. Let $Y^\varepsilon(t)$ be a nontrivial Jacobi field along γ that is perpendicular to γ and satisfies $Y^\varepsilon(0) = 0$. For $k = 1, 2, \ldots$ let $t_k(\varepsilon) > 0$ be the k'th time when $Y^\varepsilon(t) = 0$. Notice that γ is a closed geodesic with the same length for all values of ε; let L denote this common length. Since p is a simple point of γ, it suffices to choose ε so that $t_k(\varepsilon)$ is not a multiple of L for any $k \geq 1$.

It is obvious from the curvature formula (5) that the curvature of h^ε at any given point of γ is a nondecreasing function of ε, and the curvature of h^ε at x is a strictly increasing function of ε. It now follows from Sturm's First Comparison Theorem (Theorem XI.3.1 in [4]) that $t_k(\varepsilon)$ is a strictly decreasing function of ε for each $k \geq 1$. Thus any equation of the form $t_k(\varepsilon) = jL$, where j and k are positive integers, has at most one solution. Hence there are at most countably many values of ε for which p is conjugate to itself along γ. In particular there are arbitrarily small positive values of ε for which p is not conjugate to itself along γ.

$\diamond$

3   *Arbitrarily fast growth of the number of geodesics*

In this section we show how to construct a smooth metric on the sphere S^2 with a point p such that $n_T(p, p)$ is finite for all T, but $n_T(p, p)$ is as large as we wish. More precisely, suppose we are given a sequence $n_1, n_2, \ldots$ of positive integers. Then we can find a C^∞ Riemannian metric g on S^2 with a point p such that there are n_k closed geodesics of minimum period $2^{k-1} \cdot 2\pi$ that pass through p and there are few enough other geodesics from p to p that $n_T(p, p)$ is finite for all T. The latter property of g will be achieved by ensuring that p is not conjugate to itself along any geodesic.

We use an iterative procedure that begins with the standard round metric g_0 and successively constructs a sequence of metrics g_1, g_2, $\ldots$. We then show that this sequence can be chosen so that it converges in the C^∞ topology to a metric g_∞ that has the desired properties.

Pick a point $p \in S^2$ and let γ_0 be a geodesic of the standard metric g_0 passing through p. The geodesic γ_0 is the centre of a spherical strip Σ_0. By Lemma 2.1, abritrarily close to g_0 in the C^∞ topology, there exists a smooth metric g_0' with the following properties:

(A1) g_0' agrees with g_0 outside Σ_0;

(A2) (Σ_0, g_0') contains n_1 closed geodesics $\gamma_{1,1}, \ldots, \gamma_{1,n_1}$ of least period 2π such that p is a simple point of each $\gamma_{1,j}$ and p is not conjugate to itself along any $\gamma_{1,j}$;

(A3) (Σ_0, g_0') contains a closed geodesic γ_1 of least period 4π that is the centre of a spherical strip Σ_1 and has p as a simple point.

We show now that g_0' can be approximated arbitrarily closely by a metric g_1 that enjoys the properties (A1)–(A3) above and the additional property:

(A4) p is not conjugate to itself along any geodesic of length $< 2\pi + 1$.

Let $\mathcal{A}$ denote the set of C^∞ metrics g on S^2 with the following properties:

- g agrees with g_0' on the closure of the strip Σ_1;

- g agrees with g_0' to second order along the n_1 closed g_0'-geodesics $\gamma_{1,1}, \ldots, \gamma_{1,n_1}$ given by (A2), so that these curves are also geodesics for the metric g, and the Jacobi equations along these geodesics are the same whether we use the metric g or g_0';

- g agrees with g_0 outside Σ_0.

Any metric $g \in \mathcal{A}$ satisfies (A1)–(A3). Let $\mathcal{B}$ denote the set of metrics in $\mathcal{A}$ for which (A4) holds.

Proposition 3.1 $\mathcal{B}$ *is open and dense in* $\mathcal{A}$ *with respect to the* C^∞ *toplogy.*

Proof: It is obvious that $\mathcal{B}$ is open in the C^2 topology, and hence in the C^∞ topology. To prove that $\mathcal{B}$ is dense in the C^∞ topology, it suffices to show that $\mathcal{B}$ is dense in the C^k topology for any finite $k \geq 2$.

Note that if $k \geq 2$, then $\mathcal{A}$ with the C^k topology is isometric to an open subset of a Banach space, and is thus a C^∞ Banach manifold. Indeed, let Γ be the space of all symmetric C^∞ sections of $T^*S^2 \otimes T^*S^2$ and Γ_0 the subspace of Γ consisting of sections that vanish outside $\Sigma_1 \setminus \overline{\Sigma_0}$ and also vanish to second order on $\gamma_{1,1} \cup \cdots \cup \gamma_{1,n_1}$. Then $\mathcal{A}$ is the subset of the coset $g_0' + \Gamma_0$ in Γ consisting of those sections that are positive definite at all points of S^2. Since $k \geq 2$, the space Γ_0 is a *closed* subspace of Γ, and hence is a Banach space.

Let $B_{2\pi+1}$ denote the open ball in $T_p S^2$ with centre at the origin and radius $2\pi + 1$. Consider the map

$$e : \mathcal{A} \times B_{2\pi+1} \to S^2$$

given by

$$e(g, v) = \exp_p^g(v),$$

where $\exp_p^g$ denotes the exponential map at p of the metric g. Condition (A4) holds for g if and only if p is a regular value of the map $e(g, \cdot) : B_{2\pi+1} \to S^2$. Thus $\mathcal{B}$ is precisely the set of those $g \in \mathcal{A}$ for which the map $e(g, \cdot)$ is transversal to $\{p\}$.

14

By Lemma 3.2 below, the map e is transversal to $\{p\}$. It follows from the Transversality Density Theorem (Theorem 19.1 of [1]) that $\mathcal{B}$ is dense in $\mathcal{A}$ with the C^k topology.

$\diamond$

Lemma 3.2 *The map e is transversal to $\{p\}$.*

Proof: Let $(g, v) \in e^{-1}(p)$. We must show that the derivative map

$$D_{(g,v)}e : T_{(g,v)}(\mathcal{A} \times B_{2\pi+1}) \to T_p S^2$$

is onto. If v is a regular point of $e(g, \cdot)$, then the subspace $\{0\} \times T_v B_{2\pi+1}$ is mapped onto $T_p S^2$. Thus we shall assume that $e(g, v) = p$ and v is a singular point of $e(g, \cdot)$. Set $l = \|v\|$. Then we are assuming that the geodesic for the metric g that has $\dot{\gamma}(0) = v$ satisfies $\gamma(l) = \gamma(0) = p$ and that $\gamma(l)$ is conjugate to $\gamma(0)$ along γ.

The vector $\dot{\gamma}(l)$ lies in the image of $D_{(g,v)}e$, since

$$\gamma(l + s) = e(g, v + sv/\|v\|).$$

Thus in order to prove that $D_{(g,v)}e$ is onto, it will suffice to find a smooth one parameter family of metrics $g^\varepsilon \in \mathcal{A}$ with $g^0 = g$ such that the vector

$$\left.\frac{\partial}{\partial \varepsilon}\right|_{\varepsilon=0} e(g^\varepsilon, v)$$

has a nonzero component orthogonal to $\dot{\gamma}(l)$.

Let us define

$$W(t) = \left.\frac{\partial}{\partial \varepsilon}\right|_{\varepsilon=0} e(g^\varepsilon, tv/l).$$

Choose a parallel vector field $E(t)$ along γ that is orthogonal to $\dot{\gamma}$ and has unit norm. Set $\omega(t) = \langle W(t), E(t) \rangle$. Then $\omega(l)$ is the component that we wish to make nonzero.

As shown in the appendix, the variation vector field W satisfies the generalized Jacobi equation

$$\frac{D^2 W}{dt^2}(t) + R(W(t), \dot{\gamma}(t))\dot{\gamma}(t) = -\left.\frac{D}{d\varepsilon}\right|_{\varepsilon=0} \left(\frac{D_\varepsilon \dot{\gamma}}{dt}(t)\right),$$

in which D_ε/dt denotes covariant differentiation in the metric g^ε. Moreover $W(t)$ satisfies the initial conditions

$$W(0) = 0 = \frac{DW}{dt}(0).$$

Let $K(t)$ denote the sectional curvature (in the metric g) of the plane spanned by $\dot{\gamma}(t)$ and $E(t)$. We see that $\omega(t)$ is the solution of the scalar equation

$$\omega'' + K(t)\omega = \left\langle -\left.\frac{D}{d\varepsilon}\right|_{\varepsilon=0} \left(\frac{D_\varepsilon \dot{\gamma}}{dt}(t)\right), E(t) \right\rangle \stackrel{\text{def}}{=} f(t). \tag{6}$$

15

that satisfies the initial conditions $\omega(0) = 0 = \omega'(0)$.

Let y_1 and y_2 denote the solutions of the corresponding homogeneous equation that satisfy: $y_1(0) = 0$, $y_1'(0) = 1$ and $y_2(0) = 1$, $y_2'(0) = 0$. Solving (6) by variation of parameters gives (as is easily verified by differentiation)

$$\omega(t) = \left(-\int_0^t f(u)y_2(u)\,du \right) y_1(t) + \left(\int_0^t f(u)y_1(u)\,du \right) y_2(t). \tag{7}$$

Recall that we assumed that $\gamma(l)$ is conjugate to $\gamma(0)$ along γ. This means that $y_1(l) = 0$ and $y_2(l) \neq 0$; and thus (7) implies that $\omega(l) \neq 0$ if and only if

$$\int_0^l f(u)y_1(u)\,du \neq 0. \tag{8}$$

We must now prove that the one parameter family g^ε can be chosen so that (8) holds. The first step is to show that γ contains a point $x \in \Sigma_0 \setminus \overline{\Sigma_1}$ that does not lie on any of the geodesics $\gamma_{1,j}$ defined by (A2). Since γ starts at p, it certainly contains points of Σ_0. The construction of Σ_1 ensures that if γ did not leave Σ_1, it would be part of a closed geodesic of length 4π for which p is a simple point. This is impossible, since $\gamma(0) = p = \gamma(l)$ and $l = \|v\| \leq 2\pi + 1$ because $v \in B_{2\pi+1}$. If every point of γ lay on $\gamma_{1,j}$, we would have $\gamma(t) = \gamma_{1,j}(t)$, which is impossible, since we have assumed that p is conjugate to itself along γ and (A2) tells us that p is not conjugate to itself along $\gamma_{1,j}$. We see that we can find a point x on γ with the desired properties. It follows immediately that γ contains an open interval of such points.

As in Lemma 2.2, we take Fermi coordinates σ and τ about the unit speed geodesic γ in a neighbourhood V of x. We may assume that V lies in $\Sigma_0 \setminus \overline{\Sigma_1}$ and none of the geodesics $\gamma_{1,j}$ enters V. We may also assume that the points where $\sigma = 0$ are the only points in V that lie on $\gamma([0,l])$ and that the function $y_1(t)$ that appears in (8) is nonzero when $\gamma(t) \in V$.

Define a new metric g^ε by requiring that $g^\varepsilon_{\sigma,\sigma} = g_{\sigma,\sigma} = 1$, $g^\varepsilon_{\sigma,\tau} = g_{\sigma,\tau} = 0$ and

$$g^\varepsilon_{\tau,\tau}(\sigma,\tau) = g_{\tau,\tau}(\sigma,\tau) + 2\varepsilon\psi(\sigma)\psi(\tau)\sigma,$$

where the function ψ is as in Lemma 2.2. Since g^ε agrees with g outside V, the 1-parameter family g^ε belongs to $\mathcal{A}$. Let

$$I = \{ t \in [0,l] : \gamma(t) \in V \}.$$

It is obvious that I is an interval, $y_1(t) \neq 0$ if $t \in I$, and $f(t) = 0$ if $t \notin I$.

Lemma 3.3 *If $t \in I$, then $f(t) = \psi(\tau(\gamma(t)))$.*

Proof: If $t \in I$, we have

$$f(t) = \left\langle -\frac{D}{d\varepsilon}\bigg|_{\varepsilon=0}\left(\frac{D_\varepsilon\dot\gamma}{dt}(t)\right), E(t) \right\rangle = -\frac{d}{d\varepsilon}\bigg|_{\varepsilon=0}\left\langle \left(\frac{D_\varepsilon\dot\gamma}{dt}(t)\right), E(t) \right\rangle.$$

16

A simple computation in coordinates shows that

$$\left\langle \left(\frac{D_\varepsilon \dot{\gamma}}{dt}(t)\right), E(t) \right\rangle = \Gamma^\sigma_{\tau,\tau}(0, \tau(\gamma(t))) = -\frac{1}{2}\frac{\partial g^\varepsilon_{\tau,\tau}}{\partial \sigma}(0, \tau(\gamma(t))).$$

Using the definition of g^ε, we obtain

$$-\frac{1}{2}\frac{\partial g^\varepsilon_{\tau,\tau}}{\partial \sigma}(0, \tau(\gamma(t))) = -\frac{1}{2}\frac{\partial g_{\tau,\tau}}{\partial \sigma}(0, \tau(\gamma(t))) - \varepsilon\psi(\tau(\gamma(t))) = -\varepsilon\psi(\tau(\gamma(t))).$$

Therefore

$$f(t) = -\left.\frac{d}{d\varepsilon}\right|_{\varepsilon=0}[-\varepsilon\psi(\tau(\gamma(t)))] = \psi(\tau(\gamma(t))).$$

$\diamond$

Lemma 3.3 and the preceding remarks reduce (8) to

$$\int_I \psi(\tau(\gamma(t)))y_1(t)\,dt \neq 0 \tag{9}$$

and imply that $y_1(t)$ is nonzero and has constant sign on I. It is obvious that we can choose the function ψ so that (9) holds. This completes the proof of Lemma 3.2.

$\diamond$

Hence we have constructed an arbitrarily small perturbation g_1 of g_0 with properties (A1), (A2), (A3) and (A4). It is clear that by recursively applying Lemma 2.1 and the transversality theorem as above we can obtain a sequence of metrics $g_1, g_2, \ldots$ such that g_k is arbitrarily close to g_{k+1} and the following properties hold:

(A) there are n_k closed geodesics $\gamma_{k,1}, \ldots, \gamma_{k,n_k}$ of minimal period $2^{k-1} \cdot 2\pi$ having p as a simple point and the metrics g_i agree with g_k on these geodesics to infinite order for $i \geq k$;

(B) all of the g_i-geodesics $\gamma_{i,j}$ for $i < k$ are also g_k-geodesics and the Jacobi equation along each $\gamma_{i,j}$ is the same in the metric g_k as it was in the metric g_i;

(C) p is not conjugate to itself along any geodesic of length $< 2^{k-1} \cdot 2\pi + 1$.

We can choose our sequence $\{g_k\}$ so that it converges to a metric g_∞ in the C^∞ topology (i.e. in the C^r tolopology for every $r \geq 1$). By (A) and (B) the limit metric g_∞ possesses, for each k, n_k closed geodesics of minimal period $2^{k-1} \cdot 2\pi$ having p as a simple point.

Since condition (C) is open in the C^2 topology on metrics, it follows from (C) that for each k there exists $\varepsilon(k)$ such that if

$$\|g_k - g\|_{C^2} \leq \varepsilon(k),$$

then p is not conjugate to itself along any geodesic of g of length $\leq 2^{k-1} \cdot 2\pi$. We may assume that our sequence $\{g_k\}$ was chosen so that

$$\|g_k - g_i\|_{C^2} \leq \varepsilon(k)$$

whenever $i \geq k$. Then the limit metric g_∞ satisfies

$$\|g_k - g_\infty\|_{C^2} \leq \varepsilon(k), \tag{10}$$

for all $k \geq 1$. By (10) and the choice of $\varepsilon(k)$, we know for each $k \geq 1$ that the point p is not conjugate to itself along any geodesic of length $\leq 2^{k-1} \cdot 2\pi$. Thus p is not conjugate to itself along any geodesic of the metric g_∞.

4   Appendix: the generalized Jacobi equation

Suppose that g^ε is a 1-parameter family of metrics and γ^ε is 1-parameter family of curves such that γ^ε is a geodesic for the metric g^ε. We make the convention that if the superscript ε is omitted we are considering $\varepsilon = 0$; in particular γ will mean γ^0. Let $D_\varepsilon V/dt$ denote the covariant derivative of a vector field $V(t)$ in the metric g^ε.

Proposition 4.1 *The variation vector field*

$$W(t) = \left. \frac{\partial}{\partial \varepsilon} \right|_{\varepsilon=0} \gamma^\varepsilon(t)$$

satisfies the equation

$$\frac{D^2 W}{dt^2}(t) + R(W(t), \dot{\gamma}(t))\dot{\gamma}(t) = - \left. \frac{D}{d\varepsilon} \right|_{\varepsilon=0} \left(\frac{D_\varepsilon \dot{\gamma}}{dt}(t) \right). \tag{11}$$

Remarks (i) We are using the convention that

$$R(X, Y)Z = \nabla_X \nabla_Y Z - \nabla_Y \nabla_X Z - \nabla_{[X,Y]} Z.$$

(ii) The right hand side of (11) depends only on the metrics g^ε and the initial geodesic γ, and is otherwise independent of the choice of variation.

(iii) In the case when g^ε is independent of ε, equation (11) reduces to the usual Jacobi equation.

Proof: Set $\Gamma(\varepsilon, t) = \gamma^\varepsilon(t)$. Using the symmetry of the Levi-Cività connection and the above definition of the curvature tensor, we obtain

$$\frac{D^2}{\partial t^2} \frac{\partial \Gamma}{\partial t} = \frac{D}{\partial t} \frac{D}{\partial \varepsilon} \frac{\partial \Gamma}{\partial t} = \frac{D}{\partial \varepsilon} \frac{D}{\partial t} \frac{\partial \Gamma}{\partial t} - R(\frac{\partial \Gamma}{\partial \varepsilon}, \frac{\partial \Gamma}{\partial t})\frac{\partial \Gamma}{\partial t}.$$

Evaluating at $(\varepsilon, t) = (0, t_0)$ gives us

$$\frac{D^2 W}{dt^2}(t_0) + R(W(t), \dot{\gamma}(t_0))\dot{\gamma}(t_0) = \frac{D}{d\varepsilon}\bigg|_{\varepsilon=0} \left(\frac{D\dot{\gamma}^\varepsilon}{dt}(t_0) \right). \tag{12}$$

Let β_ε be the geodesic for the unperturbed metric g^0 that satisfies $\dot{\beta}_\varepsilon(t_0) = \dot{\gamma}^\varepsilon(t_0)$. Observe that if β and γ are any two curves with $\dot{\beta}(t_0) = \dot{\gamma}(t_0)$, then

$$\frac{D_\varepsilon \dot{\gamma}}{dt}(t_0) - \frac{D_\varepsilon \dot{\beta}}{dt}(t_0)$$

is independent of ε. (Changing the connection only changes what we think of as a constant vector field, it doesn't change differences.) It follows that

$$\frac{D\dot{\gamma}^\varepsilon}{dt}(t_0) = \frac{D\dot{\gamma}^\varepsilon}{dt}(t_0) - \frac{D\dot{\beta}_\varepsilon}{dt}(t_0) = \frac{D_\varepsilon \dot{\gamma}^\varepsilon}{dt}(t_0) - \frac{D_\varepsilon \dot{\beta}_\varepsilon}{dt}(t_0) = -\frac{D_\varepsilon \dot{\beta}_\varepsilon}{dt}(t_0). \tag{13}$$

Now let $F(r, s) = D_r \dot{\beta}_s / dt(t_0)$. Then

$$\frac{D}{d\varepsilon}\bigg|_{\varepsilon=0} \left(\frac{D_\varepsilon \dot{\beta}_\varepsilon}{dt}(t_0) \right) = \frac{D}{d\varepsilon}\bigg|_{\varepsilon=0} F(\varepsilon, \varepsilon) = \frac{D}{d\varepsilon}\bigg|_{\varepsilon=0} F(0, \varepsilon) + \frac{D}{d\varepsilon}\bigg|_{\varepsilon=0} F(\varepsilon, 0)$$

$$= \frac{D}{d\varepsilon}\bigg|_{\varepsilon=0} \left(\frac{D_0 \dot{\beta}_\varepsilon}{dt}(t_0) \right) + \frac{D}{d\varepsilon}\bigg|_{\varepsilon=0} \left(\frac{D_\varepsilon \dot{\beta}_0}{dt}(t_0) \right). \tag{14}$$

Since each β_ε is a geodesic in the metric g^0 and β_0 is the original geodesic γ, the first term on the second line of (14) vanishes and the second term is

$$\frac{D}{d\varepsilon}\bigg|_{\varepsilon=0} \left(\frac{D_\varepsilon \dot{\gamma}}{dt}(t_0) \right).$$

Combining this with (12), (13) and (14) completes the derivation of (11).

$\diamond$

References

[1] R. Abraham, J. Robbin, *Transversal mappings and flows*, W.A. Benjamin, Inc., N.Y. 1967.

[2] M. Berger, R. Bott, *Sur les variétés à courbure strictement positive*, Topology **1** (1962), 302–311.

[3] K. Burns, G. P. Paternain, *Counting geodesics on Riemannian manifolds and topological entropy of geodesic flows*, to appear in Ergod. Th. and Dyn. Syst.

[4] P. Hartman, *Ordinary Differential Equationns*, John Wiley & Sons, N. Y. 1964.

[5] R. Mañé, *On the topological entropy of geodesic flows*, preprint, IMPA, 1994.

[6] G. P. Paternain, *On the topology of manifolds with completely integrable geodesic flows*, Ergod. Th. and Dyn. Syst. **12** (1992), 109–121.

[7] G. P. Paternain, M. Paternain, *Topological entropy versus geodesic entropy*, International Journal of Math. **2** (1994) 213–218.

[8] M. Pollicott, *Entropy and geodesic arcs on surfaces*, preprint, University of Warwick, 1995.

[9] J. P. Serre, *Homologie singulière des espaces fibrés*, Ann. Math. **54** (1951), 425–505.

Keith Burns
Mathematics Department
Northwestern University
Evanston IL 60208
U.S.A.
E-mail: burns@math.nwu.edu

Gabriel P. Paternain
Centro de Matemática
Facultad de Ciencias
Eduardo Acevedo 1139
Montevideo CP 11200
Uruguay
E-mail: gabriel@cmat.edu.uy

EUGENE GUTKIN[*] AND SERGE TROUBETZKOY[†]

Directional flows and strong recurrence for polygonal billiards

1 Introduction

Let P be a *rational polygon*, i. e., the angles between the sides of P are rational multiples of π. Then any billiard trajectory in P assumes, as time runs from $-\infty$ to ∞, only a finite number of distinct directions. Elaborating on this observation, Katok and Zemlyakov have associated with P a *Riemann surface*, S (with singularities), that carries a one-parameter family of *directional billiard flows*, yielding a decomposition of the full billiard flow in P [26]. The surface S is tiled by a finite number of copies of P, providing S with a flat riemannian metric which is singular at the vertices, where the total angle is a multiple of 2π. Although the geometric ideas underlying the correspondence $P \to S$ have been around for a long time (see [10, 11] and the references there), the paper [26] gave their first application to the billiard dynamics. By now, there are many results of this nature in the literature (see, e. g., [18, 19, 23, 24], and the surveys [11, 12]). But so far the correspondence, $P \to S$, between polygons and surfaces has been restricted to the rational polygons.

In this work we associate a "flat surface" S with any polygon P, extending the construction above to arbitrary, irrational, polygons. The surface S provides a natural tool to study the *directional billiard flows* in P. However, if P is irrational, then S is not compact, and the directional flows do not decompose the full billiard flow in P. Nevertheless, in §3 we derive a few applications of the new technique to the general billiard dynamics (see, in particular, Corollary 3.1).

---

[*]Mathematics Department, University of Southern California, Los Angeles, CA 90089-1113. Email:<egutkin@math.usc.edu>. Partially supported by NSF Grants DMS-9013220, DMS-9400295

[†]Mathematics Department, University of Alabama at Birmingham, Alabama 35294-1170. Email: <troubetz@vorteb.math.uab.edu>.

Our main applications are to the periodic billiard orbits. We consider the *perpendicular billiard trajectories*: those that hit P at the right angle at least once. If P is rational, every nonsingular perpendicular orbit is periodic [9, 1], and there is at most a finite number of singular ones. We prove an analog of this result for a special class of irrational polygons, *generalized parallelograms* (Definition 4.4). For such polygons a perpendicular orbit is periodic, with probability one (see §4.1 for precise definitions). This follows from a strong recurrence property of the generalized parallelograms (Theorem 4.1), whose proof is based on the "flat surface realization" of the directional flows (Proposition 2.2).

More generally, the periodicity, with probability one, of perpendicular orbits is a consequence of the strong recurrence property (Proposition 4.1). The class of strongly recurrent irrational polygons is wider than the class of generalized parallelograms (Theorem 4.2), and includes the right triangles. The periodicity almost surely of perpendicular orbits in right triangles was proved in [4]. Their result served as a catalizator for the present work, and we take this opportunity to thank A. Kolan for sending us the preprint [4]. We also note that the idea of treating the directional billiard flows for certain irrational polygons as flows on an appropriate flat surface is implicit in [7], where Galperin has constructed examples of directional flows with remarkable topological properties.

The billiard in a right triangle is equivalent to the mechanical system of two elastic point masses in one dimension (Proposition 5.1). The ratio of the masses is determined by the acute angle of the right triangle. Therefore the periodicity of perpendicular orbits in right triangles has direct applications to classical mechanics (Corollaries 5.1 - 5.3). There are many open problems about periodic trajectories in polygons (see the survey [12]). It is not even known if an arbitrary triangle has a periodic orbit, although this is so for a large set of parameters [9]. We recommend the books [3, 21] for a discussion of periodic orbits and other material on billiards in polygons.

It is a pleasure to acknowledge stimulating electronic correspondence with Amy Kolan, and equally stimulating discussions with Anatole Katok (see, in particular, Theorem 3.1). Parts of this work were done while the authors were visiting the Mathematics Institute of the University of Warwick, the Institute for Mathematical Sciences at Stony Brook, the Mathematisches ForschungsInstitut at ETH in Zurich, the Forschungszentrum BiBoS at the Universitat Bielefeld (S. T.), and the Physics Institute in Technion, Haifa (E. G.). We are grateful to these institutions for their hospitality and financial support. We thank Elsa Troubetzkoy for drawing the figures. The work of E. G. was also partially supported by NSF Grants DMS-9013220, DMS-9400295.

2 Directional billiard flows for polygons

2.1 Surface associated with a polygon

Let P be a (closed, connected) Euclidean polygon, and let $a_1, \ldots, a_p$ be its sides. Let G be any group with a set, $\Sigma = \{\sigma_1, \ldots, \sigma_p\}$, of generators, satisfying $\sigma_i^2 = e, 1 \leq i \leq p$. A general construction associates with the pair (P, G) an oriented surface, S, and we use notation $S = G \times P$ (see, e. g., [5]). The group G acts on S as a discrete group of diffeomorphisms, and S is tiled by the "polygons" $P_g = gP, g \in G$. We identify P with the *fundamental domain*, P_e. Thus $S/G \simeq P$, a natural isomorphism.

We denote by $O(2)$ the group of isometries of the unit circle, S^1. Let $\sigma_i \in O(2)$ be the linear part of the reflection of $\mathbf{R}^2$ about the side a_i. The group G generated by $\sigma_i, 1 \leq i \leq p$, is a subgroup of $O(2)$. The subgroup $G_0 = G \cap SO(2)$ has index two in G, and G_0 is generated by rotations by $2\alpha_k$, where $\alpha_k, 1 \leq k \leq q$, are the angles of P. Recall that a polygon P is *rational* if all angles α_k are rational multiples of π. Otherwise we say that P is *irrational*. The surface $S = S(P)$ is determined by the intrinsic geometry of P. We do not dwell on the details of the correspondence $P \to S$ here. The following proposition will suffice.

Proposition 2.1 *Let P be an arbitrary polygon, and let $S = S(P)$ be the corresponding surface.*

The surface S is closed if and only if P is a rational polygon. Then S is the conventional surface associated with the rational polygon P [26, 10].

Proof. If P is rational, our definition of $S(P)$ reduces to the conventional one (see, e. g., [10]). In general, $S = G \times P$ is compact if and only if G is finite. The latter happens if and only if there exists $N > 1$, such that $G_0 \subset (2\pi/N)\mathbf{Z}/2\pi\mathbf{Z}$. The group G_0 is generated by the angles $2\alpha_k$ of P. Hence G is finite if and only if $\alpha_k \in (\pi/N)\mathbf{Z}$ for all k. ∎

2.2 Flat structure and directional geodesic flows

We keep the notation of §2.1. The tiling, $S = \cup_{g \in G} P_g$, partitions S into *faces,*
edges, and *vertices.* The projection, $\pi : S \to P$, lifts the Euclidean structure
on P to the subset $S_2 \subset S$ of interior points of the faces. This structure
uniquely extends to a flat Riemannian metric on S, with singularities, in
general, on the set, $S_0 \subset S$, of vertices. On $S \backslash S_0$ (the set of *regular points*)
there is a special a (G, X)-structure, a la Thurston [22], with $G = X = \mathbf{R}^2$.
See also [24, 25, 15] for a theory of a closely related class of surfaces (in a
somewhat different setting). In what follows we recap the properties of flat
surfaces that we will need.

On the unit tangent bundle, $T_1(S \backslash S_0)$, there is a natural map,
$\theta : T_1(S \backslash S_0) \to S^1$, which assigns to a vector, $v \in T_1(S \backslash S_0)$, its *direction,*
$\theta(v) \in S^1$. The mapping θ is invariant (locally) under the geodesic flow on
$T_1(S)$. Let $F_\theta \subset T_1(S)$ be set of vectors with direction θ. Then F_θ is invariant
under the geodesic flow, and identifying F_θ with S, we obtain the family of
directional geodesic flows, T_θ^t, on S. The flowlines of T_θ^t are the *geodesics in
direction* θ. We denote by F_θ the corresponding foliation of S; its *singular
leaves* begin and/or end at the vertices. Since S_0 is (at most) countable, so
is the set of singular leaves of F_θ.

The geodesic flow of S preserves the Lebesgue measure, μ, on $T_1(S)$, and
the flows T_θ^t preserve the area form, ω, on S. Note that if P is irrational, then
$\omega(S) = \infty$. Let λ be the Lebesgue measure on S^1. The family $T_\theta^t, \theta \in S^1$, of
directional flows decomposes the geodesic flow of S.

2.3 Directional billiard flows of a polygon

The *billiard flow,* B^t, in a polygon, P, is modeled on the frictionless motion
of an elastic billiard ball (point mass) inside P [11]. The phase space, W, of
the flow consists of unit vectors, v, whose footpoints, $x(v)$, are in P, with the
usual identifications on the boundary, ∂P. We set $Int(P) = P \backslash \partial P$. Thus
$V = Int(P) \times S^1$ is open and dense in W. The standard (Liouville) invariant
measure, μ, on W decomposes as the product of Lebesgue measures, ω on
$Int(P) \subset \mathbf{R}^2$, and λ on S^1.

The identifications on ∂P are not, in general, defined at the vertices of P, which causes the singularities of the billiard flow. For $\theta \in S^1$ let $V_\theta \subset V$ be the set of unit vectors with direction θ. Then $V = \coprod_{\theta \in S^1} V_\theta$, a disjoint union. For $\theta \in S^1$ set $\{\theta\} = G\theta$, using the natural action of $G \subset O(2)$ on S^1, and let G_θ be the stabilizer of θ. Let $W_\theta \subset W$ be the image of $\cup_{\alpha \in \{\theta\}} V_\alpha$ under the identifications on ∂P.

Proposition 2.2 *Let $\theta \in S^1$, and let the notation be as above.*
1. The set $W_\theta \subset W$ is invariant under the billiard flow.
2. The projection $\pi : S \to P$ induces a mapping onto $\pi_\theta : S \to W_\theta$ which semiconjugates the flows T_θ^t with $B^t|_{W_\theta}$.
3. The semiconjugacy $\pi_\theta : S \to W_\theta$ is a normal covering with the group G_θ of deck transformations.

Proof. Let $v = (x, \theta) \in V$ Then the set of directions of the billiard orbit through v is contained in $G\theta$. This proves 1). Assertions 2), 3) are of geometric nature, and are special cases of geometric relations for translation surfaces [15]. They are proved using the standard facts about groups generated by reflections [2]. ∎

Definition 2.1 *Let the notation be as above. The flow on W_θ induced by the billiard flow of P is the* directional billiard flow, *in direction θ. We denote it by B_θ^t. Then W_θ is the phase space of B_θ^t.*

We say that a direction, $\theta \in S^1$, is *regular* if $G_\theta = \{e\}$. We denote by $\Theta_P \subset S^1$ the set of irregular directions. If $\alpha \in \Theta_P$, then $\alpha = g\theta_k$ where θ_k is the direction of a side of P. Since Θ_P consists of a finite number of G-orbits, it is at most countable. The following is immediate from Proposition 2.2.

Corollary 2.1 *For all regular directions $\theta \in S^1$, the map $\pi_\theta : S \to W_\theta$ conjugates the flows T_θ^t and B_θ^t.*

Thus for all but a countable (at most) set of directions θ there is an isomorphism between the directional geodesic flow on S, and the directional billiard flow on P, in direction θ. For this reason, in what follows we simply speak of the *directional flows*, and use notation T_θ^t.

Let P be a rational polygon, and let N be the least common denominator of the angles of P. Then $|\Theta_P| = 2N$, and the directional flows $T_\theta^t, 0 <$

$\theta < \pi/N$, yield a decomposition of the billiard flow of P. Accordingly, the phase space W of the billiard flow is the disjoint union of closed surfaces $W_\theta, 0 \leq \theta \leq \pi/N$, and $W_\theta \simeq S$ for $0 < \theta < \pi/N$. These facts are well known [10], see [18] for the ergodic theory of the directional flows in a rational polygon. Let now P be irrational. Then $|\Theta_P| = \infty$, and for $\theta \in S^1 \backslash \Theta_P$ the noncompact surface $W_\theta \simeq S$ is dense in W. The directional flows $T_\theta^t, \theta \in S^1$, do not provide a decomposition of the billiard flow B^t of P, and the relation between the dynamics of T_θ^t and B^t is more complicated. The following section contains a few results in this direction.

3   Dynamics of the directional flows

To simplify the exposition below we adopt the following conventions. Let (X, μ) be a measure space, possibly with $\mu(X) = \infty$. For measurable sets $Y_1, Y_2 \subset X$ we write $Y_1 = Y_2$ if they differ by a set of measure zero. A measurable set $Y \subset X$ is nontrivial if $Y \neq \emptyset, X$. A decomposition $X = \coprod_{i=1}^k Y_i$ is nontrivial if at least one set Y_i is nontrivial. A measure preserving flow T^t on (X, μ) is ergodic if X does not admit a nontrivial invariant set.

Let P be an arbitrary polygon. If P is rational, then T_θ^t is ergodic for Lebesgue almost all θ [18], hence $B^t = \int_{0<\theta<\pi/N} T_\theta^t d\lambda(\theta)$ is the ergodic decomposition. In the irrational case the situation is quite different.

Proposition 3.1 *Let P be an irrational polygon. 1. If the billiard flow of P is not ergodic, then the directional billiard flows T_θ^t are not ergodic for Lebesgue almost all θ. 2. If the billiard flow of P is ergodic, then the set $\Theta \subset S^1$ of directions θ, such that T_θ^t is ergodic, is either trivial, or nonmeasurable.*

Proof. Let $X \subset W$ be a nontrivial invariant set. For any $\alpha \in S^1$ the set $X \cap V_\alpha$ is naturally identified with a subset of P. By Fubini's theorem, for almost all α the set $X \cap V_\alpha$ is ω-measurable, the function $\omega(X \cap V_\alpha)$ on S^1 is λ-measurable, and

$$\mu(X) = \int_{S^1} \omega(X \cap V_\alpha) d\lambda(\alpha). \tag{1}$$

The function $\alpha \to \omega(X \cap V_\alpha)$ satisfies $0 \leq \omega(X \cap V_\alpha) \leq \omega(P)$, where $\omega(P)$ is the area of P. Set $A = \{\alpha : \omega(X \cap V_\alpha) = 0\}, B = \{\alpha : \omega(X \cap V_\alpha) =$

$\omega(P)\}$, and $C = \{\alpha : 0 < \omega(X \cap V_\alpha) < \omega(P)\}$. Then $A, B, C \subset S^1$ are measurable, and $S^1 = A \amalg B \amalg C$. By eq.(1), $A, B \neq S^1$. Suppose first the decomposition $S^1 = A \amalg B \amalg C$ is trivial. Then $C = S^1$. Recall that $G \subset O(2)$ contains irrational rotations. Thus for almost every $\theta \in S^1$ we have $\{\theta\} \subset C$. Hence $X_\theta = \cup_{\alpha \in \{\theta\}} X \cap V_\alpha \subset W_\theta$ is a nontrivial T_θ^t-invariant set. If the decomposition $S^1 = A \amalg B \amalg C$ is nontrivial, then at least two of the sets A, B, C are nontrivial. Suppose A, B are nontrivial. Then for almost all θ the orbit $\{\theta\}$ intersects both A and B (by infinite sets). Therefore X_θ is a nontrivial T_θ^t-invariant set in W_θ. The other possible cases are disposed of in the same way, proving assertion 1.

To prove assertion 2, let $\theta \in \Theta$ be arbitrary, and let $\eta = g\theta, g \in G$. By the material of §2, g induces an isomorphism of directional flows T_θ^t, T_η^t. Therefore the set Θ is G-invariant. Since G contains irrational rotations, Θ is either trivial or nonmeasurable. ∎

Proposition 3.1 has an obvious but useful Corollary.

Corollary 3.1 *Let P be an irrational polygon. If the directional flows T_θ^t are ergodic for $\theta \in \Theta$, a measurable set, $\mu(\Theta) \neq 0$, then the billiard flow of P is ergodic.*

For many questions of billiard dynamics it is convenient to pass from the billiard flow to the *billiard map* (see [11, 12, 3, 21]). It is the first return map induced by the billiard flow on the standard cross-section, $V \subset W$. (Warning: the symbol V previously had a different meaning!). We use notation $b : V \to V$ for the billiard map. The cross-section V has two representations, and we use both in what follows. The *ray representation* identifies V with the space of oriented lines (rays for brevity) in the plane, intersecting the billiard table P. Let $\rho \in V$, and let $\{s, s_1\} = \rho \cap \partial P$ be a pair of consecutive intersection points. Reflecting ρ about ∂P at s_1 by the usual law of the geometric optics, we obtain $\rho_1 = b(\rho) \in V$. In the *vector representation*, elements of V are the unit vectors, v, with footpoints in ∂P, directed inward. Parametrizing ∂P by the arclength, $0 \leq s < |\partial P|$, we introduce coordinates, $v = v(s, \theta)$, on V. Here $0 \leq \theta \leq \pi$ is the angle between v and ∂P. Thus $V = \partial P \times [0, \pi]$. The natural isomorphism of the two representations of V defines the billiard map, $b(s, \theta) = (s_1, \theta_1)$ in the vector realization. The Lebesgue measure on W induces the invariant measure $d\mu = \sin \theta ds d\theta$ on V.

Let now P be a polygon, and let $a_1, \ldots, a_p$ be the sides of P. We code a billiard trajectory by the sequence of sides it hits [11]. More precisely, let $\Sigma = \{(i_0, i_1, \ldots, i_n, \ldots) : 1 \leq i_k \leq p\}$ be the one-sided shift on p symbols. For $v \in V$, let $a_{i_0}, a_{i_1}, \ldots, a_{i_n}, \ldots$ be the sequence of sides the forward billiard trajectory of v hits. The (forward) coding map, $\xi : V \to \Sigma$, is given by $\xi(v) = (i_0, i_1, \ldots, i_n, \ldots) \in \Sigma$. The map ξ sends V onto a *subshift*, $\Sigma_P \subset \Sigma$, and semiconjugates the billiard map with the *left shift transformation*, $\tau(i_0, i_1, \ldots, i_n, \ldots) = (i_1, \ldots, i_n, \ldots)$, on Σ. Note that the coding is not defined on $v \in V$ whose forward trajectory hits a corner of P. This is a countable union of intervals in V, in particular, it has measure zero.

The coding map, $\xi : V \to \Sigma_P$, is, essentially, one-to-one [8], but not much is known about Σ_P, for general polygons. Let $\Sigma' \subset \Sigma$ be an arbitrary subshift. Denote by $\Sigma'(n)$ the set of words, $w = (i_0, i_1, \ldots, i_n)$, of length n in Σ'. The standard measure of the *complexity* of Σ' is the *growth rate* of the sequence $|\Sigma'(n)|$, as $n \to \infty$. This growth rate is subexponential for any P [17, 13, 14], which has a number of implications for polygonal billiards, in particular, *zero topological entropy* [17, 13, 14, 8].

A long standing conjecture says that for any polygonal billiard the growth rate above is at most polynomial. This is known only for rational polygons [11], and in fact, that growth rate is quadratic [19]. The framework of directional billiard flows suggests the notion of *directional complexity* for polygonal billiards. Choosing a reference direction, we identify the circle of directions with $[0, 2\pi)$. For $\theta \in [0, 2\pi)$ let $V_\theta \subset V$ be the set of rays in direction θ. Set $\xi(V_\theta) = \Sigma_P^\theta \subset \Sigma_P$. The growth rate of the sequence $|\Sigma_P^\theta(n)|$, as $n \to \infty$, measures the complexity of the billiard in P, in direction θ. A. Katok has pointed out the following fact to us.

Theorem 3.1 (A. Katok) *Let P be an arbitrary polygon. For any direction θ the sequence $|\Sigma_P^\theta(n)|$ grows at most polynomially in n.*

Theorem 3.1 is a direct consequence of Proposition 3.2 below. It gives an explicit bound on $|\Sigma_P^\theta(n)|$, which is better then the bound outlined by A. Katok. Let $a_1, \ldots, a_p$ be the sides of P. The angle, $0 < \alpha(a_i, a_j) < \pi$, between two nonparallel sides is the amount of rotation in the positive direction that makes the line through a_j parallel to the line through a_i. Thus $\alpha(a_j, a_i) = \pi - \alpha(a_i, a_j)$. Let $q \leq p(p-1)$ be the number of distinct angles thus obtained. Let r be the number of rational angles ($\alpha = \pi m/n$) among them, and let s be the number of irrational angles, $q = r + s$. Let $N = N(P)$ be the least common denominator of the rational angles above.

Proposition 3.2 *Let the notation be as above. Then for any direction θ, we have*

$$|\Sigma_P^\theta(n)| \leq pNn(1 + \frac{n}{2})^s. \tag{2}$$

Proof. Let $G \subset O(2)$ be the group generated by the reflections $\sigma_i, 1 \leq i \leq p$ (see §2). Any $g \in G$ has a representation $g = \sigma_{i_1} \cdots \sigma_{i_k}, k \geq 0$. Set $G(n) = \{g = \sigma_{i_1} \cdots \sigma_{i_k}, 0 \leq k \leq n\}$. Then $\{e\} = G(0) \subset G(1) \subset \cdots \subset G(n) \subset \cdots$, and $G = \cup_{n=0}^\infty G(n)$. We will estimate $|G(n)|$.

The group $H = G \cap SO(2)$ is a normal subgroup of index 2 in G. Let $\alpha_1, \ldots, \alpha_q$ be the angles of P introduced above, where $\alpha_1, \ldots, \alpha_s$ are irrational, and $\alpha_{s+1}, \ldots, \alpha_q$ are rational. For each α_k in this collection we choose an element $\rho_k = \sigma_i \sigma_j$ so that $\rho_k \in H$ is the rotation by $2\alpha_k$. Then $\rho_1, \ldots, \rho_q$ generate the abelian group H. The group I generated by $\rho_1, \ldots, \rho_s$ is infinite, the group F, generated by $\rho_{s+1}, \ldots, \rho_q$, is finite cyclic of order N, and $H = F \times I$. Set $H(n) = H \cap G(n), I(n) = I \cap G(n)$.

Any element $g \in I(2m)$ has a representation (nonunique, in general) $g = \rho_1^{k_1} \cdots \rho_s^{k_s}$, with $0 \leq k_i \leq m$. Thus $|I(2m)| \leq (1 + m)^s$. Since $H(n) \subset F \times I(n)$, we have

$$|H(2m)| \leq N(1 + m)^s. \tag{3}$$

Let σ be any of the generators $\sigma_1, \ldots, \sigma_p$. Then $G = H \amalg \sigma H$, and $\sigma H \cap G(n) \subset \sigma H(n + 1)$, yielding $|G(n)| \leq 2|H(n + 1)|$. Since $|H(2m + 1)| = |H(2m)|$, the inequality above and eq.(3) yield

$$|G(n)| \leq 2N(1 + \frac{n+1}{2})^s. \tag{4}$$

We consider T_θ^t as a flow on the surface S (see §2), tiled by the polygons $P_g, g \in G$. The billiard trajectories defined by $v \in V_\theta$ correspond to the flow lines of T_θ^t, emanating from P_e. Each flow line, ℓ, is divided into "links", the intersections of ℓ with the "polygons" P_g. The coding of ℓ corresponds to the sequence of edges of the tiling that ℓ crosses along the way. This establishes a one-to-one correspondence between the codes, $w \in \Sigma_P^\theta(n)$, and the flow tubes formed by the n-link flow lines above. The boundary lines of these flow tubes correspond to the flow lines that run into a vertex of the tiling on

k-th link, $k \leq n$, for the first time after leaving P_e. The number of available vertices is bounded by $p|G(n-1)|$, because each tile P_g has p vertices.

For each vertex v there is a unique flow line of T_θ^t emanating from P_e, and winding up at v, before returning to P_e. Therefore to compute the contribution of the flow line ℓ that winds up at v, we examine the last n links of ℓ, and sum up the number of times ℓ has returned to P_e before hitting v. Since it takes at least two links to return to P_e, each v contributes at most $n/2$ flow tubes. Thus the number of elements in $\Sigma_P^\theta(n)$ is bounded above by $\frac{1}{2}pn|G(n-1)|$. It remains to substitute this into eq.(4). ∎

Corollary 3.2 *Let P be a rational polygon with p sides, and let $G \subset O(2)$ be the finite group generated by reflections in the sides. Then for any direction θ we have*

$$|\Sigma_P^\theta(n)| \leq \frac{p|G|}{2}n. \tag{5}$$

Proof. We specialize eq.(2). In this case $s = 0$, and $|G| = 2N$ [10]. ∎

Remarks. With additional assumptions on the angles of (irrational) P, the estimate (2) can be improved. For rational polygons the subshifts Σ_P^θ are generated by *interval exchange transformations*. These have been extensively studied [23, 12]. For convex rational polygons P. Hubert has found an explicit formula for a quantity closely related to $|\Sigma_P^\theta(n)|$ [16]. Hubert's formula implies for these polygons:

$$|\Sigma_P^\theta(n)| \leq \frac{(p-2)|G|}{2}n + |G|.$$

4 Strong recurrence and periodicity

4.1 Directional recurrence

Various versions of the notion of recurrence are used in dynamics [6]. The billiard flow, $B^t : W \to W$, of a polygon preserves the Lebesgue measure, μ. Since $\mu(W) < \infty$, the Poincare recurrence theorem [6] insures that with probability one an element $w \in W$ returns under $B^t, t > 0$, arbitrarily close to itself. Set $w = (z, \theta)$, where $z = (x, y) \in P$, and $\theta \in S^1$ is the direction of w. Let $B^t w = w(t) = (z(t), \theta(t))$. Thus, for almost all w the distances $|z(t) - z|, |\theta(t) - \theta|$ are arbitrarily small, for sufficiently large t.

Let P be a rational polygon, and let $T_\theta^t, \theta \in S^1$, be the directional flows. By the same argument, applied to T_θ^t, we obtain that for t sufficiently large, $|z(t) - z|$ is arbitrarily small, and $\theta(t) = \theta$. Thus for any direction θ, an element $w = (z, \theta) \in W$ returns with probability one arbitrarily close to itself, with exactly the same direction. This simple fact has remarkable consequences for the billiard dynamics in rational polygons [1, 9]. The argument above fails for irrational polygons, hence the definition below.

Definition 4.1 *Let P be an arbitrary polygon, and let $\omega = dxdy$ be the Lebesgue measure on P. A direction $\theta \in S^1$ is recurrent, if under B^t a point $w = ((x, y), \theta)$ returns, with probability one (with respect to ω), arbitrarily close to $(x, y) \in P$, and in the same direction θ. If every direction is recurrent, we say that (the billiard in) P is strongly recurrent.*

By preceding remarks, any rational polygon is strongly recurrent. Let P be irrational, and let S be the corresponding surface. The directional billiard flows T_θ^t, viewed as flows on S, preserve the Lebesgue measure, ω, on S. Since $\omega(S) = \sum_{g \in G} \omega(P_g) = \infty$, Poincare recurrence theorem does not apply. In particular, the flow lines of T_θ^t, starting from P_e, may not return to P_e. If this happens with positive probability, then the direction θ is not recurrent.

Definition 4.2 *A billiard trajectory (in a polygon P), $w(t) = (z(t), \theta(t)), -\infty < t < \infty$, is perpendicular if there is a side, a, of P, and a time moment, t_0, such that $z(t_0) \in a$, and $\theta(t_0)$ is orthogonal to a.*

Let $a_1, \ldots, a_p$ be the sides of P, and let $\theta_1, \ldots, \theta_p \in S^1$ be the corresponding perpendicular directions (inward). They form a finite set, $\Theta_P \subset S^1, |\Theta_P| \leq p$, of *perpendicular directions*. Let a be an arbitrary side, and let $\theta_a \in \Theta_P$ be the corresponding direction. We identify $a \subset \partial P$ with its image, $V_a \subset V_{\theta_a}$, under the injective mapping, $s \to v(s, \theta_a) = v(s)$, and let $ds/|a|$ be the induced probability measure on V_a. Regarding vectors $v \in V_a$ as initial data for billiard trajectories, $w_v(t), w_v(0) = v$, we speak of the *probability that the perpendicular orbit, starting from $s \in a$, is periodic.*

Proposition 4.1 *Let P be a polygon, and let $a \subset \partial P$ be a side, such that its orthogonal direction, θ_a, is recurrent. Then the orbits, perpendicular to the side a, are periodic, with probability one.*

Proof. Let $\theta \in S^1$ be an arbitrary direction, and let $V_\theta \subset V$ be the set of rays in direction θ. The euclidean length in the direction, orthogonal to θ, induces a measure on V_θ. In the ray representation, this measure is given by the *width of a beam* in direction θ (fig.1). We will call it the *width measure* on V_θ, and denote this measure by w_θ, or simply by w, if there is no danger of confusion.

Recall that the standard measure on V is given, in the vector representation, by $\sin\theta ds d\theta$. The measure w on V_θ corresponds to the 1-form $\sin\theta ds$, hence it is well defined for any billiard table. Its usefulness hinges on the elementary fact, that for polygonal tables the induced measure on the sets $\cup_{n\geq0}b^n(V_\theta)$ is b-invariant [13, 14, 4]. The (σ-finite) measure w on $\cup_{n\geq0}b^n(V_\theta)$ is, in general, infinite.

Now, suppose θ is a recurrent direction. Then for w-almost all $v \in V_\theta$, the *first return map*, $r_\theta(v) = b^{n(v)}v \in V_\theta$ is defined, and $r_\theta : V_\theta \to V_\theta$ preserves the measure w. Note that $w(V_\theta) < \infty$, it depends on P and θ.

Set $\theta = \theta_a$. By Poincare recurrence theorem, the map r_θ induces a first return map, $r_a : V_a \to V_a$, preserving $w|_{V_a}$. Identifying V_a with a, and using that $w = cds, c > 0$, we regard r_a as an automorphism of the measure space (a, ds). Thus for almost all $s \in a$ the perpendicular orbit staring at s returns to a, at $s_1 = r_a(s)$, for the first time, after $n(s)$ bounces. If $s_1 = s$, then the orbit in question is periodic, of period $n(s)$ (i. e., consists of $n(s)$ links). If $s_1 \neq s$, then the orbit reflects at s_1 and "backtracks" to s after $n(s)$ bounces. Thus we get a periodic orbit of period $2n(s)$. ∎

Remarks. If P is a rational polygon, the assumption of Proposition 4.1 is satisfied for any side a. Besides, the map $r_a : a \to a$ is an interval exchange. Thus, but for a finite number of singular ones, the perpendicular orbits starting at $s \in a$ are periodic, and $n(s)$ takes on a finite number of values [1, 9].

Corollary 4.1 *Let P be a polygon, and let Θ_P be the set of perpendicular directions. 1. If the directions $\theta \in \Theta_P$ are recurrent, then the perpendicular orbits in P are periodic with probability one. 2. If P is strongly recurrent, then the perpendicular orbits are periodic with probability one.*

Proof. The latter assertion is contained in the former. That one was essentially proved by the argument of Proposition 4.1. We leave it to the reader to supply the details. ∎

Recall that the *time reversal involution, $\sigma : V \to V$,* acts in the ray representation by reversing the directions of the rays. It satisfies $\sigma^2 = id$, $\sigma b \sigma = b^{-1}$. The latter implies that σ sends billiard orbits into billiard orbits. Proposition 4.1 and Corollary 4.1 stem from the fact that σ acts trivially on perpendicular orbits. This observation leads to a simple and useful generalization of Proposition 4.1 and Corollary 4.1. The material below partially overlaps with the results of Stepin and Vorobetz, contained in a paper submitted to the Russian Mathematical Surveys [20].

Definition 4.3 *An automorphism $g : V \to V$ of the billiard map phase space is a* time reversal symmetry, *if $gbg^{-1} = b^{-1}$.*

Lemma 4.1 *Let $g : V \to V$ be a time reversal symmetry, and let $V^g = \{v \in V : gv = v\}$ be the fixed point set. Then the set $R_g = \cup_{n>0} V^g \cap b^n(V^g)$ consists of periodic points.*

Proof. For any $v_1 \in R_g$ there is $n > 0$, and $v \in V^g$, such that $v_1 \in V^g, v_1 = b^n(v)$. Then $v_1 = gv_1 = gb^n v = gb^n g^{-1} v = b^{-n} v$. Combining this with $v_1 = b^n v$, we obtain $b^{2n} v = v$. ∎

Proposition 4.2 *Let g be a time reversal symmetry, and let the notation be as above. Let μ be a probability measure on V^g, such that for μ-almost every $v \in V^g$ there is $n > 0$ satisfying $b^n v \in V^g$, and such that the first return map, $r(v) = b^n v$, preserves μ. Then the elements $v \in V^g$ are periodic, with probability one.*

Proof. By Lemma 4.1, it suffices to show $\mu(R_g) = 1$. This is clear from the assumptions. ∎

If ℓ is a line in the plane, we denote by $V_\ell \subset V$ the set of rays $v \in V$ intersecting $\ell \cap P$ at the right angle. Let w be the normalized width measure on V_ℓ. A billiard orbit is *perpendicular to ℓ* if it crosses $\ell \cap P$ at the right angle.

Corollary 4.2 *Let ℓ be an axis of symmetry of a polygon P. If the direction, orthogonal to ℓ is recurrent, then the billiard orbits perpendicular to ℓ are periodic, with probability one.*

Proof. Let $\tau : V \to V$ be the reflection about ℓ, and set $g = \sigma\tau = \tau\sigma$ (fig. 2). Since τ commutes with the billiard map, g is a time reversal symmetry. By fig. 2, $V^g = V_\ell$. Let θ be any of the two orthogonal directions to ℓ. Since θ is recurrent, Proposition 4.2 applies (with $\mu = w$). ∎

4.2 Strongly recurrent polygons

Definition 4.4 *A polygon P is a* generalized parallelogram *if there are two lines, $\lambda \neq \mu$, in the plane, such that every side of P is parallel to either λ or μ.*

Let P be a generalized parallelogram. Then there is a unique $0 < \alpha \leq \pi/2$, such that the values of the angles $\alpha(a_i, a_j)$ between the sides of P are $0, \alpha, \pi - \alpha$. In fact, this property is equivalent to Definition 4.4. Thus P is rational if and only if $\alpha(P) = \pi m/n$.

Theorem 4.1 *Any generalized parallelogram is strongly recurrent.*

Proof. If P is rational, there is nothing to prove, thus $\alpha = \alpha(P)$ is irrational in what follows. Let λ, μ be as in Definition 4.4. We divide the sides of P into two classes: a side a is in λ-class, $a \in C_\lambda$, if a is parallel to λ. Otherwise $a \in C_\mu$. We assume without loss of generality that if $a \in C_\lambda, b \in C_\mu$, then $\alpha(a,b) = \alpha, \alpha(b,a) = \pi - \alpha$. Let L (M) be the total length of the sides in C_λ (C_μ). Denote by $\sigma_\lambda, \sigma_\mu \in O(2)$ the reflections about λ, μ. Then $\sigma_\lambda\sigma_\mu \in SO(2)$ is the rotation by 2α.

Let now θ be an arbitrary direction, and let $\cup_{n \geq 0} b^n(V_\theta) \subset V$ be the corresponding invariant set. Let w be the "width" measure on $\cup_{n \geq 0} b^n(V_\theta)$. It suffices to consider the set $Z = \cup_{n \geq 0} b^{2n}(V_\theta)$, invariant under b^2. Set $Z^{(k)} = Z \cap V_{\theta+2k\alpha}$. Then $Z = \coprod_{k=-\infty}^\infty Z^{(k)}$, and $b^{2n}(V_\theta) = \coprod_{k=-n}^n b^{2n}(V_\theta) \cap Z^{(k)}$.

Let $n > 0$. The set $Z_n = Z^{(n)} \cap b^{-2}(Z^{(n+1)})$ consists of rays v, in direction $\theta + 2n\alpha$, such that $b^2(v)$ has direction $\theta + 2(n+1)\alpha$. Denote by $\lambda, \mu \in S^1$ the directions of lines λ, μ (the *mod* π ambiguity will not interfere in what follows). The rays $v \in Z_n$ are reflected, first, by the sides of class C_μ, and then by the sides of class C_λ. By elementary geometry (fig. 3)

$$w(Z_n) \leq M|\sin(\theta + 2n\alpha - \mu)|. \tag{6}$$

Let now $X \subset V_\theta$ be the set of rays that don't return to the direction θ. Note that b^2 changes directions by $\epsilon \cdot 2\alpha, \epsilon = 0, \pm 1$. Thus $X = X_+ \coprod X_-$, where X_+ (X_-) consists of $v \in X$, such that for all $n > 0$, the rays $b^{2n}(v)$ have directions $\theta + 2k\alpha$, $(\theta - 2k\alpha), k > 0$. The sets X_+, X_- are measurable, and we need to show that $w(X_-) = w(X_+) = 0$. Consider X_+. The sets $X_n = b^{2n}(X_+), n > 0$, are disjoint, and setting $X_n^{(k)} = X_n \cap Z^{(k)}$, we have $\cup_{n>0} b^{2n}(X_+) = \coprod_{n>0} \coprod_{k=1}^n X_n^{(k)}$.

Let $v \in X_+$ be such that the sequence of directions of $b^{2n}(v)$ is "bounded", i. e., $\{b^{2n}(v), n > 0\} \subset \cup_{k=1}^K Z^{(k)}, K < \infty$. Since $w(\cup_{k=0}^K Z^{(k)}) \leq \sum_{k=0}^K w(V_{\theta+2k\alpha}) < \infty$, Poincare recurrence theorem applies. Thus such $v \in X_+$ form a set of measure zero, and we disregard them in what follows.

Fix $K > 0$. For any $v \in X_+$ there is a unique "first time", $n = n(v, K)$, such that $b^{2n}(v) \in Z^{(K)}$, and $b^{2n+2}(v) \in Z^{(K+1)}$. Therefore $v \to b^{2n}(v)$ defines a measure preserving embedding of X_+ into Z_K. Hence $w(X_+) \leq w(Z_K)$.

Since α is irrational, for any $\epsilon > 0$ there exists $K \in \mathbf{Z}_+$ such that for $n = K$ the right hand side of eq. (6) is less than ϵ. By eq. (6) and preceding remarks, $w(X_+) < \epsilon$ for any $\epsilon > 0$. Thus $w(X_+) = 0$. The equality $w(X_-) = 0$ is proved the same way. ∎

Let P be any polygon, let a be a side of P, and let $s_a \in Iso(\mathbf{R}^2)$ be the reflection about the line through a. The polygon $P_1 = P \cup s_a(P)$ is the *unfolding* of P about a. If P_1 has selfintersections, then P_1 is a "two-sheeted" polygon. In any event, the polygon P_1 is *tiled* by two copies of P. To iterate the unfolding, we reflect one of these two copies about a side, and so on. This way we obtain an infinite countable set, $\mathcal{U}(P)$, of (multiple-sheeted, in general) polygons, which are tiled by isometric copies of P.

Definition 4.5 *Two polygons, Q, Q', are nearest neighbors if $Q, Q' \in \mathcal{U}(P)$, for some P, or if $\mathcal{U}(Q) \cap \mathcal{U}(Q') \neq \emptyset$. We use notation $Q \sim Q'$ for the unique equivalence relation on the set of multiple-sheeted polygons, generated by the notion of nearest neighbors. If $Q \sim Q'$, we say that Q, Q' are weakly equivalent.*

Each weak equivalence class contains a countable number of polygons. For instance, the equivalence class of the unit square contains polygons *drawn on the square lattice* [10]. It also contains all rectangles with rational side lengths.

Theorem 4.2 *Any polygon weakly equivalent to a generalized parallelogram is strongly recurrent.*

Proof. Let $P_1 \sim P_2$, and let S_1, S_2 be the corresponding flat surfaces. Then S_1, S_2 either cover the same surface, $S = S(Q)$, or they have a common covering surface, $S' = S(Q')$. In either case the coverings are finite. By Poincare recurrence theorem, directional recurrence is preserved under finite coverings. Therefore a polygon, P, is strongly recurrent if and only if all polygons $Q \sim P$ are strongly recurrent. Theorem 4.1 implies the claim. ∎

Corollary 4.3 [4] *Right triangles are strongly recurrent.*

Proof. Let P be a right triangle. Doubling P about one of the sides, we obtain an isoceles triangle, P_1. Doubling P_1 about its base, we obtain a rhombus, which is a special parallelogram. The claim follows, by Theorem 4.2. ∎

Corollary 4.4 *Let Δ be an isoceles triangle. Then almost every billiard orbit with a link, parallel to the base of Δ, is periodic.*

Proof. The direction of the base is orthogonal to the axis of symmetry. By Corollary 4.2, it suffices to show this direction is recurrent. The triangle Δ is weakly equivalent to the obvious right triangle, Δ_1, and the direction, θ, is recurrent for Δ_1. Therefore θ is a recurrent direction for Δ (see the Proof of Theorem 4.2). ∎

Remarks. Let P be a polygon, and let S be the associated "flat" surface. The notions of directional and strong recurrence can be formulated in terms of the directional flows, T_θ^t, and the geodesic flow of S. Theorem 4.2 and Corollary 4.4 are special cases of more general results on recurrence, in the framework of *coverings of translation surfaces* [15].

5   Applications: mechanical system of two elastic point masses

The system in question consists of two elastic point masses, $0 < m_1 \leq m_2 < \infty$, confined to move in the "box", $0 \leq x \leq L$, of length L. The particles collide with each other, and with the "walls", $\{x = 0, L\}$, elastically. This elementary physical system is isomorphic to the billiard in a right triangle, $\Delta(m_1/m_2, L)$ [18, 11, 21]. Hence every theorem about the billiard orbits in right triangles is equivalent to a fact about the mechanics of two colliding elastic particles. For instance, by [18], for a dense set of mass ratios, $0 < r = m_1/m_2 \leq 1$, the system is ergodic. Many questions remain open. Thus, it is not known if the set of "ergodic mass ratios", $E \subset (0,1)$, has positive Lebesgue measure.

For convenience of the reader we outline the isomorphism, mentioned above. The physical phase space of the system consists of positions, $0 \leq x_1 \leq x_2 \leq L$, and velocities, $v_1 = dx_1/dt, v_2 = dx_2/dt$. The velocities are constant in time intervals between collisions. There are 3 types of collisions, corresponding to the 3 equations: $x_1 = 0, x_2 = L$, and $x_1 = x_2$. Each type of collisions corresponds to an instantaneous change of velocities. At collisions of the first (second) type we have $v_1 \to -v_1$ ($v_2 \to -v_2$). These are the collisions between one of the particles and a wall. They change the *total momentum*, $p = m_1 v_1 + m_2 v_2$, but not the *total energy*, $E = \frac{1}{2} m_1 v_1^2 + \frac{1}{2} m_2 v_2^2$. Collisions of the third type are the collisions between particles. The velocity transformation $(v_1, v_2) \to (v_1', v_2')$ preserves both the total momentum and the total energy. By an elementary calculation, the transformation is linear: $v_1' = a v_1 + b v_2, v_2' = c v_1 + d v_2$. This linear transformation is the reflection of the (v_1, v_2)-plane about the line $p = 0$, orthogonal with respect to the quadratic form E.

We set $y_1 = \sqrt{m_1} x_1, y_2 = \sqrt{m_2} x_2$. The configuration space becomes $0 \leq y_1/\sqrt{m_1} \leq y_2/\sqrt{m_2} \leq L$. This is a right triangle, $\Delta = \Delta(m_1, m_2, L)$, in the (y_1, y_2)-plane (fig. 4), with the sides $\sqrt{m_1} L \leq \sqrt{m_2} L$, and the hypothenuse $\sqrt{m_1 + m_2} L$. The smallest angle of Δ is $\alpha = \arctan \sqrt{m_1/m_2}$. Set $u_1 = \dot{y}_1 = \sqrt{m_1} v_1, u_2 = \dot{y}_2 = \sqrt{m_2} v_2$. The collision transformations $(v_1, v_2) \to (v_1', v_2')$ become the orthogonal reflections of $\mathbf{R}^2$ (with respect to the standard metric $u_1^2 + u_2^2$) about the lines ℓ_1, ℓ_2, ℓ_3 parallel to the vertical, and the horizontal sides of Δ, and the hypothenuse, respectively. The corresponding normal directions are given by the equations $u_2 = 0, u_1 = 0$, and $\sqrt{m_1} u_1 + \sqrt{m_2} u_2 = 0$. The normal unit vectors (directed inward) are $\vec{n_1} = (1, 0)$, $\vec{n_2} = (0, 1)$, $\vec{n_3} = (-\sqrt{m_2}, \sqrt{m_1})/\sqrt{m_1 + m_2}$. For future reference we summarize the assertions above as a Proposition.

Proposition 5.1 *The transformation* $y_1 = \sqrt{m_1} x_1, y_2 = \sqrt{m_2} x_2$ *is an isomorphism of the system of two elastic masses* m_1, m_2 *in the box of length* L *onto the billiard ball dynamics in the right triangle* $\Delta(\sqrt{m_1} L, \sqrt{m_2} L)$. *Under this isomorphism the left wall* $\{x_1 = 0\}$, *and the right wall* $\{x_2 = $

$0\}$ *correspond to the longer side,* $a, |a| = \sqrt{m_2}L$, *and to the shorter side,* $b, |b| = \sqrt{m_1}L$, *respectively. Let* v_1, v_2 *be the velocities of the masses, and let* $\vec{u} = (u_1, u_2)$ *be the velocity vector of the billiard ball. Then the total momentum satisfies*

$$p = m_1 v_1 + m_2 v_2 = \sqrt{m_1} u_1 + \sqrt{m_2} u_2.$$

Now we fix the parameters $0 < m_1 \leq m_2$, $L > 0$, and consider the corresponding physical system. For a vector $\vec{v} \in \mathbf{R}^2$ let $Z_{\vec{v}}$ be the set of initial data, $((x_1, v_1), (x_2, v_2))$, with $(v_1, v_2) = \vec{v}$. The set $Z_{\vec{v}}$ is naturally isomorphic to the configuration space, $X = \{0 \leq x_1 \leq x_2 \leq L\}$, of our system. The density $2dx_1 dx_2 / L^2$ defines a probability measure on X, and, via the isomorphism $Z_{\vec{v}} \simeq X$, a probability measure on $Z_{\vec{v}}$.

Corollary 5.1 *Let* $\vec{v} \in \mathbf{R}^2$, *and let* $Z_{\vec{v}}$ *be the corresponding set of initial data. Then, with probability one, for any* $\epsilon > 0$, *and any* $T \geq 0$, *the physical trajectory* $\{z(t) : t \geq 0, z(0) = z \in Z_{\vec{v}}\}$, *starting at* z, *returns to* $Z_{\vec{v}}$ ϵ-*close to* z, *infinitely many times, for* $t \geq T$.

Proof. The assertion says that the trajectory, starting at (x_1, x_2) with velocity $(v_1, v_2) = \vec{v}$, passes by, with probability one, ϵ-close to (x_1, x_2), with the same velocity, and does it infinitely many times, after arbitrarily long time periods. By standard technique, it suffices to show the return with probability one, at least once, at some future time. This follows immediately from Proposition 5.1, and the strong recurrence of right triangles (Corollary 4.3). ∎

The trajectories of our physical system are confined to the energy manifolds, $\frac{1}{2}m_1 v_1^2 + \frac{1}{2}m_2 v_2^2 = E$, and we restrict our attention to the energy level $E = 1/2$. Let Z be the corresponding energy manifold. Up to a set of positive codimension, Z is the direct product of the configuration space X, and the velocities' ellipse, $\{\vec{v} = (v_1, v_2) : m_1 v_1^2 + m_2 v_2^2 = 1\} \subset \mathbf{R}^2$. We define 3 subsets of Z. The set $Z_1 \subset Z$ is defined by equations $x_1 = v_2 = 0$. Analogously, Z_2 is given by $x_2 = L, v_1 = 0$. The set Z_3 consists of points $((x_1, v_1), (x_2, v_2)) \in Z$, satisfying $x_1 = x_2, m_1 v_1 + m_2 v_2 = 0$. Since the velocities' ellipse intersects any line through the origin at 2 points, each Z_i is parametrized by the corresponding position variable: Z_1 by $x = x_2$, Z_2 by

$x = x_1$, and Z_3 by $x = x_1 = x_2$. The variable x runs through $(0, L)$ in each case. The density dx/L defines a probability measure on each Z_i.

The subsets $Z_i, i = 1, 2, 3$, have simple physical meanings: Z_1 (Z_2) is the set of data such that the left (right) particle is colliding with the left (right) wall, and the other particle is at rest, anywhere in the space. The data in Z_3 is as follows: the particles are colliding with each other $(x_1 = x_2)$, and the center of mass is at rest $(m_1v_1 + m_2v_2$ is the centre of mass velocity).

Corollary 5.2 *Let the notation be as above, and let $Z_i \subset Z$, $i = 1, 2, 3$, be any of the three sets. Then the physical trajectory, which passes by Z_i, is periodic with probability one.*

Proof. Immediate from the preceding discussion, Proposition 5.1, and Proposition 4.1, with Corollary 4.3. ∎

Corollary 5.3 *Through almost every point of the configuration space of the system of two elastic particles passes a periodic trajectory.*

Proof. Using Proposition 5.1, we pass to the billiard in the corresponding right triangle, $\Delta = \Delta(m_1, m_2, L)$. Let $Z_1' \subset Z_1$ be the set of periodic initial data. We identify Z_1' with a subset of the vertical side of the triangle Δ, and consider the corresponding orbits. The first link of any such orbit is horizontal (fig. 4). These links cover the set $Y_1 = \{(y_1, y_2) : y_2 \in Z_1'\} \subset \Delta$. Since Z_1' has full measure in the vertical side, Y_1 is a set of full measure in Δ. ∎

Remark. The argument above shows that, for $i = 1, 2, 3$, with probability one, there is a periodic trajectory, passing by $(x_1, x_2) \in X$, and visiting Z_i. Since a trajectory may visit all three sets Z_i, this does not imply the existence of more than one periodic trajectory passing by (x_1, x_2).

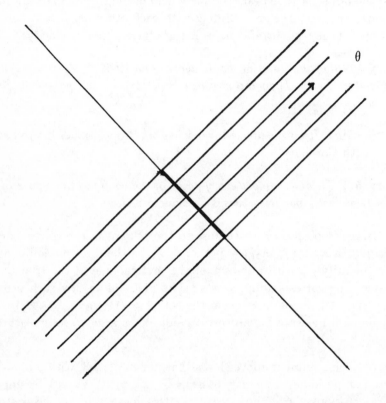

Figure 1

"Width of a beam" measure

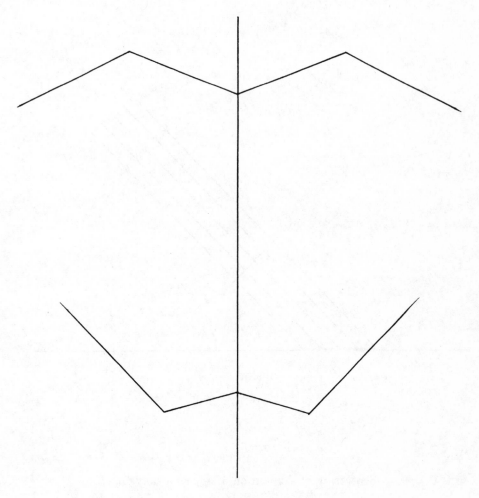

Figure 2

Polygon with an axis of symmetry

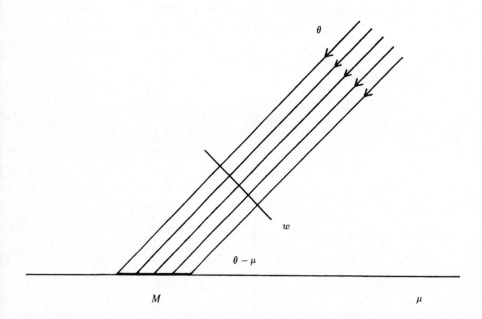

Figure 3

Scattering of a beam on a side of a polygon

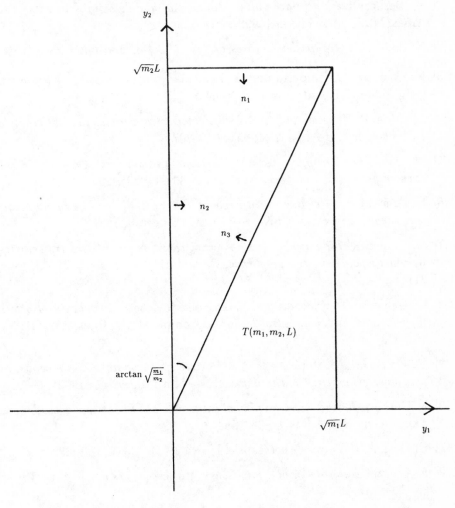

Figure 4

The right triangle corresponding to the physical system of two elastic masses

References

[1] M. Boshernitzan, *Billiards and rational periodic directions in polygons*, Amer. Math. Monthly 99 (1992), 522–529

[2] N. Bourbaki, *Groupes et algebres de Lie, Ch. 4-6*, Hermann, Paris, 1968

[3] N. Chernov, G. Galperin and A. Zemlyakov, *Mathematics of Billiards*, Cambridge University Press, to appear

[4] B. Cipra, R. Hanson, and A. Kolan, *Periodic trajectories in right triangle billiards*, Phys. Rev. E 52 (1995) 2066–2071.

[5] M. Davis, *Groups generated by reflections and aspherical manifolds not covered by Euclidean space*, Ann. Math. 117 (1983), 293–324

[6] H. Furstenberg, *Recurrence in Ergodic Theory and Combinatorial Number Theory*, Princeton University Press, Princeton, 1981

[7] G. Galperin, *Nonperiodic and not everywhere dense billiard trajectories in convex polygons and polyhedra*, Comm. Math. Phys. 91 (1983), 187–211

[8] G. Galperin, T. Krueger, and S. Troubetzkoy, *Local instability of orbits in polygonal and polyhedral billiards*, Comm. Math. Phys. 169 (1995), 463–473

[9] G. Galperin, A. Stepin, and Ya. Vorobetz, *Periodic billiard orbits in polygons; generating mechanisms*, Russ. Math. Surv. 47 (1992), 5–80

[10] E. Gutkin, *Billiard flows on almost integrable polyhedral surfaces*, Ergod. Theor. Dyn. Sys. 4 (1984), 569–584

[11] E. Gutkin, *Billiards in polygons*, Physica 19 D (1986), 311–333

[12] E. Gutkin, *Billiards in polygons: survey of recent results*, J. Stat. Phys. 83 (1996), 7–26

[13] E. Gutkin and N. Haydn, *Topological entropy of generalized polygon exchanges*, Bull. AMS 32 (1995), 50–57

[14] E. Gutkin and N. Haydn, *Topological entropy of polygon exchange transformations and polygonal billiards*, Ergod. Theor. Dyn. Sys. (1996), in press

[15] E. Gutkin and C. Judge, *Coverings of translation surfaces, prime geodesic theorems, and spectral arithmeticity*, preprint, USC (1996)

[16] P. Hubert, *Complexite des suites definies par des trajectoires de billard dans un polygone rationnel*, Bull. S. M. F. 123 (1995)

[17] A. Katok, *The growth rate for the number of singular and periodic orbits for a polygonal billiard*, Comm. Math. Phys. 111 (1987), 151–160

[18] S. Kerckhoff, H. Masur, and J. Smillie, *Ergodicity of billiard flows and quadratic differentials*, Ann. Math. 124 (1986), 293 – 311

[19] H. Masur, *The growth rate for trajectories of a quadratic differential*, Ergod. Theor. Dyn. Sys. 10 (1990), 151–176

[20] A. Stepin, private communication, June 1995

[21] S. Tabachnikov, *Billiards*, Panoramas et Syntheses, n.1, Soc. Math. France, 1995

[22] W. Thurston, *The Geometry and Topology of 3-Manifolds*, Lecture Notes, Princeton University, 1978

[23] W. A. Veech, *Gauss measures for transformations on the space of interval exchange maps*, Ann. Math. 115 (1982), 201–242

[24] W. A. Veech, *Teichmuller curves in modular space, Eisenstein series, and an application to triangular billiards*, Inv. Math. 97 (1989), 553–583

[25] W. A. Veech, *Flat surfaces*, Amer. J. Math. 15 (1993), 589–689

[26] A. Zemlyakov and A. Katok, *Topological transitivity of billiards in polygons*, Math. Notes 18 (1976), 760–764

R LABARCA

A note on one dimensional dynamics associated to singular cycles

1   Introduction

Given $0 < x_0 < 1 - \varepsilon < 1$, let $I = [0, x_0] \cup [1 - \varepsilon, 1]$. Denote by

$$S_\lambda = \{f \in C^1([0, x_0], \mathbb{R}), f'(x) > \lambda > 1, \ f(0) = 0, \ f(x_0) = 1\},$$

$$S_\epsilon = \{h \in C^1([1 - \varepsilon, 1], \mathbb{R}), h'(x) < 0, \quad h(x) > \varepsilon^\alpha\}.$$

Clearly $S_\lambda \subset C^1([0, x_0], \mathbb{R})$ and $S_\epsilon \subset C^1([1 - \varepsilon, 1], \mathbb{R})$ are C^∞-submanifolds of the Banach spaces $C^1([0, x_0], \mathbb{R})$, $C^1([1 - \varepsilon, 1], \mathbb{R})$, respectively. In both spaces we work with the C^1-uniform topology.

Let $\psi : S_\lambda \times S_\epsilon \times]0, 1] \times]0, \alpha_0[\to C^0(I, \mathbb{R})$ be the map given by

$$\psi(f, h, C, \alpha)(x) = \begin{cases} f(x), & x \in [0, x_0] \\ -(1-x)^\alpha h(x) + C, & x \in [1 - \varepsilon, 1] \end{cases}$$

We note that maps such as $\psi(f, h, C, a)$ are typically associated to the dynamics of singular cycles (see [1], [2], [5]).

It is easy to see that ψ is an injective C^1-map (see Section 1). Under this condition $M = \psi(S_\lambda \times S_\epsilon \times]0, 1] \times]0, \infty[)$ is a C^1-Banach manifold and we can use ψ as a coordinate chart.

For $g \in M$ denote by

$$\Lambda_g = \{x \in I; g^n(x) \in I, \ for \quad every \quad n \in \mathbb{N}\}.$$

We say $g_1, g_2 \in M$ are Λ - equivalent if there exists a homeomorphism $H : \Lambda_{g_1} \to \Lambda_{g_2}$ such that $H \circ g_1 = g_2 \circ H$. If there exists a neighborhood of $g \in M, \cup_g \subset M$, such that any $g' \in \cup_g$ is Λ-equivalent to g, we say that g is Λ-stable.

Our aim is to prove the following

Theorem 1 *M can be decomposed into a disjount union $M = A \cup B^- \cup B^+$, where A is an open and dense set whose elements are Λ-stable and the set B^- (resp. B^+) is laminated by C^1-codimension one submanifolds such that elements in the same leaf are Λ-equivalent.*

We can decompose M into a disjoint union $E^+ \cup E^-$. The set E^- (resp. E^+) is given by $\psi(S_\lambda \times S_\epsilon \times]0,1] \times]1, \alpha_0])$ (resp. $\psi(S_\lambda \times S_\epsilon \times]0,1] \times]0,1]))$. Elements in E^+ (resp. E^-) will be called expanding (resp. contracting) maps. Let B denote the complement of the set A in M. Here B^+(resp. B^-) denotes the intersection $B \cap E^+($ resp. $B \cap E^-$).

As we have seen in [2] the laminas in B^- are given by :

(i) Those laminas where the map $\psi(f, h, C, \alpha)$ presents a saddle-node bifurcation for periodic orbits;

(ii) Those laminas where the critical value $C = \psi(f, h, C, \alpha)$ is a periodic point;

(iii) Those laminas where the critical value $C = \psi(f, h, C, \alpha)$ is a preperiodic point, and

(iv) Those laminas where the critical value $C = \psi(f, h, C, \alpha)$ presents a recurrent behavior.

The main difference between the expanding and the contracting cases is the following : for expanding maps there are no laminas as in (i) above. Consequently when the parameter α decreases fron $\alpha \geq 1$ to the region $\alpha \leq 1$, the laminas in (i) of above vanishe. Thus, there is a transition between contracting and expanding maps. The way this transition occurs is described in [3]. The next results explains this fact in our context.

Let $L_2 \subset M$ be a lamina as in (ii) of above and let $\Phi_2 : S_\lambda \times S_\epsilon \times]1, \alpha_0[\rightarrow [0,1]$ be the C^1-map that satisfies :

$$Graph(\psi(f, h, \Phi_2(f, h, \alpha), \alpha) = L_2.$$

We have

Corollary 1 *There exists a lamina,$L_1 \subset E^-$, as in (i) above such that if $\Phi_1 : S_\lambda \times S_\epsilon \times]1, \alpha_0[\rightarrow [0,1]$ is the C^1-map that satisfies*

$$Graph(\psi(f, h, \Phi_1(f, h, \alpha), \alpha) = L_1,$$

then $\Phi_2(f, h, 1) = \lim_{\alpha \to 1} \Phi_1(f, h, \alpha)$ and $\Phi_2(f, h, 1) = \lim_{\alpha \to 1} \Phi_1(f, h, \alpha)$. That is, L_1 and L_2 are tangent at $\psi(f, h, \Phi_2(f, h, 1), 1)$.

1. Here we extend a similar result proved in [4]. The main difference is the following : In [4] the class of maps under consideration is the set of C^1–expanding map $f : I \to \mathbb{R}$ satisfying $f(0) = f(1 - \varepsilon) = 0$, $f(x_0) = 1$ and $f(1) \geq 0$.

2. As a consequence of the proof of Theorem 2 in [2] and the properties of the map ψ we obtain that the set $B(f, h, \alpha) = \{\, C \in [0, 1]; \psi(f, h, C, \alpha) \, is \, not \, \Lambda - stable \,\}$ has Lebesgue measure zero.

The organization of this paper is a follows. In Section 2 we give some preliminary results, in Section 3 we introduce the first return map and in Section 4 we prove Theorem 1.

Aknowledgments : The author wishes to thank **IMPA-Brazil** for their support and hospitality while preparing this work.

2 Preliminary Results

Let us consider the map $\psi : S_\lambda \times S_\epsilon \times]0.1] \times]0, \alpha_0[\to C^0(I, \mathbb{R})$ as in Section 1.

Lemma 1 ψ is a C^1 injective map

Proof.

Clearly ψ is a continuous map. An easy computation will show that the partial derivatives $(f, h, C, \alpha) \to \frac{\partial \psi}{\partial f}(f, h, C, \alpha)$, $(f, h, C, \alpha) \to \frac{\partial \psi}{\partial h}(f, h, C, \alpha)$ and $(f, h, C, \alpha) \to \frac{\partial \psi}{\partial C}(f, h, C, \alpha)$ are continuous maps.

Concerning the partial derivative $\frac{\partial \psi}{\partial \alpha}(f, h, C, \alpha)$ we have that

$$\left(\frac{\psi(f, h, C, \alpha + \rho) - \psi(f, h, C, \alpha)}{\rho} \right)(x) = \begin{cases} 0, & x \leq x \leq x_0 \\[2mm] \frac{-(1-x)^\alpha h(x) + (1-x)^\alpha h(x)}{\rho}, & 1 - \varepsilon \leq x \leq 1 \end{cases}$$

Therefore

$$\lim_{\rho \to 0} \frac{(\psi(f, h, C, \alpha + \rho) - \psi(f, h, C, \alpha))(x)}{\rho} = -(1 - x)^\alpha h(x) Ln|1 - x|$$

Hence

$$\frac{\partial \psi}{\partial \alpha}(f, h, C, \alpha) \cdot t(x) = \begin{cases} 0 & 0 \leq x \leq x_0 \\ -(1 - x)^\alpha h(x) Ln|1 - x| , & 1 - \varepsilon \leq x < 1 \\ 0 & x = 1 \end{cases}$$

From this equality we conclude that $(f, h, C, \alpha) \to \frac{\partial \psi}{\partial \alpha}(f, h, C, \alpha)$ is a continuous map.

Now assume $\psi(f, h, C, \alpha) = \psi(f', h', C', \alpha')$. Evaluating this map in $x = 1$ we obtain $C = C'$. Assume $\alpha < \alpha'$. Since $-(1 - x)^\alpha h(x) = -(1 - x)^{\alpha'} h'(x)$ we obtain $(1 - x)^{\alpha - \alpha'} = \frac{h'(x)}{h(x)}$, that is $\lim_{x \to 1} \frac{h'(x)}{h(x)} = \infty$. This is impossible since h, h' are bounded positive maps.

In the same way the assumption $\alpha' < \alpha$ gives a contradiction. Hence, we conclude that $\alpha = \alpha'$. This equality implies $h(x) = h'(x)$ for $x \neq 1$. Since h and h' are continuous maps we get $h(1) = h'(1)$ that is, $h = h'$.

The equality $f(x) = f'(x)$ is obvious $\diamond$

We note, for any $(f, h, C, \alpha) \in S_{x_0} \times S_\varepsilon \times]0, 1] \times]0, \infty[$ that we have

$$\psi(f, h, C, \alpha)(1 - \varepsilon) < 0.$$

In fact, otherwise $-\varepsilon^\alpha h(1 - \varepsilon) + C \geq 0$, that is,

$$C \geq \varepsilon^\alpha h(1 - \varepsilon) > \varepsilon^{\alpha - \alpha_0} > 1 \qquad (\to \leftarrow).$$

We let $\bar{x} = \bar{x}(f, h, C, \alpha) \in]1 - \varepsilon, 1[$ denote the unique solution of the equation $\psi(f, h, C, \alpha)(\bar{x}) = 0$.

It is clear that $(f, h, C, \alpha) \to \bar{x}(f, h, C, \alpha)$ is a C^1–map .

Lemma 2 *For any $\psi(f, h, C, \alpha) \in E^+$ we have*

$$\frac{\partial}{\partial x}(\psi(f, h, C, \alpha)(x)) \geq \bar{\lambda} > 1, x \in [\bar{x}, 1].$$

<u>Proof</u>. We obtain this result from the relations

$$\frac{\partial}{\partial x}(\psi(f, h, C, \alpha)(x)) = \alpha(1 - x)^{\alpha - 1} h(x) - (1 - x)^\alpha h'(x).$$

and $h(x) > \varepsilon^{-\alpha_0}$. $\diamond$

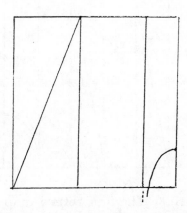

Figure 1. The map $\psi(f, h, C, \alpha), (f, h, C, \alpha) \in \psi^{-1}(E^-)$.

In a similar way we obtain

$$\frac{\partial}{\partial x}(\psi(f,h,C,\alpha))(x)|_{x=\bar{x}} > 1 \quad , any \quad (f,h,C,\alpha) \in \psi^{-1}(E^{-})$$

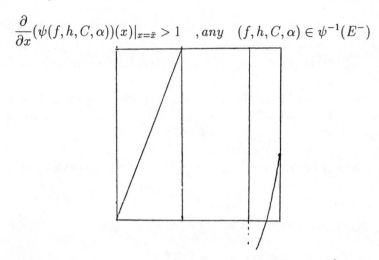

Figure 2. The map $\psi(f,h,C,\alpha), (f,h,C,\alpha) \in \psi^{-1}(E^{+})$.

3 The first return map

Let us now consider a fixed map $\psi(f,h,C,\alpha) \in M$. Let $\nabla_n(f,h,C,\alpha) = f^{-n}([\bar{x},1])$. Let $\bar{n} \in \mathbb{N}$ be an integer defined by the relations $f^{-\bar{n}}(\bar{x}) \leq C \leq f^{-\bar{n}}(1)$.

For any $n \geq \bar{n}$ there is an interval $J_n \subset [\bar{x},1]$ such that $f^n \circ \psi(J_n) = [\bar{x},1]$. Clearly $J_n \to \bar{x}$ as $n \to \infty$ and we can define a first return map $F: \bigcup_{n=\bar{n}}^{\infty} J_n \to [\bar{x},1]$ by

$$F(x) = F^n \circ \psi(x) \quad if \quad x \in J_n$$

Clearly the restriction map $F|_{J_n}$ is a C^{∞}-diffeomorphism onto its image, any $n > \bar{n}$.

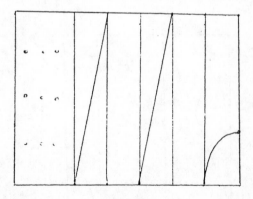

Figure 3. The first return map .

We will denote by $I_n \subset [\bar{x},1]$ the open interval between J_n and J_{n+1}.

4   Proof of the Theorem 1

Assume that there are $j \in \mathbb{N}, k \in \mathbb{N}$ such that $F^k(1) \in I_j$. Under this conditions there is an interval $L_{k,j} \subset [\bar{x}, 1], 1 \in L_{k,j}$ such that for any $x \in L_{k,j}$ we have $F^k(x) \in I_j$. In particular we obtain

Lemma 3   *There is a neighborhood $V(f, h, C, \alpha) \subset S_\lambda \times S_\epsilon \times]0, 1] \times]0, \alpha_0]$ such that, for any $(f', h', C', \alpha') \in V$, we have that there is an interval $L_{k,j}(f', h', C', \alpha') \subset [\bar{x}(f', h', C', \alpha'), 1], 1 \in L_{k,j}$ such that $F^k(x) \in I_j(f', h', C', \alpha')$. Here $F = F(f', h', C', \alpha')$.*

Let $\sum_{\bar{n}} = \{\theta : \mathbb{N} \to \{\bar{n}, \bar{n}+1, ...\}\}$ be the set of sequences $\theta : \mathbb{N} \to \{\bar{n}, \bar{n}+1, ...\}$ endowed with the topology given by the neighborhoods

$$\mathcal{V}(n, \theta) = \{\theta' \in \sum_{\bar{n}}; \quad \theta(j) = \theta'(j), \quad 1 \leq j \leq n\}$$

As usual in $\sum_{\bar{n}}$ we have defined a shift map $\sigma_{\bar{n}} : \sum_{\bar{n}} \to \sum_{\bar{n}}$ given by $(\sigma_{\bar{n}}\theta)(i) = \theta(i+1)$.

For any infinite matrix $A = \begin{pmatrix} a_{\bar{n}\bar{n}} & a_{\bar{n}\bar{n}+1} & \cdots \\ a_{\bar{n}+1\bar{n}} & a_{\bar{n}+1\bar{n}+1} & \cdots \\ \vdots & \vdots & \vdots \end{pmatrix}$ we have a set $\sum_A \subset \sum_{\bar{n}}$

given by $\{\theta \in \sum_{\bar{n}}; \quad a_{\theta(i)\theta(i+1)} = 1\}$

Clearly, $\sigma(\sum_A) \subset \sum_A$ and we have

Lemma 4   *Given (f, h, C, α) as in Lemma 3. We can find an infinite matrix $A = A(f, h, C, \alpha)$ and a homeomorphism $H_F : \Lambda_F \to \sum_A$ such that $H_F \circ F = \sigma \circ H_F$.*

As a consequence we obtain

Lemma 5   *For any $(f, h, C, \alpha) \in S_\lambda \times S_\epsilon \times]0, 1] \times]0, \alpha_0[$ such that there is $n \in \mathbb{N}$ that satisfies $\psi^n(1) \notin I$, we have that $\psi(f, h, C, \alpha)$ is Λ-stable.*

Proof.   In fact we note that $\Lambda_F = \{x \in]\bar{x}, 1]; F^n(x) \in]\bar{x}, 1], n \in \mathbb{N}\}$, is equivalent to $\sum_A$, for any $F = F(f', h', C', \alpha'), (f', h', C', \alpha') \in V$ and $A = A(f, h, C, \alpha)$ as in Lemma 4. In this way we can obtain a homeomorphism $H : \Lambda_{F(f,h,C,\alpha)} \to \Lambda_{F(f',h',C',\alpha')}$ such that $H \circ F(f, h, C, \alpha) = F(f', h', C', \alpha') \circ H$. Clearly, we can extend this homeomorphism to Λ_F. The extension of this map to $\Lambda_{\psi(f,h,C,\alpha)}$ follows easily ◇

In what follows we will assume that $\psi^n(1) \in I, n \in \mathbb{N}$.

(4.1) Suppose $\psi(f, h, C, \alpha) \in E^+$, that is, $0 < \alpha \leq 1$.

Lemma 6 *The set $\{(f,h,C,\alpha) \in \psi^{-1}(E^+); \psi^n(f,h,C,\alpha)(1) \notin I$ some $n \in \mathbb{N}\}$ is dense in $\psi^{-1}(E^+)$.*

Proof. Take $(f,h,C,\alpha) \in \psi^{-1}(E^+)$ such that $\psi^n(f,h,C,\alpha)(1) \in I$ all $n \in \mathbb{N}$. We have three possibilities:

(i) $1 \in I$ is a periodic point;

(ii) $1 \in I$ is eventually periodic, and

(iii) neither *(i)* or *(ii)*.

Let assume that *(i)* of above is the situation.

Let $k \in \mathbb{N}$ be the integer that satisfies $F^k(1) = 1$. Let $D_k \subset J_{\bar{n}}$ be the connected component of the domain of the map F^k that contains 1. This component satisfies $F^k(D_k) = [\bar{x}, 1]$.

It is clear that the equation $F^k(f,h,C,\alpha)(1) = 1$ defines a C^1-codimension one submanifold $T_k \subset E^+$. Any two elements in this submanifold are Λ_F- equivalent.

Consider the curve $C' \overset{\Gamma}{\longmapsto} \psi(f,h,C',\alpha), C' \in]C - \delta, C + \delta[$. It is easy to see that Γ is transversal to T_k at C.

Let $I^j_{k,\bar{n}}$ denote the pre-images $F^{-kj}(I_{\bar{n}}) \cap D_k$. This sequence of intervals accumulates at $1 \in D_k$. The equations: $F^k(f,h,C',\alpha)(1) \in I^j_{k,\bar{n}}, j$ big enough, define a sequence of intervals $\triangle^j_{k,\bar{n}} \subset \Gamma(]C - \delta, C + \delta[)$ that accumulates in T_k and satisfy $\triangle^j_{k,\bar{n}} \subset \{\psi(f,h,C,\alpha) \in E^+; \quad \psi^n(f,h,C,\alpha) \notin I \quad$ for some $n \in \mathbb{N}\}$.

This completes the proof of the the Lemma 6 in case *(i)*.

Let assume that *(ii)* is the situation.

Let $p(f,h,C,\alpha)$ be the periodic point that satisfies: $F^k(f,h,C,\alpha)(1) = p(f,h,C,\alpha)$ here $k \in \mathbb{N}$ is the first integer with this property. Let $n_0 \in \mathbb{N}$ be the integer defined by $p(f,h,C,\alpha) \in I_{n_0}$.

Since $p(f,h,C,\alpha)$ is a hyperbolic repelling periodic point there is neighborhood $V(\psi(f,h,C,\alpha)) \subset E^- \cup E^+$ and a C^1-map $p : V(\psi(f,h,C,\alpha)) \to J_{n_0}$ such that $p(\psi(f',h',c',\alpha'))$ is a hyperbolic repelling periodic point of the map $F(f',h',c',\alpha')$.

Clearly the equation $F^k(f',h',c',\alpha')(1) = p \circ \psi(f,h,C,\alpha)$ defines a C^1-codimension one submanifold $T_k \subset V(\psi(f,h,C,\alpha))$. As before the curve $\Gamma :]C - \delta, C + \delta[\to V(\psi(f,h,C,\alpha)), \Gamma(C') = \psi(f,h,C',\alpha)$ is transversal to T_k. Now we obtain the result as before.

Let us assume that *(iii)* is the situation.

We consider the sequence $\Theta(f,h,C,\alpha) : \mathbb{N} \to \{\bar{n}, \bar{n} + 1, \ldots\}$ defined by: $\Theta(f,h,C,\alpha)(j) = i$ if and only if $F^j(f,h,C,\alpha)(1) \in J_i = [a_i, b_i]$.

Define
$$V_j = \{(f', h', C', \alpha') \in S_\lambda \times S_\epsilon \times]0, 1] \times]0, \alpha_0] \mid F^i(f', h', c', \alpha')(1) \in J_{\theta(f, h, C, \alpha)(i)},$$
$$0 \le i \le j\}.$$

Clearly $V_1 \supset V_2 \supset V_3 \supset \ldots$ and each of the set V_j is a region whose boundary is formed by two disjoint C^1-codimension one-submanifold T_{a_j}, T_{b_j} defined by the following relations:
$$T_{a_j} = \{(f', h', c', \alpha') \in V_{j-1}; \quad F^j(f', h', c', \alpha')(1) = a_j\}$$
and
$$T_{b_j} = \{(f', h', c', \alpha') \in V_{j-1}; \quad F^j(f', h', c', \alpha')(1) = b_j\}.$$

Now, as we did in [2] Lemma 7 we can obtain:

Lemma 7 $\bigcap\limits_{i=1}^{\infty} V_i$ *defines a C^1-codimension one submanifold formed by the values* (f', h', c', α') *that satisfies* $F^i(f', h', c', \alpha')(1) \in J_{\Theta(f, h, C, \alpha)(i)};$ *$i \in \mathbb{N}$. Moreover, two elements in $\bigcap\limits_{i=1}^{\infty} V_i$ satisfies that its corresponding ψ-maps are Λ-equivalent.*

This result completes the proof of the main Theorem in the expanding case.

(4.2) Suppose $\psi(f, h, C, \alpha) \in E^-$, that is, $1 < \alpha \le \alpha_0$.

Definition: We will say $g \in C^k(I, \mathbb{R}); \quad k \ge 0$ satisfies axioma A if

(i) g has a finite number of hyperbolic, attracting periodic orbits and no other attractors,

(ii) Let $B(g)$ denote the basin of attraction of the attracting periodic orbits for g. The set $\Sigma_g = I \setminus B(g)$ is a hyperbolic set for g.

For $(f, h, C, \alpha) \in \psi^{-1}(E^-)$ let $\Lambda_{(f, h, C, \alpha)} = \{x \in I; \psi^n(f, h, C, \alpha)(x) \in I, n \ge 0\}$
and
$\Gamma_1 = \{(f, h, C, \alpha) \in \psi^{-1}(E^-); 1 \in \Lambda_{(f, h, C, \alpha)}$ and there exists a hyperbolic attracting periodic orbit for the map $\psi(f, h, C, \alpha)\}$.

Now, as we did in [2] Lemma 4 we obtain:

Lemma 8 *For $(f, h, C, \alpha) \in \Gamma_1$ we have that $\psi(f, h, C, \alpha)$ satisfies Axiom A.*

Under this situation it is not difficult to prove that $\psi(f, h, C, \alpha)$ is Λ-stable.

Now we are ready to complete the proof of the main Theorem. To do so we state the following

Lemma 9 *The set*

$$\{(f, h, C, \alpha) \in \psi^{-1}(E^-); \psi^n(f, h, C, \alpha) \notin I \text{ for some } n \in \mathbb{N} \quad or \quad (f, h, C, \alpha) \in \Gamma_1\}$$

is dense in $\psi^{-1}(E^-)$.

Proof. Take $(f, h, C, \alpha) \in \psi^{-1}(E^-)$ such that $\psi^n(f, h, C, \alpha)(1) \in I$, all $n \in \mathbb{N}$. As before we can define a first return map $F : \bigcup_{i \geq \bar{n}} J_i \to [\bar{x}, 1]$. Let define the itinerary of the map F as follows: $\Theta : \mathbb{N} \to \{\bar{n}, \bar{n} + 1, \ldots\}$ is defined by $\Theta(i) = j$ if and only if $F^i(1) \in J_j$.

For the itinerary $\Theta(f, h, C, \alpha)$ associated with (f, h, C, α) we have three possibilities:

(i) $\Theta(f, h, C, \alpha)$ is a periodic itinerary;

(ii) $\Theta(f, h, C, \alpha)$ is eventually periodic and

(iii) neither *(i)* or *(ii)*.

Assume that *(i)* is the situation.

Let $V_\Theta = \{(f', h', C', \alpha') \in \psi^{-1}(E^+) \mid \Theta(f', h', C', \alpha') = \Theta(f, h, C, \alpha)\}$.

It is easy to see that V_Θ has a boundary formed by

$T_{sa} = \{(f', h', C', \alpha') \in \psi^{-1}(E^-); \quad \Theta(f', h', C', \alpha') = \Theta(f, h, C, \alpha)$ and 1 is a periodic point for the map $\psi(f', h', c', \alpha')\}$

and
$T_{sn} = \{(f', h', c', \alpha') \in \psi^{-1}(E^+); \quad \Theta(f', h', c', \alpha') = \Theta(f, h, C, \alpha)$ and the map $\psi(f', h', c', \alpha')$ presents a saddle-node bifurcation$\}$.

Clearly T_{sa} and T_{sn} are C^1-codimension one submanifold.

These manifolds satisfy (see [3]) the following properties:

(i) for any $(f', h', C', \alpha') \in T_{sa}$ (resp. $(f', h', C', \alpha') \in T_{sn}$) the curve $C'' \longmapsto (f', h', C'', \alpha')$ is transversal to T_{sa} (resp. to T_{sn}) at C',

(ii) the curve $C'' \longmapsto (f', h', C', \alpha')$ intersects both manifolds at points $(f', h', C_1, \alpha') \in T_{sa}$ and $(f', h', C_2, \alpha') \in T_{sn}$, and we have $C_1 > C_2$;

(iii) $\lim_{\alpha' \to 1}(C_1 - C_2) = 0$, and

(iv) The tangent spaces $T_{(f', h', C_2, \alpha')}(T_{sn})$ and $T_{(f', h', C_1, \alpha')}(T_{sa})$ satisfies:

$$\lim_{\alpha' \to 1} T_{(f', h', C_2, \alpha')}(T_{sn}) = \lim_{\alpha' \to 1} T_{(f', h', C_1, \alpha')}(T_{sa})$$

Taking into account these differences the proof now follows as in the expanding case $\diamond$

This completes the proof of the main result.

54

References

[1] R. BAMÓN, R. LABARCA, R. MAÑÉ AND M.J. PACÍFICO, *The explosion of singular cycles*. Publication Mathematiques I.H.E.S. 78 pg. 207-232, 1993.

[2] R. LABARCA *Bifurcation of Contracting Singular cycles* Ann. Scient. Ec. Norm. Sup. 4e Serie, t. 28, p.705-745, 1995.

[3] R. LABARCA *Transition from contracting to expanding singular cycles: The one-dimensional case.* Proceedings of the International Workshop on Dynamical Systems and Chaos. Tokyo 1994

[4] M. J. PACÍFICO AND R. LABARCA *Dynamics of Expanding maps of the Interval* Advanced Series in Dynamical Systems, Vol. 9, World Scientific, pg. 307-315,1991.

[5] M.J. PACÍFICO AND A. ROVELLA *Unfolding contracting singular cycles.* Ann. Scient. Ec. Norm. Sup. 4^e serie, t. 26, pg. 691-700, 1993.

Departamento de Matemática y C. C.
Universidad de Santiago de Chile
Casilla 307 - Correo 2
Santiago - Chile

Partially supported by FONDECYT Grant 1941080 and the Dirección de Investigaciones of the Universidad de Santiago de Chile

CARLANGELO LIVERANI
Central limit theorem for deterministic systems

§0 Introduction.

A discrete time dynamical system consists of a measure space X (be $\mathcal{F}$ the σ-algebra), a measurable map $T : X \to X$ which describes the dynamics, and a probability measure P invariant with respect to T. This setting is particularly well suited to study problems involving statistical properties of the motion of deterministic systems.

Typically the properties of interest are ergodicity, mixing, bounds on the decay of correlations, Central Limit Theorems (CLT) and so on. Several approaches have been developed to tackle such problems at various levels. Given a system, one first explores the weaker statistical properties and then tries to investigate the stronger ones using the already obtained results plus some extra properties.

The position of this paper in the above mentioned hierarchy is between obtaining bounds on the decay of correlations and CLT. In other words we discuss a general approach that gives checkable conditions under which, in a mixing system, an observable enjoys the CLT. Many results on CLT in dynamical systems already exist but they are either limited to one dimensional systems [Ke] or relay on the existence of special partitions of the phase space [Ch], partitions whose concrete construction may be far from trivial [BSC1], [BSC2]; for a very nice review of the state of affairs up to 1989 (but still actual) see [De].

Here, I want to put forward the following point of view: the above described dynamical systems are most naturally viewed as giving rise to a (deterministic) Markov process. It is therefore tempting to think that there should exists some general probabilistic theorem that states abstract conditions for the validity of the CLT, and that all the concrete cases can simply be obtained by the direct application

---

This paper originated out of discussions with D.Szasz and A.Kramli, and was made possible by D.Szasz key suggestion to use K-partitions. I wish to thank Y.Kifer, E.Olivieri, E.Presutti, B.Tot, and L.Triolo for helpful discussions. In addition, I am indebted to S.Olla for explaining me the subtleties of the Kipnis-Varadhan approach. This work has been partially supported by grant CIPA-CT92-4016 of the Commission of the European Community. I wish also to thank ESI, where part of this work was done.

of such a theorem to the system under consideration (without having to code the system in some symbolic type dynamics). General theorems of this type are well known in probability theory but they are normally not well suited for applications to the case at hand. Two such general theorems, tailored for dynamical systems, can be found in this paper.

Attempts in this directions have been in existence for some time [Go], [IL], but they are satisfactory only in the one dimensional case (the equivalent of Theorem 1.1 in this paper). Particular mention must be given to [DG], the results obtained there are essentially comparable to the one presented here in section 1 and could be applied to the multidimensional case. Unfortunately, not much attention is given there to applications, so that the possibility to bypass a symbolic representation of the system is completely overlooked.

The approach used here is a martingale approximation inspired by [KV]. Since this is a typical probabilistic technique, I think it underlines very well the purely probabilistic nature of the result hereby clarifying which characteristics of a deterministic system yield such a drastic statistical behavior.

As we will see, a major difference with the analogous type results in probability is that the CLT holds for a much smaller class of observables than the square summable ones. This is not an artifact of the proof: it is an inevitable consequence of the deterministic nature of the systems under consideration so that only observables that operates some "coarse graining" (and therefore enjoy some degree of smoothness) can yield strong statistical behavior. Here no particular attempt is made to find the most general class of observables to which the Theorems apply; nonetheless, the technique put forward lends itself to an extension in such a direction.

The paper includes some concrete examples as well. Their aim is to show how the general theorems can be applied in special cases. The cases discussed belong to quite general classes (expanding one dimensional maps, area preserving piecewise smooth uniformly hyperbolic maps in two and more dimensions), yet no real new result is contained in such examples. This reflects the spirit of the paper intended to presenting an approach rather than new implementations. Nevertheless, the application of the present results in technically complex situations (e.g., hyperbolic billiards) greatly simplifies the proof of the validity of the CLT. In addition, it is conceivable that some new results can be obtained by this approach since the two above mentioned theorems hold in more general cases than the ones already present in the literature (a brief comparison with previously known results is inserted after the proof of each theorem).

The plan of the paper is as follows. Section 1 contains two probabilistic theorems that are well suited for the study of dynamical systems. In fact, they may seem a bit unnatural from the pure probabilistic point of view. On the one hand, both theorems deal only with functions in L^∞ instead than L^2. The reason is that normally the decay of correlations in dynamical systems can be obtained only for classes of functions with some amount of smoothness, which makes them automatically bounded. The issue is not purely a matter of taste: a look at the proofs will show that such an hypothesis has really been used and that many key estimates would not hold in L^2. On the other hand, in Theorem 2 are introduced σ-algebras $\mathcal{F}_i$ that behave nicely with respect to the dynamics. This may make little sense from the purely probabilistic point of view but it is instead a cornerstone in the treatment of hyperbolic dynamical systems and such σ-algebras are completely natural in the framework of K-systems (see footnote 4).

In the above sense the results of section 1, although purely probabilistic in nature, are expressly developed for applications to dynamical systems.

Section 2 describe how the technique applies to non-invertible maps. The case of piecewise smooth expanding maps of the interval is discussed in detail.

Section 3 deals with the most interesting applications: the multidimensional case. As an example I treat a large subclass of piecewise smooth symplectic maps. Such maps are well studied in the literature for some relevant physical models (e.g. billiards) are naturally described in their terms. It is shown that very general considerations imply the applicability of the results developed in section one.

§1 A general probabilistic result.

Let X be a complete separable metric space, $\mathcal{F}$ a σ-algebra, P a probability measure $(P(X) = 1)$ and $T : X \to X$ a measurable map.[1] We will call $\mathbb{E}$ the expectation with respect to P. In addition, we require that P is invariant with respect to T (i.e., for all $A \in \mathcal{F}$ holds $P(T^{-1}A) = P(A)$), and that the dynamical system (T, X, P) be ergodic.

For each $\phi \in L^2(X)$ define $\widehat{T} : L^2(X) \to L^2(X)$ by

$$\widehat{T}\phi = \phi \circ T,$$

and let $\widehat{T}^* : L^2(X) \to L^2(X)$ be the dual of $\widehat{T}$.

If $\mathbb{E}(f) = 0$, then by ergodicity $\lim_{n \to \infty} \frac{1}{n} \sum_{i=0}^{n-1} \widehat{T}^n f = \mathbb{E}(f) = 0$. The CLT gives us informations on the speed of convergence; namely the conditions under which there exists $\sigma \in \mathbb{R}^+$: for each interval $I \subset \mathbb{R}$

$$\lim_{n \to \infty} P\left(\left\{\frac{1}{\sqrt{n}} \sum_{i=0}^{n-1} \widehat{T}^n f \in I\right\}\right) = \frac{1}{\sqrt{2\pi}\sigma} \int_I e^{-\frac{x^2}{2\sigma^2}}\, dx;$$

this is called "convergence in law (or distribution)" to a Gaussian random variable of zero mean and variance σ.

Consider a sub-σ-algebra $\mathcal{F}_0$ of $\mathcal{F}$ and define $\mathcal{F}_i = T^{-i}\mathcal{F}_0$, $i \in \mathbb{Z}$, then the following holds.

---

[1] Actually, we assume that, for each $A \in \mathcal{F}$, not only $T^{-1}A \in \mathcal{F}$ but also $TA \in \mathcal{F}$.

Theorem 1.1. *If $\mathcal{F}_i$ is coarser than $\mathcal{F}_{i-1}$ and, for each $\phi \in L^\infty(X)$, we have*

$$\mathbb{E}(\widehat{T}\widehat{T}^*\phi|\mathcal{F}_1) = \mathbb{E}(\phi|\mathcal{F}_1),$$

then, for each $f \in L^\infty(X)$, $\mathbb{E}(f) = 0$ and $\mathbb{E}(f|\mathcal{F}_0) = f$, such that

(1) $\sum_{n=0}^{\infty} |\mathbb{E}(f\widehat{T}^n f)| < \infty$,

(2) *the series $\sum_{n=0}^{\infty} \mathbb{E}(\widehat{T}^{*n} f|\mathcal{F}_0)$ converges absolutely almost surely,[2]*

the sequence

$$\frac{1}{\sqrt{n}} \sum_{i=0}^{n-1} \widehat{T}^i f$$

converges in law to a Gaussian random variable of zero mean and finite variance σ,
$\sigma^2 \leq -\mathbb{E}(f^2) + 2\sum_{n=0}^{\infty} \mathbb{E}(f\widehat{T}^n f)$.
In addition, $\sigma = 0$ if and only if there exists a $\mathcal{F}_0$–mesurable function g such that

$$\widehat{T}f = \widehat{T}g - g.$$

Finally, if (2) converges in $L^1(X)$, then $\sigma^2 = -\mathbb{E}(f^2) + 2\sum_{n=0}^{\infty} \mathbb{E}(f\widehat{T}^n f)$.

Proof. The key idea is to use a Martingale approximation. That is, to find $Y_i \in L^2(X)$ and g $\mathcal{F}_0$–measurable, and almost everywhere finite, such that

(1.1) $$\mathbb{E}(Y_{i-1}|\mathcal{F}_i) = Y_{i-1} \; ; \quad \mathbb{E}(Y_i|\mathcal{F}_i) = 0,$$

(i.e., Y_i is a reverse Martingale difference with respect to the filtration $\{\mathcal{F}_i\}_{i=0}^{\infty}$), and

(1.2) $$\widehat{T}^i f = Y_i + \widehat{T}^i g - \widehat{T}^{i-1} g \quad \forall i > 0.$$

Accordingly,

(1.3) $$\frac{1}{\sqrt{n}} \sum_{i=0}^{n-1} \widehat{T}^i f = \frac{1}{\sqrt{n}} \sum_{i=0}^{n-1} Y_i + \frac{1}{\sqrt{n}}[\widehat{T}^n g - g].$$

Equation (1.3) shows that we can obtain the central limit theorem for our random variable provided we have the central limit theorem for the martingale difference Y_i. In fact, $\frac{1}{\sqrt{n}}[\widehat{T}^n g - g]$ converges to zero in probability when $n \to \infty$.

Note that (1.1) and (1.2) are equivalent to

$$\mathbb{E}(\widehat{T}^i f|\mathcal{F}_i) = \mathbb{E}(\widehat{T}^i g|\mathcal{F}_i) - \mathbb{E}(\widehat{T}^{i-1} g|\mathcal{F}_i) \quad \forall i > 0.$$

---

[2] As we will see in the proof, this implies that there exists an almost everywhere finite $\mathcal{F}_0$-measurable function g, such that $f = g - \mathbb{E}(\widehat{T}^* g|\mathcal{F}_0)$.

Since by the definition of $\mathcal{F}_i$ follows that, for each $\phi \in L^1(X)$,

$$\mathbb{E}(\widehat{T}^i \phi | \mathcal{F}_i) = \widehat{T}^i \mathbb{E}(\phi | \mathcal{F}_0) \quad \forall i > 0,$$

and because the invariance of $\mathbb{E}$ with respect to T implies $\widehat{T}^* \widehat{T} = \mathbb{1}$, we have

$$
\begin{aligned}
f &= \mathbb{E}(g | \mathcal{F}_0) - \widehat{T}^* \mathbb{E}(g | \mathcal{F}_1) = g - \widehat{T}^* \mathbb{E}(\widehat{T}\widehat{T}^* g | \mathcal{F}_1) \\
&= g - \mathbb{E}(\widehat{T}^* g | \mathcal{F}_0).
\end{aligned}
$$
(1.4)

It is immediate to see that $g = \sum_{n=0}^{\infty} \mathbb{E}(\widehat{T}^{*n} f | \mathcal{F}_0)$ (if the series converges in $L^1(X)$) is a solution of the above equation, and therefore of (1.2), (clearly, $Y_i = \widehat{T}^{i-1} Y_1$).[3]

In fact, setting $T_0 \phi = \mathbb{E}(\widehat{T}^* \phi | \mathcal{F}_0)$, the solution of (1.4) is given by the Neumann series $\sum_{n=0}^{\infty} T_0^n f$. But $T_0^n f = \mathbb{E}(\widehat{T}^{*n} f | \mathcal{F}_0)$ since

$$
\begin{aligned}
\mathbb{E}(\widehat{T}^* \mathbb{E}(\widehat{T}^{*n} f | \mathcal{F}_0) | \mathcal{F}_0) &= \widehat{T}^* \widehat{T} \mathbb{E}(\widehat{T}^* \mathbb{E}(\widehat{T}^{*n} f | \mathcal{F}_0) | \mathcal{F}_0) = \widehat{T}^* \mathbb{E}(\widehat{T}\widehat{T}^* \mathbb{E}(\widehat{T}^{*n} f | \mathcal{F}_0) | \mathcal{F}_1) \\
&= \widehat{T}^* \mathbb{E}(\mathbb{E}(\widehat{T}^{*n} f | \mathcal{F}_0) | \mathcal{F}_1) = \widehat{T}^* \mathbb{E}(\widehat{T}^{*n} f | \mathcal{F}_1) \\
&= \widehat{T}^* \mathbb{E}(\widehat{T}\widehat{T}^{*(n+1)} f | \mathcal{F}_1) = \mathbb{E}(\widehat{T}^{*(n+1)} f | \mathcal{F}_0).
\end{aligned}
$$

To insure that the central limit theorem for Y_i holds, we need only to show that Y_i is square summable due to the following [Ne]:

Theorem. *Let $(Y_n)_{n \geq 1}$ be a stationary, ergodic, martingale difference (or reversed martingale difference) with respect to a filtration $\{\mathcal{F}_n\}_{n \geq 1}$. If $Y_1 \in L^2(X)$, then the CLT holds and $\sigma^2 = \mathbb{E}(Y_1^2)$.*

The above theorem could apply to our case since the stationarity of (Y_n) is implied by the invariance of the measure with respect to T, while the ergodicity follows from the ergodicity of the dynamical system (X, T, P).

If the series $\sum_{n=0}^{\infty} \mathbb{E}(\widehat{T}^{*n} f | \mathcal{F}_0)$ would converge in $L^2(X)$, then $Y_1 \in L^2(X)$ would hold and the Theorem would be proven. It is however a remarkable fact that Y_i can be in $L^2(X)$ without g being even integrable [KV]. Unfortunately, the road to this result is a bit indirect and consists in carrying out an argument similar to the one above but producing a sequence of martingale differences $Y_i(\lambda)$ that approximate Y_i.

Let us look for $Y_i(\lambda)$, $\lambda > 1$, such that

$$\mathbb{E}(Y_{i-1}(\lambda) | \mathcal{F}_i) = Y_{i-1}(\lambda) \; ; \quad \mathbb{E}(Y_i(\lambda) | \mathcal{F}_i) = 0;$$
(1.5)

---

[3] It is noteworthy that, once we have g, the Y_i are defined by (1.2) itself, and will automatically satisfy (1.1).

and

(1.6) $\qquad \widehat{T}^i f = Y_i(\lambda) + \widehat{T}^i g(\lambda) - \lambda^{-1} \widehat{T}^{i-1} g(\lambda) \quad \forall i > 0, \ \lambda > 1.$

In analogy with what we have seen before $g(\lambda) = \sum_{n=0}^{\infty} \lambda^{-n} \mathbb{E}(\widehat{T}^{*n} f | \mathcal{F}_0)$, only now $g(\lambda) \in L^2(X)$ for each $\lambda > 1$. Since $\lim_{\lambda \to 1} g(\lambda) = g(1) = g$ almost surely, it follows that $\lim_{\lambda \to 1} Y_i(\lambda) = Y_i$ almost surely. In addition,

$$
\begin{aligned}
\mathbb{E}(Y_i(\lambda)^2) =& \mathbb{E}(Y_1(\lambda)^2) = \mathbb{E}([\widehat{T} f - \widehat{T} g(\lambda) + \lambda^{-1} g(\lambda)]^2) \\
=& \mathbb{E}(\widehat{T} f [\widehat{T} f - \widehat{T} g(\lambda) + \lambda^{-1} g(\lambda)]) \\
& - \mathbb{E}([\widehat{T} g(\lambda) - \lambda^{-1} g(\lambda)][\widehat{T} f - \widehat{T} g(\lambda) + \lambda^{-1} g(\lambda)]),
\end{aligned}
$$

since $\mathbb{E}(\widehat{T} f - \widehat{T} g(\lambda) + \lambda^{-1} g(\lambda) | \mathcal{F}_1) = \mathbb{E}(Y_1 | \mathcal{F}_1) = 0$. Hence,

$$
\begin{aligned}
\mathbb{E}(Y_i(\lambda)^2) =& - \mathbb{E}((\widehat{T} f)^2) + \mathbb{E}([\widehat{T} g(\lambda) - \lambda^{-1} g(\lambda)]^2) \\
=& - E(f^2) + \mathbb{E}(\widehat{T} g(\lambda) [\widehat{T} g(\lambda) - \lambda^{-1} g(\lambda)]) \\
& - \lambda^{-1} \mathbb{E}(g(\lambda) \widehat{T} g(\lambda)) + \lambda^{-2} \mathbb{E}(\widehat{T} g(\lambda)^2) \\
=& - \mathbb{E}(f^2) + 2 \mathbb{E}(\widehat{T} g(\lambda) [\widehat{T} g(\lambda) - \lambda^{-1} g(\lambda)]) - (1 - \lambda^{-2}) \mathbb{E}(g(\lambda)^2) \\
=& - \mathbb{E}(f^2) + 2 \mathbb{E}(\widehat{T} g(\lambda) \widehat{T} f) - (1 - \lambda^{-2}) \mathbb{E}(g(\lambda)^2) \\
=& - \mathbb{E}(f^2) + 2 \mathbb{E}(g(\lambda) f) - (1 - \lambda^{-2}) \mathbb{E}(g(\lambda)^2) \\
\leq& - \mathbb{E}(f^2) + 2 \sum_{n=0}^{\infty} \lambda^{-n} \mathbb{E}(f \widehat{T}^n f) \leq -\mathbb{E}(f^2) + 2 \sum_{n=0}^{\infty} |\mathbb{E}(f \widehat{T}^n f)|.
\end{aligned}
$$

The wanted estimates follows from

$$
\mathbb{E}(Y_1^2) = \mathbb{E}(\liminf_{\lambda \to 1} Y_1(\lambda)^2) \leq \liminf_{\lambda \to 1} \mathbb{E}(Y_1(\lambda)^2) \leq -\mathbb{E}(f^2) + 2 \sum_{n=0}^{\infty} \mathbb{E}(f \widehat{T}^n f).
$$

In conclusion, we have seen that the random variable under consideration converges in law to a Gaussian of variance $\sigma^2 = \mathbb{E}(Y_1^2) < \infty$. If $\sigma = 0$ then the second assertion of the statement holds since

$$
\mathbb{E}(Y_1^2) = \mathbb{E}([\widehat{T} f - \widehat{T} g + g]^2).
$$

If we assume that the series in (2) converges in $L^1(X)$, then it is possible to obtain the much sharper result

$$
\lim_{\lambda \to 1} \mathbb{E}(Y_1(\lambda)^2) = -\mathbb{E}(f^2) + 2 \sum_{n=0}^{\infty} \lambda^{-n} \mathbb{E}(f \widehat{T}^n f).
$$

61

In fact, for each $\varepsilon > 0$

$$\left| \mathbb{E}(Y_i(\lambda)^2) - \mathbb{E}(f^2) + 2 \sum_{n=0}^{\infty} \lambda^{-n} \mathbb{E}(f\widehat{T}^n f) \right| \leq \sum_{n=0}^{\infty} (1 - \lambda^{-n}) \mathbb{E}(f\widehat{T}^n f)$$

$$+ (1 - \lambda^{-2}) \mathbb{E}(g(\lambda)^2) \leq (1 - \lambda^{-M}) \sum_{n=0}^{\infty} \mathbb{E}(f\widehat{T}^n f) + \sum_{n=M}^{\infty} \mathbb{E}(f\widehat{T}^n f)$$

$$+ (1 - \lambda^{-2}) \mathbb{E}(g(\lambda)^2) \leq \varepsilon + (1 - \lambda^{-2}) \mathbb{E}(g(\lambda)^2)$$

where M has been chosen sufficiently large and λ sufficiently close to one. In order to continue we need to estimate the last term in the above expression. For further use we will deal with a more general estimate: for each $\lambda, \mu \in (1, \infty)$ holds

$$\mathbb{E}(g(\lambda)g(\mu)) = \sum_{n,m=0}^{\infty} \lambda^{-n} \mu^{-m} \mathbb{E}(\widehat{T}^{*n} f \mathbb{E}(\widehat{T}^{*m} f | \mathcal{F}_0))$$

(1.7) $$\leq \sum_{n=0}^{\infty} \lambda^{-n} \sum_{m=0}^{M-1} \|f\|_{\infty} \mathbb{E}(|\mathbb{E}(\widehat{T}^{*n} f | \mathcal{F}_0)|) + \sum_{n=0}^{\infty} \lambda^{-n} \sum_{m=M}^{\infty} \mathbb{E}(|\mathbb{E}(\widehat{T}^{*m} f | \mathcal{F}_0)|)$$

$$\leq M\|f\|_{\infty} \sum_{n=0}^{\infty} \mathbb{E}(|\mathbb{E}(\widehat{T}^{*n} f | \mathcal{F}_0)|) + \frac{\|f\|_{\infty}}{1 - \lambda^{-1}} \sum_{m=M}^{\infty} \mathbb{E}(|\mathbb{E}(\widehat{T}^{*m} f | \mathcal{F}_0)|).$$

That is, choosing again M large and λ sufficiently close to 1,

$$(1 - \lambda^{-1}) \mathbb{E}(g(\lambda)^2) \leq 2\varepsilon.$$

This is not the end of the story: it is possible to prove that Y_1 is the limit of $Y_1(\lambda)$ in $L^2(X)$. To see this it suffices to estimate

$$\mathbb{E}([Y_1(\lambda) - Y_1(\mu)]^2) = \mathbb{E}([\lambda^{-1}g(\lambda) - \mu^{-1}g(\mu)][Y_1(\lambda) - Y_1(\mu)])$$

$$= \mathbb{E}([\lambda^{-1}g(\lambda) - \mu^{-1}g(\mu)]^2) - \mathbb{E}([g(\lambda) - g(\mu)]^2)$$

$$\leq 2(1 - \mu^{-1}\lambda^{-1}) \mathbb{E}(g(\lambda)g(\mu)),$$

since no generality is lost by choosing $\lambda \geq \mu > 1$, the result follows thanks to the estimate (1.7). $\square$

Let us discuss briefly how the above result compares with the ones present in the literature. In the work of Gordin [Go], used by Keller [Ke], a very similar theorem is present. The main difference is that condition (1) and (2) are replaced by the much stronger condition

$$\sum_{n=0}^{\infty} \mathbb{E}(\mathbb{E}(\widehat{T}^{*n} f | \mathcal{F}_0)^2) < \infty.$$

A similar comment applies to [DG], where moreover there is no discussion of the case $\sigma = 0$.

Theorem 1.1 often is applicable in cases in which T is not invertible, where sometime it is possible to choose $\mathcal{F}_0 = \mathcal{F}$ (see §2).

When T is invertible the choice $\mathcal{F}_0 = \mathcal{F}$ is likely to yield $\mathcal{F}_i = \mathcal{F}$ for each $i \in \mathbb{Z}$, this would undermine the possibility of capturing any type of dynamical coarse graining effect, whereby nullifying the hope of obtaining an interesting statistical behavior. In such a case, there are situations in which a natural choice for $\mathcal{F}_0$ exists[4] (see §3), but it would be too restrictive to require f to be $\mathcal{F}_0$–measurable.

The above difficulties can be dealt with by the following Theorem.

Theorem 1.2. *Suppose T one to one and onto. If $\mathcal{F}_i$ is coarser than $\mathcal{F}_{i-1}$, then, for each $f \in L^\infty(X)$, $\mathbb{E}(f) = 0$ such that*

(1) $\sum_{n=0}^{\infty} |\mathbb{E}(f\widehat{T}^n f)| < \infty$,

(2) *the series* $\sum_{n=0}^{\infty} |\mathbb{E}(\widehat{T}^{*n} f | \mathcal{F}_0)|$ *converges in L^1,*

(3) $\exists\, \alpha > 1$: $\sup_{k \in \mathbb{N}} k^\alpha \mathbb{E}(|\mathbb{E}(f|\mathcal{F}_{-k}) - f|) < \infty$,[5]

the sequence

$$\frac{1}{\sqrt{n}} \sum_{i=0}^{n-1} \widehat{T}^i f$$

converges in law to a Gaussian random variable of zero mean and finite variance σ, $\sigma^2 = -\mathbb{E}(f^2) + 2\sum_{n=0}^{\infty} \mathbb{E}(f\widehat{T}^n f)$.

In addition, if $\sum_{n=0}^{\infty} n|\mathbb{E}(f\widehat{T}^n f)| < \infty$, then $\sigma = 0$ if and only if there exists $g \in L^2(X)$ such that

$$\widehat{T}f = \widehat{T}g - g.$$

Proof. The key idea is to first approximate f by $\mathbb{E}(f|\mathcal{F}_{-k})$ and then use the same type of Martingale approximation introduced in Theorem 1.1. That is, to find $Y_i(k, \lambda) \in L^2(X)$ and $g(k, \lambda) \in L^2(X)$ such that, given $k > 0$, for each $i > 0$ and $\lambda > 1$

(1.8) $\qquad \mathbb{E}(Y_{i-1}(k, \lambda)|\mathcal{F}_{i-k}) = Y_{i-1}(k, \lambda)$; $\quad \mathbb{E}(Y_i(k, \lambda)|\mathcal{F}_{i-k}) = 0$,

---

[4] In fact, such σ-algebras are natural for K-systems. A K-system is a dynamical system for which there exists a σ-algebra $\mathcal{F}_0$ such that $T^{-n}\mathcal{F}_0$, $n > 0$, is coarser than $\mathcal{F}_0$; the finer σ-algebra containing $\{T^{-n}\mathcal{F}_0\}_{n \in \mathbb{Z}}$ is $\{X, \emptyset\}$ and the coarsest σ-algebra contained in $\{T^{-n}\mathcal{F}_0\}_{n \in \mathbb{Z}}$ is the full σ-algebra $\mathcal{F}$. It is easy to check that if a system is K then it is mixing; $\mathbb{E}(|\mathbb{E}(\widehat{T}^{*n} f|\mathcal{F}_0)|) = \mathbb{E}(|\mathbb{E}(f|\mathcal{F}_n)|) \to 0$ as $n \to \infty$, and $\mathbb{E}(|\mathbb{E}(f|\mathcal{F}_{-n}) - f|) \to 0$ as $n \to \infty$.

Thus, the conditions of theorem 1.2 are very close to the conditions of a K-system; it is just required a stronger control on the speed of the above convergence for the given function f. In particular, the systems discussed in section 3 are K-systems, although this fact it is not explicitly mentioned there.

[5] This condition it is not optimal, as it can be seen by looking at the proof, yet I do not know of any application in which a weaker condition could be of interest.

(in other words, $Y_i(k, \lambda)$ is a reverse Martingale difference with respect to the filtration $\{\mathcal{F}_i\}_{i=-k}^\infty$) and

(1.9) $\quad \widehat{T}^i \mathbb{E}(f|\mathcal{F}_{-k}) = Y_i(k, \lambda) + \widehat{T}^i g(k, \lambda) - \lambda^{-1}\widehat{T}^{i-1} g(k, \lambda) \quad \forall i > 0, \lambda \geq 1.$

Note that (1.8) and (1.9) are equivalent to

$$\mathbb{E}(f|\mathcal{F}_{-k}) = g(k, \lambda) - \lambda^{-1}\mathbb{E}(\widehat{T}^* g(k, \lambda)|\mathcal{F}_{-k}).$$

It is immediate to see that $g(k, \lambda) = \sum_{n=0}^\infty \lambda^{-n}\mathbb{E}(\widehat{T}^{*n} f|\mathcal{F}_{-k}) \in L^2(X)$ for each $\lambda > 1$ and in $L^1(X)$ for $\lambda = 1$ (this is a consequence of hypothesis (2) in the statement of the Theorem) is a solution of the above equation (see the analogous discussion in Theorem 1.1).

Again we want to show that the $Y_i(k, 1)$ are square summable, actually, in this case, we need a uniform estimate in k. In partial analogy with Theorem 1.1, we have

$$\mathbb{E}(Y_i(k, \lambda)^2) = \mathbb{E}(Y_1(k, \lambda)^2) = -\mathbb{E}(\mathbb{E}(f|\mathcal{F}_{-k})^2) + \mathbb{E}([\widehat{T}g(k, \lambda) - \lambda^{-1}g(k, \lambda)]^2)$$
$$= -\mathbb{E}(\mathbb{E}(f|\mathcal{F}_{-k})^2) + 2\mathbb{E}(g(k, \lambda)\mathbb{E}(f|\mathcal{F}_{-k})) - (1 - \lambda^{-2})\mathbb{E}(g(k, \lambda)^2).$$

In addition, for each $\lambda > 1$,

$$\mathbb{E}(g(k, \lambda)\mathbb{E}(f|\mathcal{F}_{-k})) \leq \sum_{n=0}^\infty |\mathbb{E}(f\mathbb{E}(\widehat{T}^{*n} f|\mathcal{F}_{-k}))| = \sum_{n=0}^\infty |\mathbb{E}(\widehat{T}^{*n} f\mathbb{E}(f|\mathcal{F}_{-k}))|$$

$$\leq 2\|f\|_\infty k\mathbb{E}(|\mathbb{E}(f|\mathcal{F}_{-k}) - f|) + \sum_{n=k}^\infty \mathbb{E}(\widehat{T}^k f\mathbb{E}(\widehat{T}^{*n} f|\mathcal{F}_0))$$

$$+ \sum_{n=0}^{2k-1} \mathbb{E}(\widehat{T}^n ff) < \infty,$$

where the uniform bound follows from the hypotheses (1), (2), (3) of the Theorem. The previous estimates show that $Y_i(k, 1)$ are uniformly square integrable martingale differences. Moreover,

$$\lim_{k\to\infty} \lim_{\lambda\to 1} \mathbb{E}(Y_1(k, \lambda)^2) = -\mathbb{E}(f^2) + 2\sum_{n=0}^\infty \mathbb{E}(f\widehat{T}^n f) = \sigma^2.$$

64

To see this it, it is enough to compute

$$\mathbb{E}(g(k, \lambda)g(k, \mu)) = \sum_{n,m=0}^{\infty} \lambda^{-n}\mu^{-m}\mathbb{E}(\widehat{T}^{*n}f\mathbb{E}(\widehat{T}^{*m}f|\mathcal{F}_{-k}))$$

$$\leq \sum_{n=0}^{\infty} \lambda^{-n}M\|f\|_{\infty}\mathbb{E}(|\mathbb{E}(\widehat{T}^{*n}f|\mathcal{F}_{-k})|) + \sum_{n=0}^{\infty} \lambda^{-n}\|f\|_{\infty}\sum_{m=M}^{\infty} \mathbb{E}(|\mathbb{E}(\widehat{T}^{*m}f|\mathcal{F}_{-k})|)$$

$$\leq M\|f\|_{\infty}^2 k + M\|f\|_{\infty}\sum_{n=0}^{\infty}\mathbb{E}(|\mathbb{E}(\widehat{T}^{*n}f|\mathcal{F}_0)|)$$

$$+ (1-\lambda^{-1})^{-1}\|f\|_{\infty}\sum_{m=M-k}^{\infty}\mathbb{E}(|\mathbb{E}(\widehat{T}^{*m}f|\mathcal{F}_0)|),$$

so, since M can be chosen arbitrarily large, $\lim_{\lambda \to 1}(1-\lambda)\mathbb{E}(g(k, \lambda)^2) = 0$. Furthermore, in analogy with Theorem 1.1, easily follows that $Y_1(k, \lambda)$ converges to $Y_1(k, 1)$ in $L^2(X)$.

This implies that, defining

$$S_n = \frac{1}{\sqrt{n}}\sum_{i=0}^{n-1}\widehat{T}^i f \; ; \quad S_n^k = \frac{1}{\sqrt{n}}\sum_{i=0}^{n-1}\widehat{T}^i\mathbb{E}(f|\mathcal{F}_k),$$

the S_n^k converges in law to a gaussian with zero means and variance $\mathbb{E}(Y_1(k, 1)^2)$.

The next step is to obtain the needed convergence as k goes to infinity.

$$\mathbb{E}([S_n^k - S_n]^2) = \frac{1}{n}\sum_{i,j=0}^{n-1}\mathbb{E}(\widehat{T}^i[f - \mathbb{E}(f|\mathcal{F}_{-k})]\widehat{T}^j[f - \mathbb{E}(f|\mathcal{F}_{-k})])$$

$$\leq \mathbb{E}([f - \mathbb{E}(f|\mathcal{F}_{-k})]^2) + 2\sum_{i=1}^{n-1}|\mathbb{E}([f - \mathbb{E}(f|\mathcal{F}_{-k})]\widehat{T}^i[f - \mathbb{E}(f|\mathcal{F}_{-k})])|$$

$$\leq 2\|f\|_{\infty}\mathbb{E}(|f - \mathbb{E}(f|\mathcal{F}_{-k})|) + 2\sum_{i=1}^{n-1}|\mathbb{E}(\widehat{T}^i f[f - \mathbb{E}(f|\mathcal{F}_{-k})])|$$

$$= 2\|f\|_{\infty}\mathbb{E}(|f - \mathbb{E}(f|\mathcal{F}_{-k})|) + 2\sum_{i=1}^{n-1}|\mathbb{E}(\widehat{T}^{*i}f[f - \mathbb{E}(f|\mathcal{F}_{-k-i})])|$$

$$\leq 2\|f\|_{\infty}\sum_{i=k}^{\infty}\mathbb{E}(|f - \mathbb{E}(f|\mathcal{F}_{-i})|),$$

which it is smaller than ε uniformly in n, since (3) implies the convergence of the series $\sum_{i=0}^{\infty}\mathbb{E}(|f - \mathbb{E}(f|\mathcal{F}_{-i})|)$.

65

Collecting the previous estimates follows that S_n converges to a Gaussian of zero mean and variance σ^2.

Next, suppose that $\sigma^2 = 0$ and $\sum_{n=0}^{\infty} n|\mathbb{E}(f\widehat{T}^n f)| \leq \infty$, then

$$(1 - \lambda^{-2})\mathbb{E}(g(k, \lambda)^2) + \mathbb{E}(Y_1(k, \lambda)^2) = -\mathbb{E}(\mathbb{E}(f|\mathcal{F}_{-k})^2) + 2\mathbb{E}(g(k, \lambda)f)$$

$$= \mathbb{E}(f^2) - \mathbb{E}(\mathbb{E}(f|\mathcal{F}_{-k})^2) - 2\sum_{n=1}^{\infty}(1 - \lambda^{-n})\mathbb{E}(f\widehat{T}^n f)$$

$$+ 2\sum_{n=0}^{\infty}\lambda^{-n}\left[\mathbb{E}(\mathbb{E}(\widehat{T}^{*n}f|\mathcal{F}_{-k})f) - \mathbb{E}(f\widehat{T}^n f)\right]$$

$$\leq \|f\|_{\infty}\mathbb{E}(|f - \mathbb{E}(f|\mathcal{F}_{-k})|) + 2(1 - \lambda^{-1})\sum_{n=0}^{\infty} n|\mathbb{E}(f\widehat{T}^n f)|$$

$$+ 2\|f\|_{\infty}\left(\sum_{n=0}^{2k-1}\mathbb{E}(|\mathbb{E}(f|\mathcal{F}_{-k}) - f|) + \sum_{n=k}^{\infty}\mathbb{E}(|\mathbb{E}(\widehat{T}^{*n}f|\mathcal{F}_0)|)\right) + 2\sum_{n=2k}^{\infty}\mathbb{E}(f\widehat{T}^n f).$$

Accordingly, it is possible to define $\phi : (0, 1) \to \mathbb{N}$, $\lim_{\lambda \to 1}\phi(\lambda) = \infty$, such that

$$\mathbb{E}(g(\phi(\lambda), \lambda)^2) \leq M \quad \forall \lambda > 1$$

$$\lim_{\lambda \to 1}\mathbb{E}(Y_1(\phi(\lambda), \lambda)^2) = 0,$$

where M is some fixed positive number.

Since $L^2(X)$ is a Hilbert space, and therefore reflexive, the unit ball is compact in the weak topology, so $\{g(\phi(\lambda), \lambda\}_{\lambda > 1}$ is a weakly compact set and we can extract a subsequence $\{\lambda_j\}$, $\lim_{j \to \infty}\lambda_j = 1$, such that $\{g(\phi(\lambda_j), \lambda_j\}$ converges weakly to a function $g \in L^2(X)$. In addition, (1.9) implies, for each $\varphi \in L^2(X)$,

$$\mathbb{E}(\widehat{T}^*\varphi\mathbb{E}(f|\mathcal{F}_{-k})) = \mathbb{E}(Y_1(\phi(\lambda_j), \lambda_j)\varphi) + \mathbb{E}(\widehat{T}^*\varphi g(\phi(\lambda_j), \lambda_j)) - \lambda_j^{-1}\mathbb{E}(\varphi g(\phi(\lambda_j), \lambda_j)),$$

and taking the limit $j \to \infty$ yields

$$\mathbb{E}(\widehat{T}^*\varphi f) = \mathbb{E}(\widehat{T}^*\varphi g) - \mathbb{E}(\varphi g) \quad \forall \varphi \in L^2(X).$$

That is

$$\widehat{T}f = \widehat{T}g - g.$$

$\square$

This theorem is rather similar to Theorem 4.4 in [DG], the main difference is the absence, in [DG], of a discussion of the degenerate case $\sigma = 0$. The only other results known to the author having a breath similar to Theorem 2.1 are contained in [Ch]. The comparison it is not so easy because the results in [Ch] are stated directly in the language of special families of finite partitions. This language it is well suited for applications to the case in which the system is studied by the type of coding called Markov sieves, but it is not so transparent in an abstract context. At any rate, an evident difference is that Chernov's result requires the existence of the first moment of the correlations (i.e., $\sum_{n=0}^{\infty} n\mathbb{E}(ff \circ T^n) < \infty$) in order to obtain the CLT while in Theorem 1.2 such a condition is not necessary, unless one wants the coboundary characterization of the functions that yield a degenerate limit.

§2 Non invertible maps.

In this section we will see how the results of the previous section apply to the case in which T is onto but not one to one.

We choose $\mathcal{F}_0 = \mathcal{F}$, so $\mathcal{F}_i = \mathcal{F}$ for all $i \leq 0$. Note that if $\mathbb{E}(\phi|\mathcal{F}_1) = \phi$, then $g(x) = \phi(T^{-1}x)$ is well defined, hence Range(T) is exactly the $\mathcal{F}_1$-measurable functions. Moreover, $\widehat{T}\widehat{T}^*$ is an orthogonal projection onto Range(T), while $\mathbb{E}(\cdot|\mathcal{F}_1)$ is an orthogonal projection onto the $\mathcal{F}_1$-mesuarable functions. That is, for each $\phi \in L^1(X)$

$$\widehat{T}\widehat{T}^*\phi = \mathbb{E}(\phi|\mathcal{F}_1).$$

Therefore, the first condition of Theorem 1.1 is satisfied quite generally. To see how the theorem works let us apply it to the case of one dimensional maps (i.e. $X = [0, 1]$).

Let us consider a partition of $[0, 1]$ into finitely many intervals $\{I_k\}_{k=1}^p$. And $T : [0, 1] \to [0, 1]$ such that

(1) $T|_{\overline{I}_k} \in \mathcal{C}^{(2)}$ for each $k \in \{1, ..., p\}$
(2) $\inf\limits_{x\in[0, 1]} |D_x T| \geq \lambda > 1$.

That is a piecewise smooth expanding map. If the reader wants to consider a concrete example, here is a very simple one: the piecewise linear map $T : [0, 1] \to [0, 1]$ define by

$$T(x) = \begin{cases} \dfrac{9}{2}\left(\dfrac{1}{9} - x\right) & x \in \left(0, \dfrac{1}{9}\right) \\[2mm] \dfrac{9}{2}\left(x - \dfrac{1}{9}\right) & x \in \left(\dfrac{1}{9}, \dfrac{3}{9}\right) \\[2mm] \dfrac{9}{2}\left(\dfrac{5}{9} - x\right) & x \in \left(\dfrac{3}{9}, \dfrac{5}{9}\right) \\[2mm] \dfrac{9}{2}\left(x - \dfrac{5}{9}\right) & x \in \left(\dfrac{5}{9}, \dfrac{7}{9}\right) \\[2mm] \dfrac{9}{2}(1 - x) & x \in \left(\dfrac{7}{9}, 1\right) \end{cases}$$

The map satisfies our assumptions since $|DT| = \frac{9}{2} > 1$.
The following result is well known [HK]:

Theorem 2.1. *There exists a unique probability measure μ, absolutely continuous with respect to Lebesgue, which is invariant with respect to the map T. In addition, there exist[6] $\Lambda \in (0, 1)$ and $K > 0$ such that, for each $f \in BV([0, 1])$ (the space of functions of bounded variation), and $g \in L^1([0, 1], \mu)$*

$$\left| \int_0^1 fg \circ T^n d\mu - \int_0^1 f d\mu \int_0^1 g d\mu \right| \leq K\Lambda^n \|f\|_{\mathrm{BV}} \|g\|_1.$$

Since μ is absolutely continuous with respect to the Lebesgue measure m, the Radon–Nicodým derivative $h = \frac{d\mu}{dm}$ is in $L^1([0, 1], m)$. For simplicity assume $h \geq \varepsilon > 0$,[7] then it follows

$$\widehat{T}^* f(x) = h(x)^{-1} \sum_{y \in T^{-1}(x)} h(y) f(y) |D_y T|^{-1}.$$

Such a representation implies that the last statement of Theorem 2.1 can be rephrased as follows: for each $f \in BV([0, 1])$, $\int_0^1 f d\mu = 0$

$$\|\widehat{T}^{*n} f\|_\infty \leq K\Lambda^n \|f\|_{\mathrm{BV}}.$$

It is then immediate to see that Theorem 1.1 applies to this situation yielding the central limit theorem for all functions of bounded variation. The reader can easily see that such a result can be improved obtaining the central limit theorem for functions with less regularity (e.g., by an approximation argument) but this is not the main focus here. In addition, similar results can be obtained for several cases in which the map T consists of infinitely many smooth pieces.

It is also immediate to verify that the theorem will yield the CLT for BV functions also for some non-hyperbolic maps (such as the quadratic family [Y]) or maps that are non-uniformly hyperbolic ([LSV]).

§3 Invertible maps.

In this case it would be useless to choose $\mathcal{F}_0 = \mathcal{F}$: typically this would yield $\mathcal{F}_i = \mathcal{F}$ for each $i \in \mathbb{Z}$. So the choice of $\mathcal{F}_0$ must be motivated by dynamical considerations. Here we will discuss a general class of systems for which such a choice is quite natural: the hyperbolic systems.[8]

For simplicity I will confine the discussion to the case in which X is as compact symplectic manifold with a Riemannian structure that yields a volume form equivalent to the symplectic one and T a piecewise $\mathcal{C}^2$ symplectic map, but see [KS] and [LW] for more general possibilities. By hypothesis the symplectic (or Riemannian)

---

[6] In fact, it is possible to obtain explicit estimates of K and Λ; see [L2] for details.

[7] This is always verified if T is continuous, like in our example; but see [L2] for a discussion of the general case.

[8] More generally this strategy can be applied to K-systems.

volume μ is invariant. (The more general case of dissipative systems can also be treated by the same arguments, again the details are left to the reader).

We will assume T uniformly hyperbolic, since almost nothing is known on the decay of correlations for non-uniformly hyperbolic systems. By this we mean that at each point $x \in X$ there exists two subspaces $E^u(x)$, $E^s(x) \in T_x X$, $E^u(x) \cap E^s(x) = \{0\}$ and $E^u(x) \oplus E^s(x) = T_x X$, invariant (i.e., $D_x T E^{u,s}(x) = E^{u,s}(Tx)$), and there exists $\lambda > 1$ such that for each $x \in X$,

$$\|D_x T v\| \geq \lambda \|v\| \quad \forall v \in E^u(x)$$
$$\|D_x T v\| \leq \lambda^{-1} \|v\| \quad \forall v \in E^s(x).$$

Also, we assume that $E^{u,s}(x)$ depends continuously with respect to x (the above systems are called Anosov, in the smooth case). In the smooth case such systems are known to be ergodic (in fact, Bernoulli), one can see [LW] for sufficient conditions that insure ergodicity also in the non smooth case. To help the reader in better visualizing the following discussion let us consider the simplest possible non-trivial example.

We consider a family of linear maps of the plane defined by

$$x_1' = x_1 + a x_2$$
$$x_2' = x_2,$$

where a is a real parameter. We use these linear maps to define (discontinuous if $a \notin \mathbb{N}$) maps of the torus by restricting the formulas to the strip $\{0 \leq x_2 \leq 1\}$ and further taking them modulo 1. In this way we define a mapping T_1 of the torus $\mathbb{T}^2 = \mathbb{R}^2/\mathbb{Z}^2$ which is discontinuous on the circle $\{x_2 \in \mathbb{Z}\}$ (except when a is equal to an integer) and preserves the Lebesgue measure μ.

Similarly we define another family of maps depending on the same parameter a by restricting the formulas

$$x_1' = x_1$$
$$x_2' = a x_1 + x_2$$

to the strip $\{0 \leq x_1 \leq 1\}$ and then taking them modulo 1. Thus for each a we get a mapping T_2 of the torus which is discontinuous on the circle $\{x_1 \in \mathbb{Z}\}$ (except when a is equal to an integer) and preserves the Lebesgue measure μ.

Finally we introduce the composition of these maps $T = T_2 T_1$ which depends on one real parameter a. An alternative way of describing the map T is by introducing two fundamental domains for the torus $\mathcal{M}^+ = \{0 \leq x_1 + a x_2 \leq 1, 0 \leq x_2 \leq 1\}$ and $\mathcal{M}^- = \{0 \leq x_1 \leq 1, 0 \leq -a x_1 + x_2 \leq 1, \}$.

The linear map defined by the matrix

$$\begin{pmatrix} 1 & a \\ a & 1+a^2 \end{pmatrix} = \begin{pmatrix} 1 & 0 \\ a & 1 \end{pmatrix} \begin{pmatrix} 1 & a \\ 0 & 1 \end{pmatrix}$$

takes $\mathcal{M}^+$ onto $\mathcal{M}^-$ thus defining a map of the torus which is discontinuous at most on the boundary of $\mathcal{M}^+$ and preserves the Lebesgue measure. This is the map T that constitute our toy model.

Let us go back to the more general case, according to [KS] such systems have a natural measurable partition (in fact a K-partition): the partition into stable manifolds.

Such a partition $\mathcal{P}$ can be constructed as to satisfy the following requirements:

(1) there exists a finite number of codimension one smooth manifolds $\{S_i\}_{i=1}^{m_0}$, transversal to the stable direction, such that each $p \in \mathcal{P}$ has the boundaries points belonging to the set $\cup_{j=1}^{m_0} \cup_{n=0}^{\infty} T^{-n}S_i$;[9]

(2) for each $p \in \mathcal{P}$ diam$(p) \le 2\delta_0$;[10]

(3) for each $p \in \mathcal{P}$ there exists $\{p_i\}_{i=1}^k \subset \mathcal{P}$ such that $T^{-1}p = \cup_{i=1}^k p_i$.

According to above properties, if one chooses as $\mathcal{F}_0$ the σ-algebra generated by the partition $\mathcal{P}$, then $\{\mathcal{F}_i\}_{i=0}^{\infty}$ has the dynamical properties requested in the hypotheses of Theorem 1.2.

To make the previous statement more clear let us see how such a partition looks like in the concrete example mentioned above.

The map T is piecewise linear and it has constant contracting direction v. Let us call S the discontinuity set of T^{-1} and $S_{\infty} = \cup_{n=0}^{\infty} T^{-n}S$. Then the stable partition is made of segments along the direction v with the endpoints belonging to S_{∞}.[11] Since S_{∞} is an invariant set, properties (1)-(3) are readily verified.

Further, we will assume that the manifolds $\{S_i\}$ satisfy the following property:

Property 0. *For each $i \ne j$ $\overline{S}_i \cap \overline{S}_j$ is either empty or consists of smooth submanifolds I_{ij} of codimension at least two. Moreover, setting*[12]

$$M \equiv \sup_{ij} \sharp\{k \in \{1, ..., m_0\} \mid \overline{S}_k \cap I_{ij} \ne \emptyset\},$$

we require

$$\nu \equiv \lambda^{-1}M < 1.$$

Note that Property 0 may not be satisfied by T but may be enjoyed by T^q, for some $q > 1$. In fact, it is not so hard to see that "generically" this will be the case (i.e., Property 0 will hold for some iterate of the map). In such a situation we can apply all the following to the dynamical system (X, μ, T^q) obtaining the same conclusions as far as the CLT is concerned. Here, for simplicity, we restrict ourselves to the case $q = 1$.

---

[9] In the discontinuous case such manifolds can be simply chosen as the set of points at which T is not $\mathcal{C}^{(2)}$.

[10] δ_0 is some previously fixed number.

[11] See [LW] for the details of such a construction and the proof that almost every point belongs to one such segment.

[12] Here by $\sharp B$ is meant the cardinality of the set B.

If we think to our model example we see that $M = 2$, so that $\lambda^{-1}M < 1$ if $|a| > \frac{1}{\sqrt{2}}$. The reader can easily compute M for powers of T and see that Property 0 is satisfied for smaller and smaller values of a. Of course, $a = 0$ corresponds to the identity, for which no hyperbolicity is present.

For the systems under consideration holds the following (see [KS] for details)

Property 1. *For each $p \in \mathcal{P}$ define the measure μ_p by*

$$\int_p g \, d\mu_p \equiv \mathbb{E}(g|\mathcal{F}_0)(x),$$

for $g \in \mathcal{C}^{(0)}(X)$, and $x \in p$.

Then, calling m_p the measure induced by the Riemannian structure on p, and $\phi_p = \frac{d\mu_p}{dm_p}$ the Radon–Nicodým derivative, there exist c_0 such that $\sup_p \|\phi_p\|_\infty \leq c_0$.

For our simple example we see that $\mu_p = \frac{1}{m_p(p)} m_p$.

The map is invertible, thus $\widehat{T}^* f = f \circ T^{-1}$. A very important consequence of Property 1 is that, if $p \in \mathcal{P}$ and $\mathcal{P}' \subset \mathcal{P}$ is such that $\bigcup_{q \in \mathcal{P}'} q = T^{-n}p$, then for each $f \in L^1(X, \mu)$

$$\int_p f \circ T^{-n} d\mu_p = \sum_{q \in \mathcal{P}'} \mu_p(T^n q) \int_q f \, d\mu_p.$$

In addition, one can prove the following (see [L1] for a complete discussion of the two-dimensional case).

Property 2. *For each $\delta \leq \delta_0$, there exists $K \in \mathbb{R}$ and $\Lambda > 1$, such that, for each $x \in X$ that belongs to a $p \in \mathcal{P}$ with $diam(p) \geq \delta$, and for each $g, f \in \mathcal{C}^\alpha(X)$ (Hölder continuous of class $\alpha > 0$), $\int_X g = 0$,[13]*

$$|\mathbb{E}(f\widehat{T}^{*n}g|\mathcal{F}_0)(x)| \leq K\Lambda^{-n}\|g\|_\alpha^s \|f\|_\alpha^u$$

In the rest of the section we will see that Properties 0-2 imply, for the systems under consideration, the hypotheses of Theorem 1.2.

Lemma 3.1. *Calling $\mathcal{A}_\varepsilon = \{x \in X \mid diam(p(x)) \leq \varepsilon\}$ we have[14]*

$$m(\mathcal{A}_\varepsilon) \leq C\varepsilon$$

---

[13] By $\|f\|_\alpha$ we mean the usual $\mathcal{C}^{(\alpha)}$ norm, while $\|f\|_\alpha^s = \sup_{p \in \mathcal{P}} \sup_{x, y \in p} \frac{|f(x)-f(y)|}{\|x-y\|^\alpha} + \|f\|_\infty$; and $\|f\|_\alpha^u$ is defined analogously by using the unstable partition. Essentially, This norms measure the Hölder derivative in the stable (or unstable) direction only.

[14] By $m(\cdot)$ we mean the symplectic or Riemannian metric that, according to our hypotheses, is the invariant measure of the system.

for some fixed $C \in \mathbb{R}^+$.

Proof. Since ∂p is made up of points belonging to the preimages of the manifolds S_i, it follows that if $\text{diam}(p) \leq \varepsilon$ then there exists $z \in \partial p$ and $n \in \mathbb{N}$, $i \in \{1, ..., m_0\}$ such that $T^n z \in S_i$. Accordingly, $T^n p$ must lie in a $\lambda^{-n} \varepsilon$ neighborhood of S_i. Such a neighborhood has measure $c_1 \lambda^{-n} \varepsilon$, for some fixed c_1. It is then clear that

$$m(\mathcal{A}_\varepsilon) \leq \sum_{n=0}^{\infty} m_0 c_1 \lambda^{-n} \varepsilon = \frac{m_0 c_1}{1 - \lambda} \varepsilon.$$

$\square$

The problem in applying our theorem comes from the possible presence in $\mathcal{P}$ of very small elements. On such elements Property 2 does not provide any direct control. To our advantage instead works Lemma 3.1 that informs us that the total measure of the very small pieces is small.

Yet, small pieces may be present.[15] The idea to deal with them consists in iterating them: if $T^{-n}|_p$ is smooth, then $\text{diam}(T^{-n}p) \geq \lambda^n \text{diam}(p)$. Unfortunately, in general T is not smooth so we have to handle the iterations with more care.

Fix $p \in \mathcal{P}$, by construction there exists $\mathcal{P}_1 \subset \mathcal{P}$ such that $T^{-1}p = \bigcup_{q \in \mathcal{P}_1} q$. Call $\mathcal{P}_1^- = \{q \in \mathcal{P}_1 \mid \text{diam}(q) \leq \delta\}$ and $p_1 = \bigcup_{q \in \mathcal{P}_1^-} Tq \subset p$. In other words p_1 consists of the portion of p that, under the actions of T^{-1}, does not give rise to sufficiently large elements of the partition. The process can obviously be iterated: let $\mathcal{P}_2$ be the collection such that $T^{-2}p_1 = \bigcup_{q \in \mathcal{P}_2} q$, $\mathcal{P}_2^- = \{q \in \mathcal{P}_2 \mid \text{diam}(q) \leq \delta\}$, $p_2 = \bigcup_{q \in \mathcal{P}_2^-} T^2 q \subset p_1$ and so on.

Lemma 3.2. *If δ is sufficiently small and $p \in \mathcal{A}_\varepsilon$, then for $n \geq \frac{\log \varepsilon^{-1} \delta}{\log \nu^{-1}} + m$,*

$$m_p(p_n) \leq \varepsilon \nu^m.$$

Proof. Thanks to Property 0, the choice of δ sufficiently small insures that each element with diameter less than δ can intersect at most M manifolds S_i. Since the S_i describe all the possible discontinuities in our system, it follows that $\sharp \mathcal{P}_1 \leq M$. But the same argument applies to each connected piece of p_j: since the diameter of $T^{-l}p_j$ is, by definition, less than δ, for $l < j$, it follows that $T^{-l-1}p_j$ can consists of, at most, M elements of the partition. In conclusion, $\sharp \mathcal{P}_n^- \leq M^n$, and

$$m_p(p_n) \leq M^n \lambda^{-n} \delta \leq \nu^m \varepsilon.$$

---

[15] In fact, this is certainly the case in the non-smooth case. If T is smooth, then it is possible to construct $\mathcal{P}$ in such a way that $\text{diam}(\mathcal{P}) \geq \delta$ for some fixed δ, by using Markov partitions. When finite Markov partitions are available the present method boils down to a repackaging of well known facts.

□

Using the above estimates, for each $n, k \in \mathbb{N}$, $k < n$, and calling $\chi_{\mathcal{B}}$ the characteristic function of the set $\mathcal{B}$, we have

$$\mathbb{E}(|\mathbb{E}(g\widehat{T}^{*n}f|\mathcal{F}_0)|) \leq \sum_{m=0}^{\infty} \mathbb{E}(\chi_{\mathcal{A}_{\nu m \delta}}|\mathbb{E}(g\widehat{T}^{*n}f|\mathcal{F}_0)|)$$

$$= \sum_{m=0}^{n-k} \mathbb{E}(\chi_{\mathcal{A}_{\nu m \delta}}|\mathbb{E}(\widehat{T}^{*k}[\widehat{T}^k g\widehat{T}^{*(n-k)}f]|\mathcal{F}_0)|)$$

$$+ \sum_{m=n-k+1}^{\infty} \|f\|_{\infty}\|g\|_{\infty}\mu(\mathcal{A}_{\nu m \delta}).$$

Next, for each $x \in p \subset \mathcal{A}_{\nu m \delta}$, using the notations of the Lemma 3.2 and setting $\phi = \widehat{T}^k g\widehat{T}^{*(n-k)}f$,

$$|\mathbb{E}(g\widehat{T}^{*n}f|\mathcal{F}_0)| = \left|\int_p \phi \circ T^{-k}d\mu_p\right| \leq \left|\int_{p_k} \phi \circ T^{-k}d\mu_p\right| + \left|\int_{p \setminus p_k} f \circ T^{-k}d\mu_p\right|$$

$$\leq c_0 m_p(p_k)\|\phi\|_{\infty} + \sum_{\substack{q \in \mathcal{P} \\ T^k q \subset p \setminus p_k}} \mu_p(T^k q)\left|\int_q \phi d\mu_p\right|$$

$$\leq c_0\|f\|_{\infty}\|g\|_{\infty}\delta\nu^k + \sum_{\substack{q \in \mathcal{P} \\ T^k q \subset p \setminus p_k}} \mu_p(T^k q)|\mathbb{E}(\phi|\mathcal{F}_0)|$$

$$\leq c_0\|f\|_{\infty}\|g\|_{\infty}\delta\nu^k + K\Lambda^{-n+k}\|\widehat{T}^k g\|_{\alpha}^s\|f\|_{\alpha}^u$$

$$\leq \|f\|_{\alpha}\|g\|_{\alpha}[c_0\delta\nu^k + K\Lambda^{-n+k}\lambda^{-k\alpha}].$$

Using the above estimates yield

$$\mathbb{E}(|\mathbb{E}(g\widehat{T}^{*n}f|\mathcal{F}_0)|) \leq \|f\|_{\alpha}\|g\|_{\alpha}\left[(n-k)(c_0\nu^k\delta + K\Lambda^{-n+k}) + \frac{C\delta}{1-\nu}\nu^{n-k+1}\right].$$

Hence, by choosing $k = [\frac{n}{2}]$, it follows that there exists $c_1 \in \mathbb{R}^+$ such that

$$\mathbb{E}(|\mathbb{E}(gT^{*n}f|\mathcal{F}_0)| \leq c_1 n \max\{\nu, \Lambda^{-1}\}^{\frac{n}{2}}.$$

This verifies the hypotheses 1–2 of Theorem 1.2. Hypothesis 3 follows trivially from the assumption $f \in \mathcal{C}^{(\alpha)}$ (or f piecewise Hölder). To see this, consider that $\mathcal{F}_{-k}$ is the σ-algebra associated to the partition $T^k\mathcal{P}$. But, if $p \in T^k\mathcal{P}$, then

73

diam$(p) \leq 2\lambda^{-k}\delta_0$. Thus, for each $x \in p \in T^k\mathcal{P}$, let $q \in \mathcal{P}$ be the unique element such that $p \subset q$, then

$$|\mathbb{E}(f|\mathcal{F}_{-k})(x) - f(x)| = |\mu_q(p)^{-1} \int_p f d\mu_q - f(x)| \leq \sup_{y \in p} |f(y) - f(x)|$$
$$\leq \|f\|_\alpha \text{diam}(p)^\alpha \leq \|f\|_\alpha 2^\alpha \lambda^{-\alpha k}\delta_0^\alpha.$$

Hence Theorem 1.2 applies to the class of systems under consideration and yields the CLT for all Hölder continuous functions (in fact, the above considerations can easily be pushed to obtain the CLT for much larger classes of functions; e.g. piecewise continuous functions with a logarithmic modulus of continuity).

References

[BSC1]   L.A.Bunimovich, Ya.G. Sinai, N.I. Chernov, *Markov partitions for two dimensional billiards*, Rus. Math. Surv. **45** (1990), 105–152.

[BSC2]   L.A.Bunimovich, Ya.G. Sinai, N.I. Chernov, *Statistical properties of two dimensional hyperbolic billiards*, Rus. Math. Surv. **46** (1991), 47–106.

[Ch]   N.I.Chernov, *Limit Theorems and Markov approximations for chaotic dynamical systems*, Probability Theory and Related Fields **101** (1995), 321–362.

[De]   M. Denker, *The Central Limit theorem for dynamical systems*, Dynamical Systems and Ergodic Theory, Banach Center Publications, **23**, Czeslaw Olech editor, PWN–Polish Scientific Publisher, Warsaw, 1989.

[DG]   D.Dürr, S.Goldstein, *Remarks on the Central Limit Theorem for weakly dependent random variables*, Stochastic Processes–Mathematics and Physics, S.Albeverio, Ph.Blanchard and L.Streit editors, Lecture Notes in Mathematics, vol. **1158**, Springer-Verlag, 1986.

[Go]   M.I.Gordin, *The Central Limit Theorem for stationary processes*, Soviet.Math.Dokl. **10** n.5 (1969), 1174–1176.

[HK]   F. Hofbauer, G. Keller, *Ergodic properties of invariant measures for piecewise monotone transformations*, Math. Zeit. **180** (1982), 119–140.

[IL]   I.A.Ibragimov, Y.V.Linnik, *Independent and stationary sequences of random variables*, Wolters-Noordhoff, Groningen, 1971.

[Ke]   G. Keller, *Un théorem de la limite centrale pour une classe de transformations monotones par morceaux*, Comptes Rendus de l'Académie des Sciences, Série A **291** (1980), 155–158.

[KS]   A.Katok, J.M. Strelcyn with collaboration of F. Ledrappier and F. Przytycki, *Invariant Manifolds, Entropy and Billiards, Smooth Maps with Singularities*, Lectures Notes in Mathematics **1222** (1986).

[KV]   C.Kipnis, S.R.S.Varadhan, *Central Limit Theorem for addictive functions of reversible Markov process and applications to simple exclusions*, Commun.Math.Phys. **104** (1986), 1–19.

[L1]   C. Liverani, *Decay of correlations*, Annals of Mathematics **142** (1995), 239–301.

[L2]   C. Liverani, *Decay of Correlations in Piecewise Expanding maps*, Journal of Statistical Physics **78**, 3/4 (1995), 1111–1129.

[LW]   C. Liverani, M.P.Wojtkowski, *Ergodicity in Hamiltonian Systems*, Dynamics Reported **4** (1995), 130–202.

[LSV]   A.Lambert, S.Siboni, V.Vaienti, *Statistical properties of a non-uniformly hyperbolic map of the interval*, Journal of Stat.Phys. **72** (1993), 1305.

[Ne]   J.Neveu, *Mathematical foundations of the calculus of probability*, Holden-Day, San Francisco, 1965.

[Y]   L.-S.Young, *Decay of correlations for certain quadratic maps*, Comm.Math.Phys. **146** (1992), 123–138.

LIVERANI CARLANGELO, MATHEMATICS DEPARTMENT, UNIVERSITY OF ROME *Tor Vergata*, 00133 ROME, ITALY.

E-mail address: liverani@mat.utovrm.it

RAFAEL DE LA LLAVE
On necessary and sufficient conditions for uniform integrability of families of Hamiltonian systems

To the memory of Ricardo Mañé

In [Po] Ch. V, specially §81, H. Poincaré discussed an obstruction to uniform integrability of families of Hamiltonians. (That is, the existence of changes of variables analytic in the parameter ϵ and in the variables that make the family of Hamiltonians a function of only action variables). We examine his proof and discover that, for non-degenerate systems, this condition is also sufficient for the integrability to first order in the parameter (That is, there exist analytical changes of variables, analytic in ϵ so that the family in these new variables depends only on the action variables up terms which are $o(\epsilon)$.) This leads to the existence of obstructions in higher order. We show that the vanishing of the obstructions to order n is sufficient for the existence of analytic and symplectic changes of variables analytic in the parameter ϵ that reduce the system to integrable up to errors of order ϵ^{n+1}. Moreover, we show that the vanishing of all the obstructions means that the system is uniformly integrable. We note that these obstructions have a geometric meaning and they are cohomology obstructions computed on periodic orbits of the system.

1. Introduction.

Many problems in Mechanics can be reduced to the study of a family of Hamiltonian systems H_ϵ where H_0 is an integrable system. (By an integrable system we mean one whose phase space admits action-angle variables and such that the Hamiltonian depends only on the actions.)

At the early times in Mechanics — and more surprisingly, up to recently — there was widespread hope that almost all families H_ϵ could be systematically reduced to integrable by canonical changes of variables. (Found, e.g., solving the Hamiltonian-Jacobi equations.)

That is, there was hope that one could find a family of canonical transformations g_ϵ in such a way that

$$(1.1) \qquad\qquad\qquad H_\epsilon \circ g_\epsilon = I_\epsilon$$

where I_ϵ is only an analytic function of the actions. This situation, where whole families can be reduced is refered to as *"uniform integrability"*. The "uniform" emphasizing the fact that there is a parameter involved and that all the constructions can be performed in ways that depend analytically on the parameter.

The fact that these hopes can not be fulfilled – even in the very weak sense of formal power series – was demonstrated very eloquently in [Po]. The whole of chapter V is devoted to finding obstructions for uniform integrability. In particular, in §81 one can find obstructions for the existence of analytic solutions of (1.1) that depend analytically on ϵ (uniform integrability) and in §84, the conditions are extended slightly and verified for the three-body problem. At the end of [Po] Chapter V (p. 259) the question is raised of whether these obstructions are sufficient. Of course, the more recent K. A. M. theorem shows that provided some non-degeneracy conditions are met, one can find solutions of (1.1) but only on a Cantor set.

The goal of this paper is to reexamine in more modern language the circle of ideas introduced in [Po]. We will show that the criterion of non-integrability introduced in [Po] can be strengthened and that this strengthening is necessary and sufficient for uniform integrability when the unperturbed system is non-degenerate.

We point out that a thorough analysis of the degenerate situation was undertaken in [Ga]. There, the situation is quite different and there are examples with systems integrable to all orders but not uniformly integrable. In [Ga], one can also find sufficient conditions for convergence of perturbation expansions of some degenerate situations.

The obstructions to uniform integrability we will describe are integrals over periodic orbits. They are equivalent to those of [Po] even if there, they were stated in terms of Fourier series coefficients. We try to write everything in terms of periodic orbits because the geometric meaning is clearer. In particular, the obstructions can be interpreted as cohomology obstructions for existence of cocycles. We will also discuss a further geometrical interpretation.

The main technical tool to prove the converse of the criterion will be a result that states that if the obstructions vanish on periodic orbits, a cohomology equation has a solution. This is somewhat reminiscent of the celebrated Livsic theorem [Li] and its use in rigidity results for Anosov diffeomorphisms. Of course, in our case there is no hyperbolic behavior, hence, the proofs are very different. In that respect we call attention to [Ve], where in is shown that cohomology equations for linear maps on the torus can be solved using only obstructions on periodic orbits. The methods, however are very algebraic and do not generalize to our situations.

We will prove two types of results: some that have to do with asymptotic expansions and their feasibility and some more complicated ones that discuss the convergence of these expansions.

Even if the obstructions and asymptotic results need only a finite number of derivations and simple minded calculations, the converse of the full set of obstructions — that is, the theorem that shows that if all the obstructions vanish, the system is uniformly integrable — does require analyticity and, as far as we know, an iterative process similar to those K. A. M. theorem.

The same circle of ideas applies to other related problems that have appeared in the literature. We will also discuss the problem of relative integrability. That is, the existence of a family of canonical transformations g_ϵ and a function of a real variable F_ϵ such that:

$$(1.2) \qquad\qquad H_\epsilon \circ g_\epsilon = F_\epsilon(H_0)$$

This means that H_ϵ describes the same motions as H_0 up to a change of variables and a reparameterization of the orbits. That is, the orbits of H_ϵ are the same as that of H_0 under a change of variables, even if they may be transversed at different speed. The change of speed depends only on the energy surface.

We emphasize that the results we present here deal exclusively with the problem of uniform integrability. In particular, we do not discuss related problems such as the integrability for open domains of the parameter. To the best of our knowledge failure of uniform integrability does not exclude that the system is integrable in a whole neighborhood of parameters. Even if it seems unlikely, we cannot exclude that there exist integrals that are not analytic with respect to parameters. Much less we can exclude integrability for dense set of parameters etc. We also call attention to the fact that we use integrability in a well defined sense that involves changes of variables that are analytic and with analytic inverses. The literature abounds with weaker notions (multi-valued integrals, integrals with singularities etc.) which are sometimes also referred as integrability. We have little to say in this paper about these notions. It is clear that our methods rely on our precise definitions. When to avoid cluttering the sentences we just refer to "integrability" we mean one of our precise definitions that we hope, will be clear from the context.

2. Definitions, preliminary calculations.

Even if several of the calculations we will perform will require only a finite number of derivatives, the main results indeed require analyticity. Hence, we will assume that all the Hamiltonians are analytic in all the parameters.

We will consider a phase space of the form $M = U \times \mathbb{T}^d$ where U is an open set in $\mathbb{R}^d$. (We will also use complexifications of this phase space.) Given a function $\eta : U \times \mathbb{T}^d \to \mathbb{C}$ we will denote the partial Fourier expansion as

$$\eta(A, \varphi) = \sum_{k \in \mathbb{Z}^d} \hat{\eta}_k(A) e^{2\pi i k \varphi}.$$

We denote the symplectic form by γ (and reserve ω for the frequencies). In the action-angle variables we have $\gamma = \sum_{i=1}^d dA_i \wedge d\varphi_i$.

Note that if the phase space admits action-angle variables we have $\gamma = d\theta$ where $\theta = \sum_i A_i d\varphi_i$. We will refer to θ as the symplectic potential. (Not all symplectic manifolds have a symplectic potential, but we are only concerned in this paper with those having action angle variables as above.)

A transformation f is called symplectic when $f_* \gamma = \gamma$. Equivalently, if $d(f_* \theta - \theta) = 0$. If, furthermore, we have $f_* \theta - \theta = dS$, we say that f is exact symplectic. Note that the notion of exact symplectic maps only makes sense for symplectic manifolds in which the symplectic form admits a potential.

We say that a Hamiltonian is integrable if it is a function only of the action variables. Perhaps it would me more appropriate to say that a system is integrable when it can be reduced to one of the above form by a canonical change of variables. However, the above notation is quite extended and we will follow it.

We will also be concerned with integrable systems that are non-degenerate in the following sense:

Definition 2.1. *We say that a Hamiltonian $H_0 : M \to \mathbb{R}$ is non degenerate if the mapping $\omega : U \to \mathbb{R}^d$ defined by $\omega = \frac{\partial H_0}{\partial A}$ is a diffeomorphism in its range and $\omega^{-1} : \omega(U) \to U$ is a uniformly analytic map.*

We will find it convenient to use the deformation calculus for families. If $\mathcal{G}_\epsilon$ is a family of diffeomorphisms we write:

$$\frac{d}{d\epsilon} g_\epsilon = \mathcal{G}_\epsilon \circ g_\epsilon$$

and we will refer to $\mathcal{G}_\epsilon$ as the generator of the family. Note that, for C^1 vector fields, $\mathcal{G}_\epsilon$ and g_0 determine uniquely g_ϵ.

The transformations are symplectic if and only if $g_{\epsilon *}\gamma = \gamma$. This happens if and only if $g_{0 *}\gamma = \gamma$ and $\frac{d}{d\epsilon}(g_{\epsilon *}\gamma) = 0$. Using the formulas for the Lie derivative and the fact that $d\gamma = 0$, we see that a family is symplectic if and only if

$$g_{0 *}\gamma = \gamma; \quad d(i(\mathcal{G}_\epsilon)\gamma) = 0$$

This motivates the following well known definitions:

Definition 2.2. *We say that a C^1 vector field $\mathcal{F}$ is locally Hamiltonian if and only if*

$$d(i(\mathcal{F})\gamma) = 0$$

We say that a C^1 vector field $\mathcal{F}$ is globally Hamiltonian if there exist a function F such that

$$i(\mathcal{F})\gamma = dF$$

Similarly, we say that g_ϵ is a locally (globally) Hamiltonian isotopy if its generator $\mathcal{G}_\epsilon$ is locally (globally) Hamiltonian. It is a well known result that g_ϵ is a family of symplectic mappings if and only if g_0 is symplectic and is a locally Hamiltonian isotopy. Similarly, it is a family of exact symplectic mappings if g_0 is exact symplectic and g_ϵ is a globally Hamiltonian isotopy.

Note that given a function F, we can compute dF and, because γ is non-degenerate, we can find a unique $\mathcal{F}$ such that $i(\mathcal{F})\gamma = dF$. Conversely, given $\mathcal{F}$, $i(\mathcal{F})\gamma = dF$ determines F up to an additive constant on every connected component of its domain of definition.

Definition 2.3. *Let H_0 be an integrable Hamiltonian, H_ϵ be an analytic family agreeing with H_0 when $\epsilon = 0$. We say that H_ϵ is a uniformly integrable family if we can find an analytic family of canonical transformations g_ϵ and a family of integrable Hamiltonians I_ϵ such that*

(2.1)
$$H_0 = I_0 , \qquad g_0 = Id$$
$$H_\epsilon \circ g_\epsilon = I_\epsilon$$

In the literature, one sometimes finds the notion of relatively integrable family.

Definition 2.4. *We say that H_ϵ is H_0-integrable if we can find a family of canonical transformations g_ϵ and a C^r-family of real valued functions of a real variable F_ϵ in such a way that*

(2.2)
$$g_0 = Id , \quad F_0 = Id$$
$$H_\epsilon \circ g_\epsilon = F_\epsilon(H_0)$$

We will refer to the g_ϵ as the integrating transformations.

The reason to introduce the concept of relative integrability is that if (2.2) is met, we can understand the dynamics of all the Hamiltonians H_ϵ in terms of the dynamics of H_0. This is advantageous when the dynamics of H_0 is "well understood." Of course, when H_0 is integrable, the dynamics is well understood because of the explicit solution. Nevertheless, a system may have dynamics for which a great deal is known even if there are no explicit formulas.

Note also that the problem of relative integrability is geometrically more natural than the problem of integrability. It does not use the fact that there exist action and angle variables and, hence, it makes sense to pose the problem in any manifold. In particular, there are less stringent notions of action-angle variables (see e.g. [Du]) where much of the ideas here could carry over. In this paper, however, we will only consider the problem of relative integrability with respect to an integrable system. The papers [CEG] and [LMM] consider the problem of relative integrability of Anosov systems, which certainly do not have action angle variables.

In the relatively integrable case, we will use some more geometric structures. Note that for E real – or complex – we can define the energy surface $\Sigma_E = H_0^{-1}(E)$. Recall also that the symplectic form γ defines a volume form γ^d. If the energy surface does not contain critical points of the Hamiltonian, we can introduce a natural $2d-1$ form μ on the energy surface Σ_E determined by by $\gamma^d = d\mu_E \wedge dH_0$. Since both H_0 and γ are invariant under the flow of H_0, so is μ. We will assume that the Hamiltonians we consider are defined on domains $H_0^{-1}(\Omega)$ where Ω is an interval in $\mathbb{R}$ or a complex domain containing a real interval. We will furthermore assume that there are no critical points in this set and, moreover that the energy surfaces are compact for real values of the energy.

Note that the I_ϵ, g_ϵ that integrate a family of Hamiltonians are not unique. In effect, if $h_\epsilon(A, \varphi) = (A + \Delta_\epsilon, \varphi)$ then,

$$H_\epsilon \circ g_\epsilon \circ h_\epsilon = I_\epsilon \circ h_\epsilon = \widetilde{I}_\epsilon$$

where $\widetilde{I}_\epsilon$ defined by $\widetilde{I}_\epsilon(A) = I_\epsilon(A + \Delta_\epsilon)$ depends only on the actions.

Similarly if $K_\epsilon(A)$ is the generator of k_ϵ, $k_0 = Id$

$$k_\epsilon(A, \varphi) = (A, \varphi + \Gamma_\epsilon(A))$$

(where $\Gamma_\epsilon(A) = \int_0^\epsilon \frac{\partial K_\sigma}{\partial A}(A)\, d\sigma$) and again, $H_\epsilon \circ g_\epsilon \circ k_\epsilon = I_\epsilon \circ k_\epsilon = \widetilde{\widetilde{I}}_\epsilon$ where $\widetilde{\widetilde{I}}_\epsilon$ depends only on the actions.

Similarly, for relatively integrable systems, we observe that if g_ϵ integrates the system and k_ϵ has Hamiltonian $K_\epsilon = K_\epsilon(H_0)$ then $g_\epsilon \circ k_\epsilon$ also integrates the system.

We will use this elementary observations to show that, without loss of generality the integrating transformations may be assumed to satisfy some extra normalizations. This will simplify some of the calculations. Note also that, unless some

82

normalizations are imposed, it will be hard to perform any analysis since these trivial factors may ruin convergence of any limiting process.

Related to the concept of the integrability but better adapted to perturbation expansions is the concept of asymptotic integrability.

Definition 2.5. *We say that an analytic system H_ϵ is "asymptotically integrable to order n", if we can find a g_ϵ and an I_ϵ as before such that:*

$$(2.3) \qquad\qquad H_\epsilon \circ g_\epsilon = I_\epsilon(A) + o(\epsilon^n)$$

Similarly we say that it is "asymptotically relatively integrable up to order n" if we can find g_ϵ, F_ϵ in such a way that

$$(2.4) \qquad\qquad H_\epsilon \circ g_\epsilon = F_\epsilon(H_0) + o(\epsilon^n)$$

Note that, if a H_ϵ is asymptotically integrable to order n, and $\tilde{H}_\epsilon$ is another system such that $H_i = \tilde{H}_i$ for $i = 0, \ldots, n$ then $\tilde{H}_\epsilon$ is also integrable and we can use the same g_ϵ. Similarly, if g_ϵ integrates a system asymptotically to order n and $\tilde{g}_\epsilon$ agrees with g_ϵ up to order $n + 1$ then $\tilde{g}_\epsilon$ also integrates the system. That is, asymptotic integrability up to order n is a property of the Taylor expansion in ϵ up to order n of H and can be verified using only the Taylor expansion to order n of g.

Since we will have to deal often with truncated Taylor expansions, we introduce the notation for Hamiltonians

$$H_\epsilon^{[\leq n]} = \sum_{i=0}^{n} H_i \epsilon^n$$

and, similarly, $g^{[\leq n]}$ will be the globally Hamiltonian isotopy starting at g_0 and of Hamiltonian $G_\epsilon^{[\leq n]}$. We will also use the notation with other subscripts with inequalities. They are meant to describe the range of the sum in a power series and, for deformations the deformation where the Hamiltonian is a sum of powers of ϵ similarly restricted.

Note that if a system is integrable, then it is asymptotically integrable to order n for all n. The converse is certainly not automatic — it is clearly false for families

C^∞ in ϵ. Take $\widetilde{H}_\epsilon = H_{\exp 1/\epsilon^2}$ where H_ϵ is not integrable. Note also that one system could be integrable to all orders in n but that the transformations that achieve this integrability could be quite different – they are not unique, as we have shown – . Nevertheless one of the main results of this paper is that analytic Hamiltonians satisfying Definition 2.1 that are integrable to all orders are uniformly integrable.

One motivation to consider this theory of necessary and sufficient conditions for asymptotic integrability is the problem of water waves in a gravitational field. It is a recent discovery [DZ] (see also [CW]) that this important example is asymptotically integrable to a high order but that it, seemingly is not to all orders. Of course, since this a field theory – a free boundary value problem – for a PDE. the Hamiltonian flow occurs in an infinite dimensional space and the results of this paper do not apply (even existence of solutions for all time seems not to have been stablished). Nevertheless, they can serve as heuristic gude.

The following proposition will be important to address the non-uniqueness question.

Proposition 2.6. *Assume that H_ϵ is integrable (resp. relatively integrable) to order n. then, it admits an integrating family such that*

$$\int_{\mathbf{T}^d} G^i = 0; \quad 0 \le i \le n \quad \left(\text{resp.} \int_{\Sigma_E} G^i = 0; \quad 0 \le i \le n\right)$$

Proof. For the uniformly integrable case, note that if $\tilde{g}_\epsilon$ integrates the problem to order n then for any k_ϵ such that $K_\epsilon = K_\epsilon(A)$, $g_\epsilon = \tilde{g}_\epsilon \circ k_\epsilon$ also integrates the problem. We claim that it is possible to choose K_ϵ in such a way that the normalization is satisfied.

Note that the Hamiltonian of g_ϵ is $G_\epsilon = \tilde{G}_\epsilon + K_\epsilon \circ \tilde{g}_\epsilon$. If we expand in powers of ϵ, we obtain the the coefficient of ϵ^i has the form $G^i = \tilde{G}^i + K^i + R^i$ where R^i is an expression involving only terms of lower order. Therefore, we can recursively pick $K^i = -\int_{\mathbf{T}^d} \tilde{G}^i + R^i$. In the relatively integrable case, the same argument tells us that it suffices to pick recursively $K^i = -\int_{\Sigma_E} \tilde{G}^i + R^i$. ∎

3. Obstructions to uniform integrability and statement of results.

Since the solutions of (2.1), if they exist, are highly non-unique (as illustrated by the remarks after the definition) to derive obstructions, it will be useful to show that if there exist solutions, they can be found in a restricted class.

Proposition 3.1. *Assume that $H_\epsilon \circ g_\epsilon = I_\epsilon$ on $U \times \mathbb{T}^d$, $g_0 = Id$ then, we can find a family t_ϵ of the form*

$$t_\epsilon(A, \varphi) = (A + \Delta_\epsilon, \varphi) , \qquad \Delta_0 = 0$$

in such a way that $H_\epsilon \circ g_\epsilon \circ t_\epsilon = I_\epsilon \circ t_\epsilon$ and $g_\epsilon \circ t_\epsilon$ is a globally Hamiltonian isotopy.

In other words, if there is an integrating family defined in a certain domain, we can find a globally symplectic transformation defined in a slightly different domain. Hence, to exclude the existence of integrating transformations, it suffices to exclude the existence of globally Hamiltonian ones.

Proof. We denote by [ ] the cohomology class of a form and by $\#$ the operator induced on cohomology by a transformation.

Since

$$\frac{d}{d\epsilon} g_\epsilon \circ t_\epsilon = (\mathcal{G}_\epsilon + g_{\epsilon *}\mathcal{T}_\epsilon) \circ g_\epsilon \circ t_\epsilon$$

the family $g_\epsilon \circ t_\epsilon$ will be globally Hamiltonian if and only if

(3.1) $$[i(\mathcal{G}_\epsilon)\gamma] + g_{\epsilon \#}[i(\mathcal{T}_\epsilon)\gamma] = 0$$

Since $g_0 = Id$, g_ϵ is isotopic to the identity and, therefore, $g_{\epsilon \#}$ is the identity.

Moreover, $\mathcal{T}_\epsilon = (\dot{\Delta}_\epsilon, 0)$ and

$$i(\mathcal{T}_\epsilon)\gamma = \sum_i \dot{\Delta}_{\epsilon, i} \, d\varphi_i$$

since $[d\varphi_i]$ are a basis for the first cohomology in our phase space, we can compute $-\dot{\Delta}_{\epsilon,i}$ as the component of $[d\varphi_i]$ of $[i(\mathcal{G}_\epsilon)\gamma]$. Hence, we can compute Δ_ϵ by integration.

Note that, if $\mathcal{G}_\epsilon$ is analytic in ϵ, so is Δ_ϵ.

■

To obtain obstructions for asymptotic integrability to order 1 — a fortiori obstructions to uniform integrability —, following [Po] §81, we take derivatives with respect to ϵ on (2.1) and obtain

$$(3.2) \qquad (\dot{H}_\epsilon + \{G_\epsilon, H_\epsilon\}) \circ g_\epsilon = \dot{I}_\epsilon + o(\epsilon)$$

where $\{\ \}$ denotes Poisson brackets and $\dot{H}$ denotes derivatives with respect to ϵ.

Evaluating at $\epsilon = 0$ we obtain $H_1 + \{G_0, H_0\} = I_1$. Furthermore, we note that $\{G_0, H_0\} = L_{\mathcal{G}_0} H_0 = -L_{\mathcal{H}_0} G_0$.

Hence, if β is a periodic orbit of the flow generated by $\mathcal{H}_0$,

$$\int_\beta H_1 + \{G_0, H_0\} = \int_\beta H_1 - L_{\mathcal{H}_0} G_0 = \int_\beta H_1$$

because the integral along a periodic orbit of the derivative along the flow of a function vanishes.

Hence, we have proved, following [Po] §81

Theorem 3.2. *A necessary condition for the family H_ϵ to be asymptotically integrable to order 2 is that*

$$\frac{1}{|\beta|} \int_\beta H_1$$

is the same for all periodic orbits of H_0 with the same action variables.

Moreover, since I_1 is a function of the actions only and the actions are conserved along the flow of H_0, if we denote by $A(\beta)$ the values of the actions and by $|\beta|$ the length of the periodic orbit $\int_\beta I_1 = I_1(A(\beta))|\beta|$. Hence, we should have $I_1 = \frac{1}{|\beta|} \int_\beta H_1$

86

Note that $\{A\} \times \mathbb{T}^d$ contains a periodic orbit if and only if $\omega(A) = \frac{1}{T}p$ with $p \in \mathbb{Z}^d$. In that case $\{A\} \times \mathbb{T}^d$ is foliated by periodic orbits that can be parameterized by $\mathbb{T}^{d-1}$. If $\frac{1}{|\beta|} \int_\beta H_1$ is independent of all the orbits by averaging over the extra variables

$$I_1(A) = \int_{\mathbb{T}^d} H_1(A, \varphi) \, d\varphi$$

on a torus where there are periodic orbits. We will generalize this observation later.

Similarly, we have

Lemma 3.3. *A necessary condition for relative integrability up to first order is that, for any periodic orbit*

$$\frac{1}{|\beta|} \int_\beta H_1 = F_1(H_0(\beta))$$

Theorem 3.2 is the main result of §81 of [Po]. Even if the ideas presented here are the same, it is quite interesting to compare the notation there. In particular, we call attention to the fact that Poincaré phrases the problem as finding uniform integrals, that is analytic first integrals that depend analytically on the parameter. Of course, due to Liouville-Arnol'd theorem, the existence of integrals in involution is the same as integrability. Even if the formulation in terms of the number of integrals is more general, it seems that it is less suited for the derivation of converses in the category of analytic functions and not just power series. In this paper we establish convergence of the formal procedures, by using a KAM method. This seems to require that the procedure has a "group structure".

An obstruction for the existence of a uniform integral A_ϵ that starts in a known one A_0 can be obtained easily by noting that if $\{H_\epsilon, A_\epsilon\} = 0$, then, equating terms of order ϵ we obtain $\{H_0, A_1\} + \{H_1, A_0\} = 0$. The second term is known and we obtain proceeding in the same way that the integrals over periodic orbits have to vanish.

We also note that the integrals we consider in the previous two lemmas are Melnikov integrals. The constancy of them can be interpreted as stating that the torus moves together in first order perturbation theory.

In Chapter V of [Po] we find many variants of this idea and applications to several problems. At the end of §86, commenting on a model for which the obstructions vanish, we find

Les conditions énoncées dans ce Chapitre étant neccesaires mais non suffisantes, rien ne prouve que cette troisième intégral existe; il convient avant de se prononcer, d'attendre la publication complète des résultats de Mme de Kowaleski.

It is not clear to us what Poincaré had in mind at this time since, as we have emphasized, the conditions are for analytic integrals depending analytically on parameters and the claims of Kowaleskaia seem to be only for special values of the parameters. The footnote accompanying this paragraph is a one of the few paragraphs in a Mathematics book that are drama.

Depuis que ces lignes ont été écrites le monde savant a eu à déplorer la morte prématurée de Mme de Kowaleski. Les notes q'on a retrouvées chez elle sont malhereusement insuffisantes pour permettre de reconstituer ses démonstrations et ses calculs.

In this paper we will be concerned with stating and proving converses to this obstruction to uniform integrability and a similar one for relative integrability. The first of the main results of this paper is:

Theorem 3.4. *Assume that H_0, H_1 are defined on $U \times \mathbb{T}^d$ and that, H_0 is non-degenerate, H_1 satisfies the necessary conditions if Theorem 3.2. Then, we can find G_0 defined also in $U \times \mathbb{T}^d$ in such a way that*

$$H_\epsilon^{[\leq 1]} \circ g_\epsilon^{[\leq 0]} = I_\epsilon^{[\leq 1]} + o(\epsilon^2)$$

In other words, when H_0 is non-degenerate, the conditions in Theorem 3.2 are not only necessary but also sufficient for asymptotic integrability to order 2.

The proof of this result will be postponed till section 4, when we have developed some technical results needed in the proof. In that section, we will also show that it is possible to obtain estimates of the analyticity properties of G_0 in terms of the analyticity properties of H_1.

An interesting example that shows that the non-degeneracy conditions in Theorem 3.4 are necessary appears in the thesis of E. Meletlidou [Me].

Example 3.5. *Consider* $H_\epsilon = H_0 + \epsilon H_1$, *with* $H_0 = \frac{1}{2}(A_1^2 + A_2^2)$, $H_1 = \cos(2\varphi_1 - \varphi_2)$.

One can verify that this system has the integrals H_ϵ and $A_1 + 2A_2$, hence, one would be tempted to call it integrable. On the tori invariant by H_0 for which $2A_1 = A_2$, $\frac{1}{|\beta|}\int H_1 = \cos(2\varphi_1(0) - \varphi_2(0))$ which is not independent of the starting point.

■

The reason why this example is not a contradiction with Theorem 3.4 is that precisely at the tori with $2A_1 = A_2$ the gradient of the second integral fails to be independent of the gradient of the first and, indeed, one can check that the integrating transformation cannot be analytic there.

This example points to the fact that some of the assumptions in the theorem are really necessary and, hence, that talking about "integrability" without attaching it a precise meaning can lead to confusion.

Once that we have necessary and sufficient conditions for asymptotically integrability up to order two, we can investigate integrability up to higher orders.

Clearly, the higher the order, the more necessary conditions will have to be met. It can happen that a system is asymptotically integrable up to a certain order but not to a higher one. (For example, if we take $H_{\epsilon^7} \circ g_\epsilon$ with H_ϵ not integrable to order 2, we obtain a family integrable to order 6 but not to order 7.) As in the criterion to exclude integrability to first order, we can derive necessary conditions for integrability up to high orders which are related to periodic orbits.

Lemma 3.6. *Let* n *be a natural number bigger than one,* H_ϵ *a family of Hamiltonians. Assume that there exists a* $g_\epsilon^{[\leq(n-1)]}$, $I_\epsilon^{[\leq n]}$ *in such a way that:*

$$H_\epsilon \circ g_\epsilon^{[\leq(n-1)]} = I_\epsilon^{[\leq n]} + o(\epsilon^{n+1})$$

A necessary condition for the existence of G^n in such a way that

$$H_\epsilon \circ g_\epsilon^{[\leq n]} = I_\epsilon^{[\leq(n+1)]} + o(\epsilon^{n+2})$$

is that, for any periodic orbit β of H_0 we have:

$$\int_\beta H_{n+1} + \int_\beta R_n = |\beta| I_{n+1}(A(\beta))$$

where R_n is an algebraic expression — described in the proof — involving $H_0, \ldots, H_n$, $G_0, \ldots, G_n$, Poisson-brackets, sums and products.

Proof. Since for any family of functions A_ϵ,

$$\frac{d}{d\epsilon} A_\epsilon \circ g_\epsilon = (\dot{A}_\epsilon + \{G_\epsilon, A_\epsilon\}) \circ g_\epsilon$$

proceeding by induction, we obtain

(3.3) $$\left(\frac{d}{d\epsilon}\right)^{n+1} H_\epsilon \circ g_\epsilon = \left[\left(\frac{d}{d\epsilon}\right)^{n+1} H_\epsilon + \left\{\left(\frac{d}{d\epsilon}\right)^n G_\epsilon, H_\epsilon\right\} + R_\epsilon^n\right] \circ g_\epsilon$$

where R_ϵ^n is an expression involving derivatives up to order $n-1$ of G and of H up to order n of H_ϵ, Poisson brackets, sums and products.

Evaluating (3.3) at $\epsilon = 0$ we obtain

$$\left(\frac{d}{d\epsilon}\right)^{n+1} H_\epsilon \circ g_\epsilon \bigg|_{\epsilon=0} = (n+1)! H_{n+1} + n! \{G_n, H_0\} + R_0^n$$

If (2.3) is to hold this should equal to $(n+1)! I_{n+1}(A)$.

As in the proof of Theorem 3.2 we observe that if β is a periodic orbit for H_0

$$\int_\beta \{G_n, H_0\} = 0$$

and hence, integrating over a periodic orbit, we obtain the statement of the claim.

■

Remark. Note that the proof gives an explicit expression for the R_n. It suffices to evaluate the derivatives in (3.3). For example:

$$R_\epsilon^0 = 0$$
$$R_0^2 = 2\{G_0, H_1\} + \{G_0, \{G_0, H_0\}\}$$

Note that as we have expressed it, the obstruction to integrability to order $n - 1$ depends on the integrating transformations up to order n we have selected. Nevertheless, it can be shown by an straightforward – albeit tedious – computation that this is not so. More interesting is to remark that it follows from our results that the vanishing of the obstructions does not depend on the choices. The notion of asymptotic integrability to order n is independent of any choices, but we show in this paper that it is equivalent to the vanishing of the obstructions.

Similarly, we have an statement for relative integrability.

Lemma 3.7. *Let n be a natural number bigger than two, H_ϵ a family of Hamiltonians. Assume that there exists a $g_\epsilon^{[\leq(n-1)]}$, $F_\epsilon^{[\leq n]}$ in such a way that:*

$$H_\epsilon \circ g_\epsilon^{[\leq(n-1)]} = F_\epsilon^{[\leq n]}(H_0) + o(\epsilon^{n+1})$$

A necessary condition for the existence of G^n in such a way that

$$H_\epsilon \circ g_\epsilon^{[\leq n]} = F_\epsilon^{[\leq n+1]}(H_0) + o(\epsilon^{n+2})$$

is that, for any periodic orbit β of H_0 we have:

$$\int_\beta H_{n+1} + \int_\beta R_n = |\beta| F_{n+1}(H_0(\beta))$$

where R_n is the same algebraic expression as in Lemma 3.6

The proof of this lemma is identical to that of Lemma 3.6.

■

By an argument very similar to the one that we will use in the proof of Theorem 3.4 we will obtain that, when H_0 is non-degenerate, the necessary conditions in Lemma 3.6 are not only necessary but also sufficient.

Theorem 3.8. *Assume that $H_0, \ldots, H_{n+1}, G_0, \ldots, G_{n-1}$ are defined on $U \times \mathbb{T}^d$ and that then they satisfy the necessary conditions in Lemma 3.6,(resp.Lemma 3.7) then, we can find G_n defined also on $U \times \mathbb{T}^d$ in such a way that*

$$H_\epsilon^{[\leq n+1]} \circ g_\epsilon^{[\leq n]} = I_\epsilon^{[\leq n+1]} + o(\epsilon^{n+2})$$

(resp.

$$H_\epsilon^{[\leq n+1]} \circ g_\epsilon^{[\leq n]} = F_\epsilon^{[\leq n+1]}(H_0) + o(\epsilon^{n+2}) \ )$$

Moreover, the G_n that makes the above equation true is unique up to addition of a function of A (resp. a function of H_0).

Note that with these results provide us with an algorithm to show that a system is not uniformly (rep. relatively) integrable:

Starting with $n = 1$ compute the obstruction for asymptotic integrability to order $n + 1$. If the obstruction does not vanish, we conclude that the system is not integrable up to order $n + 1$. If it is, compute the Hamiltonian that integrates it up to order $n + 1$ and start checking again.

Our final result shows that this algorithm is optimal for analytic systems. That is, that if it does not fail to exclude a family, then the family is uniformly integrable. We again emphasize that in the definition of family we have included analyticity with respect to the parameters.

Theorem 3.9. *Let H_ϵ be a family defined on $U \times \mathbb{T}^d$. Assume*

i) *H_0 is non-degenerate as in Definition 2.1*

ii) *For every n, we can find a family $g_\epsilon^{(n)}$ and a family of functions of the actions $I_\epsilon^{(n)}(A)$ such that*

$$H_\epsilon \circ g_\epsilon^{(n)} = I_\epsilon^{(n+1)}(A) + o(\epsilon^{n+2})$$

Then, we can find an open set $\widetilde{U}$, $\emptyset \neq \widetilde{U} \subset U$ and an analytic family g_ϵ defined in $\widetilde{U} \times \mathbb{T}^d$ and such that

$$H_\epsilon \circ g_\epsilon = I_\epsilon(A) \qquad |\epsilon| \leq \epsilon^*$$

with $\epsilon^* > 0$.

We point out that the condition in Theorem 3.9 that H_0 satisfies Definition 2.1 really does belong. In [Ga] p. 369 one can find examples of systems that fail to satisfy Definition 2.1 and are not integrable. A simple example is

$$H_\epsilon(A_1, A_2, \varphi_1, \varphi_2) = \alpha A_1 + A_2 + \epsilon(A_2 + f(\varphi_1)f(\varphi_2))$$

where α is the golden mean. The same idea can be used to produce examples where rank$\frac{\partial^2 H_0}{\partial A \partial A}$ is $1, 2, \cdots, d-1$.

The paper [Ga] studies the problem of finding analogues to Theorem 3.9 when rank$\frac{\partial^2 H_0}{\partial A \partial A}$ is $1,2$. There, one can find conditions – that go beyond those found in [Po] and discussed here – that are also necessary and sufficient.

Two questions that we have not been able to explore are the converses for existence or a reduced number of first integrals. We also do not know whether existence of analytic integrating transformations for all values of ϵ in a complex neighborhood of zero imply that the system is uniformly integrable.

4. Existence and regularity for solutions of cohomology equations.

We have shown already that the existence of asymptotic integrating transformations can be reduced to solving a hierarchy of equations of the form

$$(4.1) \qquad \frac{-1}{n+1}\{G_n, H_0\} + I_{n+1} = H_{n+1} + \frac{1}{(n+1)!}R_n = \Gamma_n$$

where H_0, H_{n+1}, R_n are known and we are supposed to find G_n and I_{n+1}. We have already shown that a necessary condition for the existence of (4.1) is that $\int_\beta \Gamma_n$ depends only on the action.

Similarly, the problem of relative integrability reduces to the study of a hierarchy of equations of the form

$$\{G_n, H_0\} + F_{n+1}(H_0) = \Gamma_n$$

To prove Theorem 3.8, and the similar result for relative integrability, it suffices to show that when the obstructions found in Lemma 3.6, (resp. Lemma 3.7) are met, we can find analytic G_n and I_{n+1}. (resp. analytic G_n, F_{n+1}.)

We first argue that, we can determine F_n, I_n in such a way that we reduce the problem to considering the equation

$$\{G_n, H_0\} = \Gamma_n$$

with the compatibility conditions $\int_\beta \Gamma_n = 0$ for every periodic orbit of H_0.

We first observe that finding the I_n (for (2.1)) and the F_n's is easy.

In effect, if we consider $\Gamma_n^t(A, \varphi) = \frac{1}{t} \int_0^t \Gamma_n(A, \varphi + \omega(A)t)$ they are equicontinuous functions. Moreover, if $\omega(A)$ is of the form $\frac{2\pi}{T}(p_1, \ldots, p_d)$ with $p_1, \ldots, p_d \in \mathbb{Z}$ they converge to the average over the periodic orbit. If the system is non-degenerate, the set of periodic orbits is dense. By Ascoli-Arzelá theorem, this sequence of functions is converging uniformly to a limit. By our hypothesis on periodic orbits, the limit depends only on the actions.

Hence, if the average on the periodic orbits depends only on the actions, (resp. on the energy) of a periodic orbit, we obtain that $\Gamma_n^t(A, \varphi)$ converges uniformly as $t \to \infty$ to a function of A (resp. the energy) which, of course, agrees with the averages over periodic orbits. (Hence, we conclude, in particular, that the averages over periodic orbits of Γ_n have to be continuous.)

Now, we proceed to compute a more convenient form of these averages.

If $\Gamma_n^t(A, \varphi)$ converges as above, then,

$$\int_{\mathbb{T}^d} d\varphi \, \Gamma_n^t(A, \varphi) = \frac{1}{t} \int_0^t \int_{\mathbb{T}^d} d\varphi \, \Gamma_n(A, \varphi + \omega(A)t)$$

$$= \frac{1}{t} \int_0^t dt \int d\varphi \, \Gamma_n(A, \varphi) = \int_{\mathbb{T}^d} d\varphi \, \Gamma_n(A, \varphi)$$

$$\int_{\Sigma_E} d\mu_E \, \Gamma_n^t(A, \varphi) = \frac{1}{t} \int_0^t \int_{\Sigma_E} d\mu_E \, \Gamma_n(A, \varphi + \omega(A)t)$$

$$= \frac{1}{t} \int_0^t \int_{\Sigma_E} d\mu_E \, \Gamma_n = \int_{\Sigma_E} d\mu_E \, \Gamma_n$$

Hence, if Γ_n^t is converging uniformly to a function I_n only of A (resp. F_n only of H_0) we see that this function has to be

$$(4.2) \qquad I_n(A) = \int_{\mathbb{T}^d} d\varphi \, \Gamma_n(A, \varphi) \qquad (\text{resp.} \quad F_n(E) = \int_{\Sigma_E} d\mu_E \, \Gamma_n \quad).$$

These functions have to agree with the average on periodic orbits.

Note that both $I_n(A)$, $F_n(E)$ given by (4.2) are analytic functions of the position if Γ_n is.

Therefore, if we set I_n, (resp. F_n) to be given by (4.2) we reduce our problem to

$$\{G_n, H_0\} = \Gamma_n - I_n \qquad (\text{resp.} \quad \{G_n, H_0\} = \Gamma_n - F_n \quad)$$

where, in both cases the R.H.S. have average zero over periodic orbits.

Remark. Since the problem of relative integrability is geometrically natural, the above reduction can be worked out in much larger generality. This is done in [LMM], pp. 549–553.

Remark. The I_n's and the F_n's play a role very similar to the averages that need to be taken off in the proofs of K. A. M. theory.

Therefore Theorem 3.4, Theorem 3.8 are reduced to the proof of:

Theorem 4.1. *Let H_0 be an integrable system satisfying Definition 2.1. Let Γ be an analytic function on $U \times \mathbb{T}^d$ satisfying:*

$$\int_\alpha \Gamma = 0$$

for all α periodic orbit of H_0. Then, there exists an analytic function G on $U \times \mathbb{T}^d$ such that

$$\{G, H_0\} = \Gamma.$$

This solution is unique up to addition of a function of the actions.

Proof. Since we will study the cohomology equation

$$\{G, H_0\} \equiv \omega(A)\frac{\partial}{\partial \varphi}G = \Gamma$$

using Fourier analysis, we first work out the implication of the compatibility conditions for the Fourier coefficients.

We observe that if $\omega(A_0) = \frac{1}{T}p$ with $T \in \mathbb{R}$, $p \in \mathbb{Z}^d$, $\{A_0\} \times \mathbb{T}^d$ consists of periodic orbits of period T and, for all $\varphi \in \mathbb{T}^d$,

$$0 = \frac{1}{T} \int_0^T \Gamma(A_0, \omega(A_0)t + \varphi) \, dt .$$

Using the partial Fourier expansion, we have

$$0 = \frac{1}{T} \int_0^T \sum_k \hat{\Gamma}_k(A_0) e^{2\pi i \frac{1}{T}pok + k\varphi} \, dt = \sum_{k \cdot p = 0} e^{2\pi i k\varphi} \hat{\Gamma}_k(A_0)$$

We denote by $\mathcal{P}$ the set of ω's that can be written as $\frac{1}{T}p$ for some $T \in \mathbb{R}^+$, $p \in \mathbb{Z}^d$. That is, the ω's that give rise to a periodic orbit.

Hence if the compatibility conditions are met,

$$\hat{\Gamma}_k(A_0) = 0 \text{ if } k \cdot \omega(A_0) = 0 \text{ and } \omega(A_0) \in \mathcal{P}$$

Now, we argue that the last condition $\omega(A_0) \in \mathcal{P}$ is superfluous. In effect, note that if $k \in \mathbb{Z}^d$ and $k \neq 0$ — say $k_d \neq 0$ — $\omega \cdot k = 0$ if and only if ω can be written as $(\omega_1, \ldots, \omega_{d-1}, -\frac{1}{k_d}(k_1\omega_1 + \cdots + k_{d-1}\omega_{d-1}))$. Since given any real numbers $\omega_1, \ldots, \omega_{d-1}$ we can approximate then by rationals, we conclude that any $\omega \in \mathbb{R}^d$ with $\omega \cdot k = 0$ can be approximated by one with rational components, which, taking T as the lowest common denominator, is of the form $\frac{1}{T}p$ with $p \in \mathbb{Z}^d$.

We conclude that if $k \in \mathbb{Z}^d$, $k \neq 0$, $\mathcal{P} \cap \{\omega \in \mathbb{R}^d \mid \omega \cdot k = 0\}$ is dense on $\{\omega \in \mathbb{R}^d \mid \omega \cdot k = 0\}$. (Of course the result is also true when $k = 0$.) Since $\hat{\Gamma}_k$ is continuous it vanishes when $\{\omega(A) \cdot k = 0\}$ even if $\omega(A) \notin \mathcal{P}$.

We observe that, in terms of Fourier coefficients, the equation reads

(4.3)
$$\left(\omega(A)\frac{\partial}{\partial\varphi}G\right)^{\wedge}_k = 2\pi i(\omega(A) \cdot k)\hat{G}_k(A) = \hat{\Gamma}_k(A)$$

The regularity of the $\hat{G}_k$ and the $\hat{\Gamma}_k$'s is easier to study when we consider them as functions of ω. Since ω^{-1} is uniformly analytic, we can consider $\tilde{\Gamma}_k = \hat{\Gamma}_k \circ \omega^{-1}$, $\tilde{G}_k = \hat{G}_k \circ \omega^{-1}$. Then (4.3) reads

(4.4)
$$\tilde{G}_k(\omega)(\omega \cdot k)2\pi i = \tilde{\Gamma}_k(\omega)$$

but we have that $\tilde{\Gamma}_k(\omega) = 0$ when $\omega \cdot k = 0$. Hence, the solution

$$(4.5) \qquad \tilde{G}_k(\omega) = \frac{1}{2\pi i} \begin{cases} (\omega \cdot k)^{-1} \tilde{\Gamma}_k(\omega) , & \omega \cdot k \neq 0 \\ \frac{1}{|k|^2} (k \cdot \partial_\omega) \tilde{\Gamma}_k(\omega) , & \omega \cdot k = 0 \end{cases}$$

is the only continuous solution. Certainly $\tilde{G}_k(\omega)$ is analytic if $\tilde{\Gamma}_k$ is (and hence $\hat{G}_k(A)$ is analytic if $\hat{\Gamma}_k$ is).

Recall that, by Cauchy inequalities we can bound

$$\left| (\partial_\omega)^n \tilde{\Gamma}_k(\omega) \right| \leq n! K_\omega a_\omega^n e^{-\delta |k|} \quad \text{where} \quad a < 1, \ \delta > 0$$

It follows from (4.5) that $\tilde{G}_k$ satisfies similar bounds and hence, is analytic.

This finishes the proof of Theorem 4.1 and, hence the proof of Theorem 3.4, Theorem 3.8.

$\blacksquare$

The proof of convergence will be obtained, not by estimating the recursion outlined before, but by using a quadratically convergent scheme of the K. A. M. type. This convergent scheme will require quantitative versions of Theorem 4.1 that not only claim the existence of the solutions.

As it turns out, this convergent scheme will need to be implemented on spaces of families of functions. Hence, our first task is to introduce appropriate spaces of families of functions and formulate quantitative versions of Theorem 4.1 for them.

Definition 4.2. *Given an open set $U \subset \mathbb{R}^d$ we can consider it also as a subset of $\mathbb{C}^d$ and define, for $\delta > 0$*

$$U^\delta = \{ z \in \mathbb{C}^d \mid d(z, U) \leq \delta \}$$

Given a non-degenerate frequency function we will denote $V = \omega(U)$ and

$$\|\eta\|^{\omega, U, \delta} = \sup_{\substack{z \in V^\delta \\ |\operatorname{Im} \varphi| \leq \delta, |\epsilon| \leq e^\delta}} |\eta(\omega^{-1} z, \varphi)|$$

The set of functions for which the above norm is finite forms a Banach space which we will denote by $\chi^{\omega,U,\delta}$. Similarly, given an analytic family of non-degenerate frequency functions ω_ϵ we define for

$$\|\Gamma\|^{\omega,U,\delta} = \sup_{\substack{z \in V^\delta \\ |\operatorname{Im}\varphi| \le \delta, |\epsilon| \le e^\delta}} |\Gamma_\epsilon(\omega_\epsilon^{-1} z, \varphi)|$$

and denote by $\Phi^{\omega,U,\delta}$ the Banach space of families for which the above norm is finite.

The reason to introduce the norms in Definition 4.2 is that, for the solution of the cohomology equation,(4.5) the variables ω are more natural than the angle variables. This has the drawback that the norms depend on the frequency ω, but the dependence is slight as we will prove later.

Note that we denote the radius of the balls of the parameter ϵ by e^δ, where δ is the same parameter that measures the complex component in the complex extension of the torus. The exponential notation is justified so that the Cauchy inequalities look the same. The use of the same parameter for the two variables is justified just to keep down the number of parameters even it it leads to wasteful estimates. We also introduce the norms

$$\|\eta\|_\infty^{U,\delta} = \sup_{\substack{z \in U^\delta \\ |\operatorname{Im}\varphi| \le \delta \\ |\epsilon| \le e^\delta}} |\eta_\epsilon(z, \varphi)|; \qquad \|\Gamma\|_\infty^{U,\delta} = \sup_{\substack{z \in U^\delta \\ |\operatorname{Im}\varphi| \le \delta \\ |\epsilon| \le e^\delta}} |\Gamma_\epsilon(z, \varphi)|$$

and denote the corresponding Banach spaces by $\chi_\infty^{U,\delta}$, $\Phi_\infty^{U,\delta}$.

When the functions take values on matrices or on linear spaces we substitute for the absolute value in the R.H.S. of the above definitions the appropriate norms.

Note that if $V = \omega(U)$ we have

(4.6) $$\|R\|^{\omega,U,\delta} = \|R \circ \omega\|_\infty^{V,\delta}$$

We clearly have that

$$\|\eta\|^{\omega,U,\delta} \le \|\eta\|^{\omega,U,\delta+\sigma} \quad \text{for} \quad \sigma > 0$$

since we are taking the supremum over a larger set.

Note that $d(z, U) \leq \alpha$ implies

$$d(\omega(z), V) \leq \alpha \|D\omega^{-1}\|_\infty^{U,\alpha}$$

Hence,

$$\|\eta\|_\infty^{U,\delta} \leq \|\eta\|^{\omega,U,\delta} \|D\omega\|_\infty^{U,\alpha}$$

and since $d(w, V) \leq \alpha$ implies $d(\omega^{-1}(w), U) \leq \alpha \|D\omega\|_\infty^{V,\alpha}$ we have

$$\|\eta\|^{\omega,U,\delta} \leq \|\eta\|_\infty^{U,\delta} \|D\omega^{-1}\|_\infty^{\omega,U,\delta}$$

The following are well known Cauchy bounds applied to the angular variables and the parameters. (We do not state the much better known inequalities for derivatives with respect to the A.)

Lemma 4.3.

$$\|\hat{\eta}_k\|^{\omega,U,\delta} \leq e^{-|k|\delta} \|\eta\|^{\omega,U,\delta}$$

$$\|\hat{\eta}_k\|_\infty^{U,\delta} \leq e^{-|k|\delta} \|\eta\|_\infty^{U,\delta}$$

$$\left\| \frac{\partial^{|j|}}{\partial \varphi^j} \eta \right\|^{\omega,U,\delta-\sigma} \leq K\sigma^{-|j|} \|\eta\|^{\omega,U,\delta}$$

$$\|\eta^n\|^{\omega,U,\delta} \leq K e^{-n\delta} \|\eta\|^{\omega,U,\delta}$$

$$\|\eta^{[\leq N]}\|^{\omega,U,\delta-\sigma} \leq K\sigma^{-1} \|\eta\|^{\omega,U,\delta}$$

$$\|\eta^{[\geq N]}\|^{\omega,U,\delta-\sigma} \leq K\sigma^{-1} e^{-N\delta} \|\eta\|^{\omega,U,\delta}$$

Before we embark in the proof of Theorem 3.9, let us state another technical lemma that will control the dependence of the norm $\| \ \|^{\omega,U,\delta}$ on ω. The main point of the lemma is that the change of the norms, when we change ω is controlled by the change in ω. We need, however to make sure that the domains match and that the change is not so drastic as to make the new frequency function non-invertible. In view of the observation (4.6) we see that the estimates that we need are just the customary estimates for the composition of functions that appear in almost all the proofs of K. A. M. theorems based in successive transformations.

Lemma 4.4. *Assume that $\omega, \tilde{\omega}$ are non-degenerate frequency functions.*

Assume moreover

i) $\tilde{U}$ *is such that* $d(\tilde{U}, \mathbb{C}^d - U) > K||\omega - \tilde{\omega}||_\infty^{U,\delta} + K\sigma$

ii) $||\omega - \tilde{\omega}||_\infty^{U,\delta} < 1/K$

where K *depends only on* $||\omega||_\infty^{U,\delta}$, $||\tilde{\omega}||_\infty^{U,\delta}$ $||\omega^{-1}||_\infty^{U,\delta}$, $||\tilde{\omega}^{-1}||_\infty^{U,\delta}$ *and can be chosen uniformly when these norms can be bounded uniformly.*

Then,
$$\left| ||R||^{\omega,\tilde{U},\delta} - ||R||^{\omega,\tilde{U},\delta} \right| \le C\sigma^{-1}||R||^{\omega,U,\delta}||\omega - \tilde{\omega}||_\infty^{U,\delta}$$

Analogous results hold for families.

Proof. We have:
$$\left| ||R||^{\omega,\tilde{U},\delta} - ||R||^{\omega,\tilde{U},\delta} \right| \le ||R \circ \omega^{-1} - R \circ \tilde{\omega}^{-1}||_\infty^{V,\delta}$$

where V is any domain such that $\omega(U) \subset V$, $\tilde{\omega}(U) \subset V$. If $\tilde{V}$ is such that $\lambda\omega^{-1}(x) + (1 - \lambda)\tilde{\omega}^{-1}(x) \in \tilde{V}$ for $x \in V$ and $\lambda \in [0, 1]$ we can bound this by

$$||\nabla R||_\infty^{\tilde{V},\delta}||\omega^{-1} - \tilde{\omega}^{-1}||_\infty^{V,\delta}$$

Finally, if $\tilde{\tilde{U}}$ is such that $d(\tilde{U}, \mathbb{C}^d - \tilde{\tilde{U}}) \ge \sigma$ we can use Cauchy bounds to bound the gradient. Moreover, given uniform bounds on the functions we can bound the difference of the inverses by the difference of the functions using the implicit function theorem. We therefore obtain the desired result.

The result for families is obtained by just applying this result for each value of ϵ and then taking suprema. ∎

Since in the proof of the theorem we will have to give up domains in a controlled way repeatedly, we adopt the convention that a domain with a ~ means a domain related to the original one by a relation such as *i*) above. The following is the main result of this section. It provides quantitative estimates for the solutions that were shown to exist in Theorem 4.1

Theorem 4.5. *Let I_ϵ be an analytic family of integrable systems with non-degenerate frequency function on U, with U a complex extension of a domain in the reals. Let R_ϵ be an analytic family in $\Phi^{\omega, U, \delta}$. Assume that δ is sufficiently small and that U is sufficiently close to the real space and that R satisfies the compatibility conditions $\int_{\beta_\epsilon} R_\epsilon = 0$ for every β_ϵ periodic orbit of I_ϵ.*

Let G_ϵ be the solution of

$$\{I_\epsilon, G_\epsilon\} = R_\epsilon$$

which also satisfies $\int_{\mathbb{T}^d} G_\epsilon(A, \varphi) = 0$. Then,

$$\|G\|^{\omega, \tilde{U}, \delta - \sigma} \leq \|R\|^{\omega, U, \delta} \sigma^{-d} K$$

where K depends only on the dimension.

Remark. Note that a non-degenerate system will not have periodic orbits outside of the real part of U if the extension is small enough. Clearly a frequency with an imaginary frequency will not lead to a periodic orbit. Once we fix a neighborhood, there is an open set of Hamiltonians with the same property. We will assume that all the perturbations are small enough that all our constructions do not leave this neighborhood.

Proof. By Cauchy estimates we have

$$\|\hat{R}_k\|^{\omega, U, \delta} \leq \|R\|^{\omega, U, \delta} e^{-|k|\delta}$$

We also recall that the solution G was given by (4.5)

$$\hat{G}_{k\epsilon} \circ \omega_\epsilon^{-1}(\alpha) = \frac{1}{2\pi i} \begin{cases} (\alpha \cdot k)^{-1} \hat{R}_{k,\epsilon} \circ \omega_\epsilon^{-1}(\alpha) , & \alpha \cdot k \neq 0 \\ \frac{1}{|k|^2}(k \cdot \partial_\alpha) \hat{R}_{k,\epsilon} \circ \omega_\epsilon^{-1}(\alpha) , & \alpha \cdot k = 0 \end{cases}$$

Hence, the supremum of $\hat{G}_k \circ \omega^{-1}$ is bounded by a constant times the supremum of $\frac{1}{|k|} \nabla(\hat{R}_k \circ \omega^{-1})$.

This can be bounded as follows. If $\tilde{U}$ is a domain in $\mathbb{C}^d$ such that $\tilde{U} \subset U$ and $d(\tilde{U}, \mathbb{C}^d - U) > \sigma$ we have:

$$\|\hat{G}_k\|^{\omega, \tilde{U}, \delta - \sigma} \leq K \frac{1}{|k|} \|\nabla \hat{R}_k\|^{\omega, \tilde{U}, \delta - \sigma} \leq K\sigma^{-1} \frac{1}{|k|} \|\hat{R}_k\|^{\omega, U, \delta} \leq K\sigma^{-1} \frac{e^{-\delta|k|}}{|k|} \|R\|^{\omega, U, \delta}.$$

Then

$$\|G_k\|^{\omega,\tilde{U},\delta-\sigma} \leq \sum_{k\in\mathbb{Z}^d-\{0\}} e^{(\delta-\sigma)|k|}\|\hat{G}_k\|^{\omega,\tilde{U},\delta} \leq \sum_{k\in\mathbb{Z}^d-\{0\}} e^{-\sigma|k|}\frac{1}{|k|}\|R\|^{\omega,U,\delta}K\sigma^{-1}$$

$$\leq K\sigma^{-1}\sum_{k\in\mathbb{N}-\{0\}} e^{-\sigma k}k^{d-2}\|R\|^{\omega,U,\delta} \leq K\sigma^{-d}\|R\|^{\omega,U,\delta}$$

■

Remark. It is quite possible that the previous estimates are rather wasteful and that the negative power of σ can be greatly reduced. It seems that since (4.5) only loses one derivative we should only have a factor σ^{-1} in the estimates above. Part of the problem is due to the use of Fourier series and supremum norms in the way that we have done it. Note that if we wanted to estimate the identity operator using the sloppy bounds that we have employed, we would have obtained a $\sigma^{-(d-1)}$ factor, which, obviously does not belong. These issues are important to improve the smoothness conditions in the usual K. A. M. theorem. Since the main results in this paper have no hope of working in any regularity substantially smaller than analytic we will refrain from such improvements.

5. Proof of Theorem 3.9.

As we have indicated before the proof will be done using a Newton algorithm to obtain quadratic convergence.

We proceed to describe the inductive steps. At step n we will have a family of Hamiltonians $H_\epsilon^{(n)}$

(5.1)
$$H_\epsilon^{(n)} = I_\epsilon(A)^{(n)} + R_\epsilon^{(n)}$$
$$H_\epsilon^{(n)} = F_\epsilon^{(n)}(H_0) + S_\epsilon^{(n)}$$

where $R_\epsilon^{(n)}$ and $S_\epsilon^{(n)}$ are $o(\epsilon^{2^n})$. Since the splitting of the Hamiltonian in these parts would not be unique otherwise, we assume that $I_\epsilon^{(n)}$ contains only terms up to $O(\epsilon^{2^n-1})$ and that $\int_{\mathbb{T}^d} R_\epsilon^{[\leq 2^n-1]}d\varphi = 0$.

The inductive step will consist in determining

$$G_\epsilon^{(n)} = \sum_{i=2^n}^{2^{n+1}-1} G^{(n),i} \epsilon^i$$

(and $I_\epsilon^{(n+1)}$, $F_\epsilon^{(n+1)}$) in such a way that setting $H_\epsilon^{(n+1)} = H_\epsilon^{(n)} \circ g_\epsilon^{(n)}$ we can also find $R_\epsilon^{(n)}$, $I_\epsilon^{(n)}$ (resp. $S_\epsilon^{(n)}$, $F_\epsilon^{(n)}$) in such a way that they satisfy (5.1) with $n+1$ in place of n. We will describe the procedure to determine the $G_\epsilon^{(n)}$ once all the other elements are known and, then, we will develop enough bounds that imply convergence. As it is customary in K. A. M. theory, the functions $H_\epsilon^{(n)}$ will be controlled in domains $U^{(n)} \times \mathbb{T}^d$ which will be decreasing as the iteration proceeds, (but which will not become empty). We will denote by $\omega^{(n)}$ the frequency function corresponding to $I^{(n)}$.

It is quite important to remark that because of the way that $H_\epsilon^{(n+1)}$ is produced, it will be integrable to all orders (resp. relatively integrable to all orders) if $H_\epsilon^{(n)}$ is.

The first step of the iteration will be obtained by taking $H_\epsilon^{(0)} = H_{\Lambda \epsilon}$ where Λ is a sufficiently large real number. Note that such $H_\epsilon^{(0)}$ is integrable (resp relatively integrable) to all orders if H_ϵ is. Moreover, by choosing Λ large enough, we can obtain that $||R_\epsilon^{(0)}||^{\omega,U,\delta}$ (resp. $||S_\epsilon^{(0)}||^{\omega,U,\delta}$) is small enough. Of course $H_0^{(0)} = H_0$ so that it is integrable and non-degenerate.

The quadratically convergent procedure will be based in finding a g_ϵ^n in such a way that $H_\epsilon^{(n+1)} \equiv H_\epsilon^{(n)} \circ g_\epsilon^{(n)}$ is much closer to an integrable family.

First, we will derive heuristically what are the equations that the $G_\epsilon^{(n)}$ have to satisfy, show that these equations admit solutions and estimate the new error. Then, it will become quite an standard procedure in K. A. M. theory to show that the $g_\epsilon^{(0)} \circ \cdots \circ g_\epsilon^{(n)}$ converge and the limit integrates the system.

The fact that we can find our $g_\epsilon^{(n)}$ will use the fact that that $H_\epsilon^{(n)}$ is integrable to all orders. Note that, by the definition of $H_\epsilon^{(n+1)}$, it will be integrable to all orders if $H_\epsilon^{(n)}$ is. For the experts in K. A. M. theory we note that, even if it follows from our result that the set of families integrable to all orders is a manifold, we cannot use it in the proof. In particular, we cannot use the formulation of K. A. M. theory as implicit function theorems and just use the fact that all the steps of the iterative procedure respect the set of families that are integrable to all orders. This is very similar to the problem of rotations of the circle as originally solved in [Ar], where we do not know a priori that the set of maps with the same rotation number is a manifold – it is consequence of the theorem. Similar arguments happen in the papers – much closer to our problem – [CEG] and [Ga].

We will find it convenient to introduce a second auxiliary parameter μ. This may be justified by noting that we are talking about small in two different senses. One is of course, that ϵ is small, but we also have to consider distances between families whose parameter is ϵ.

If we consider the family $H_\epsilon^{(n)} \circ g_{\mu\epsilon}^{(n)}$ for $\mu \in [0,1]$ and apply the mean value theorem $F(1) = F(0) + F'(1) + E$, where $|E| \le \frac{1}{2} \sup |F''|$ we obtain – in the region where the flow does not leave the domain of definition of H_ϵ^n –

(5.2)
$$H_\epsilon^{(n)} \circ g_\epsilon^{(n)} = H_\epsilon^{(n)} + \epsilon\{G_\epsilon^{(n)}, H_\epsilon^{(n)}\} \circ g_\epsilon^{(n)} + E$$
$$= H_\epsilon^{(n)} + \epsilon\{G_\epsilon^{(n)}, H_\epsilon^{(n)}\} + [\epsilon\{G_\epsilon^{(n)}, H_\epsilon^{(n)}\} \circ g_\epsilon^{(n)} - \epsilon\{G_\epsilon^{(n)}, H_\epsilon^{(n)}\}] + E$$

where E is an error term that, by the mean value theorem can be bounded by the supremum of the second derivative along the flow. We just recall that the second derivative along the flow can be expressed in terms of Poisson brackets

(5.3)
$$\frac{d^2}{d\mu^2} H_\epsilon^{(n)} \circ g_{\mu\epsilon}^{(n)} = \{G_\epsilon^{(n)}, \{G_\epsilon^{(n)}, H_\epsilon^{(n)}\}\} \circ g_{\mu\epsilon}^{(n)} .$$

Substituting (5.1) in (5.2) we obtain for the problem of uniform integrability

(5.4)
$$H_\epsilon^{(n)} \circ g_\epsilon^{(n)} = I_\epsilon^{(n)} + R_\epsilon^{(n)[\le 2^{n+1}]} + R_\epsilon^{(n)[> 2^{n+1}]} + \epsilon\{G_\epsilon^{(n)}, I_\epsilon^{(n)}\} + \epsilon\{G_\epsilon^{(n)}, R_\epsilon^{(n)}\} +$$
$$+ [\epsilon\{G_\epsilon^{(n)}, H_\epsilon^{(n)}\} \circ g_\epsilon^{(n)} - \epsilon\{G_\epsilon^{(n)}, H_\epsilon^{(n)}\}] + E$$

and, for the problem of relative integrability:

(5.5)
$$H_\epsilon^{(n)} \circ g_\epsilon^{(n)} = I_\epsilon^{(n)} + R_\epsilon^{(n)[\le 2^{n+1}]} + S_\epsilon^{(n)[> 2^{n+1}]} + \epsilon\{G_\epsilon^{(n)}, F_\epsilon^{(n)}(H_0)\}$$
$$+ \epsilon\{G_\epsilon^{(n)}, S_\epsilon^{(n)}\} + [\epsilon\{G_\epsilon^{(n)}, H_\epsilon^{(n)}\} \circ g_\epsilon^{(n)} - \epsilon\{G_\epsilon^{(n)}, H_\epsilon^{(n)}\}] + E$$

Now, we work out the consequences of the system being integrable to all orders.

We note that if there was a g_ϵ that integrated the system, we argued in Proposition 2.6 that we could find one integrating g_ϵ such that $\int_{\mathbf{T}^d} G_\epsilon = 0$. If we apply (5.4) to this g_ϵ, substituting G_ϵ for $G_\epsilon^{(n)}$ in (5.4) we obtain:

$$\{G^0, I_0^{(n)}\} = 0$$

and, that and the normalization condition, allows us to conclude that $G^0 \equiv 0$ by the uniqueness statement in Theorem 4.1 In other words, if $H_\epsilon^{(n)}$ can be integrated – by a system with normalized Hamiltonian, as we showed without loss of generality – then $G^0 = 0$. Similarly, for terms whose order in ϵ is not more that $2^n - 1$ we obtain that

$$\{G^i, I_0^{(n)}\} = \Gamma_i$$

where Γ_i is an expression involving $G^0, \cdots, G^{i-1}$ that vanishes if $G^0, \cdots, G^{i-1}$ vanishes. (This is because the terms involving the second derivatives, looking at (5.3) have order in ϵ at least $2 + 2r$ if the G_ϵ is order r in ϵ.

That is, we have shown that if the system is integrable to all orders then, any integrating normalized transformation has to satisfy: $G^0 = G^1 = \cdots = G^{2^n - 1} = 0$.

Once we know that the coefficients in the expansion of order up to $2^n - 1$ have to be equal to zero we realize that the only terms in (5.4) that contain terms of order less or equal than 2^{n+1} are the first and second.

In particular, we obtain that $R_\epsilon^{(n)[\leq 2^{n+1}]} + \epsilon\{G, I_\epsilon^{(n)}\} = I(A) + O(\epsilon^{2^{n+1}+1})$ Proceeding as in Theorem 3.2, integrating along periodic orbits of $I_\epsilon^{(n)}$, we obtain that if the system is integrable to all orders, then, $\frac{1}{|\beta_\epsilon|} \int_{\beta_\epsilon} R_\epsilon^{(n)[\leq 2^{n+1}]}$ is only a function of A.

We note that these are the compatibility conditions for the existence of $\Delta I_\epsilon^{(n)}$ $G_\epsilon^{(n)}$ solving

$$\Delta I_\epsilon^{(n)} + R_\epsilon^{(n)[\leq 2^{n+1}]} + \epsilon\{G_\epsilon^{(n)}, I_\epsilon^{(n)}\} = 0 \tag{5.6}$$

Analogously, we find that if $H_\epsilon^{(n)}$ was relatively integrable to all orders, the compatibility conditions for the equation for the equation

$$\Delta F_\epsilon^{(n)}(H_0) + S_\epsilon^{(n)[\leq 2^{n+1}]} + \epsilon\{G_\epsilon^{(n)}, F_\epsilon^{(n)}(H_0)\} = 0 \tag{5.7}$$

This is the step that makes all the method work because those are the main terms in (5.4) and (5.5). We want to obtain estimates on the resulting $G_\epsilon^{(n)}$'s, then estimate the remainder – in an slightly smaller region – and finally prove that the whole procedure converges.

We will write the rest of the proof only for uniform integrability since the argument from now on is completely identical and it only differs in the notation. As we have shown before, both cases rely on solving the same equation and the two identities (5.4), (5.5) are completely similar.

To obtain estimates for $G_\epsilon^{(n)}$ we observe that by Cauchy estimates Lemma 4.3 we have $||R_\epsilon^{(n)[\leq 2^{n+1}]}||^{\omega, U^{(n)}, \delta-\sigma} \leq K\sigma^{-1}||R_\epsilon^{(n)}||^{\omega, U^{(n)}, \delta}$. Then, applying Theorem 4.5 we obtain:

$$(5.8) \qquad ||G_\epsilon^{(n)}||^{\omega^{(n)}, \widetilde{U^{(n)}}, \delta-2\sigma} \leq K\sigma^{-(d+1)}||R_\epsilon^{(n)}||^{\omega, U^{(n)}, \delta}$$

Also, in an slightly smaller domain we will have:

$$(5.9) \qquad ||\nabla G_\epsilon^{(n)}||^{\omega^{(n)}, \widetilde{\widetilde{U^{(n)}}}, \delta-3\sigma} \leq K\sigma^{-(d+2)}||R_\epsilon^{(n)}||^{\omega, U^{(n)}, \delta}$$

$$||\nabla^2 G_\epsilon^{(n)}||^{\omega^{(n)}, \widetilde{\widetilde{U^{(n)}}}, \delta-3\sigma} \leq K\sigma^{-(d+3)}||R_\epsilon^{(n)}||^{\omega, U^{(n)}, \delta}$$

Since $G_\epsilon^{(n)}$ is chosen in such a way that the main terms in (5.4) cancel, the new remainder will be estimated by estimating systematically all the terms in (5.4) that are not involved in (5.6).

Using (5.9), provided that $\sigma \geq Ke^\delta \sigma^{-(d+2)}||R_\epsilon^{(n)}||^{\omega, U^{(n)}, \delta}$ the solutions of the Hamilton equations corresponding to $G_\epsilon^{(n)}$ that start in $\widetilde{\widetilde{U^{(n)}}} \times |\mathrm{Im}\varphi| \leq \delta - 4\sigma$ will not leave $\widetilde{U^{(n)}}] \times |\mathrm{Im}\varphi| \leq \delta - 3\sigma$ for a "time" $\epsilon |\epsilon| \leq e^\delta$. (Without any loss in generality, we will assume that $e^\delta < 2$) Then, we can estimate as follows:

(5.10)

$$||\epsilon\{G_\epsilon^{(n)}, R_\epsilon^{(n)}\}||^{\omega^{(n)}, \widetilde{\widetilde{U^{(n)}}}, \delta-3\sigma} \leq K||\nabla G_\epsilon^{(n)}||^{\omega^{(n)}, \widetilde{\widetilde{U^{(n)}}}, \delta-3\sigma}||\nabla R_\epsilon^{(n)}||^{\omega^{(n)}, \widetilde{\widetilde{U^{(n)}}}, \delta-2\sigma} e^\delta$$

$$\leq K\sigma^{-(d+3)}\left(||R_\epsilon^{(n)}||^{\omega^{(n)}, U^{(n)}, \delta}\right)^2$$

Similarly, we have

$$||\nabla \epsilon \{G_\epsilon^{(n)}, H_\epsilon^{(n)}\}||^{\omega^{(n)}, \widetilde{\widetilde{U^{(n)}}}, \delta - 3\sigma} \leq K ||\nabla^2 G_\epsilon^{(n)}||^{\omega^{(n)}, \widetilde{\widetilde{U^{(n)}}}, \delta - 3\sigma} ||\nabla H_\epsilon^{(n)}||^{\omega^{(n)}, \widetilde{\widetilde{U^{(n)}}}, \delta - 2\sigma} e^\delta$$

$$\leq K\sigma^{-(d+5)} ||R_\epsilon^{(n)}||^{\omega^{(n)}, U^{(n)}, \delta} ||H_\epsilon^{(n)}||^{\omega^{(n)}, U^{(n)}, \delta};$$

(5.11)

$$||\epsilon \{G^{(n)}, H_\epsilon^{(n)}\} \circ g_\epsilon^{(n)} - \epsilon \{G^{(n)}, H_\epsilon^{(n)}\}||^{\omega^{(n)}, \widetilde{\widetilde{U^{(n)}}}, \delta - 4\sigma} \leq$$

$$\leq ||\nabla \epsilon \{G_\epsilon^{(n)}, H_\epsilon^{(n)}\}||^{\omega^{(n)}, \widetilde{\widetilde{U^{(n)}}}, \delta - 3\sigma} ||g_\epsilon^{(n)} - \mathrm{Id}||^{\omega^{(n)}, \widetilde{\widetilde{U^{(n)}}}, \delta - 3\sigma}$$

$$\leq K\sigma^{-(d+5)} ||R_\epsilon^{(n)}||^{\omega^{(n)}, U^{(n)}, \delta} ||H_\epsilon^{(n)}||^{\omega^{(n)}, U^{(n)}, \delta} ||\nabla G_\epsilon^{(n)}||^{\omega^{(n)}, \widetilde{\widetilde{U^{(n)}}}, \delta - 2\sigma} \leq$$

$$K\sigma^{-(2d+7)} ||H_\epsilon^{(n)}||^{\omega^{(n)}, U^{(n)}, \delta} \left(||R_\epsilon^{(n)}||^{\omega^{(n)}, U^{(n)}, \delta} \right)^2$$

$$||\epsilon^2 \{G_\epsilon^{(n)}, \{G_\epsilon^{(n)}, H_\epsilon^{(n)}\}\} \circ g_\epsilon^{(n)}||^{\omega^{(n)}, \widetilde{\widetilde{U^{(n)}}}, \delta - 4\sigma}$$

$$\leq ||\epsilon^2 \{G_\epsilon^{(n)}, \{G_\epsilon^{(n)}, H_\epsilon^{(n)}\}\}||^{\omega^{(n)}, \widetilde{\widetilde{U^{(n)}}}, \delta - 3\sigma}$$

(5.12)

$$\leq K ||\nabla^2 G_\epsilon^{(n)}||^{\omega^{(n)}, \widetilde{\widetilde{U^{(n)}}}, \delta - 3\sigma} ||\nabla^2 H_\epsilon^{(n)}||^{\omega^{(n)}, \widetilde{\widetilde{U^{(n)}}}, \delta - 3\sigma}$$

$$\leq K\sigma^{-(2d+12)} ||H_\epsilon^{(n)}||^{\omega^{(n)}, U^{(n)}, \delta} \left(||R_\epsilon^{(n)}||^{\omega^{(n)}, U^{(n)}, \delta} \right)^2$$

(5.13) $$||R_\epsilon^{(n)[\geq 2^{n+1}]}||^{\omega^{(n)}, \widetilde{\widetilde{U^{(n)}}}, \delta - 3\sigma} \leq K e^{-\delta 2^{n+1}} ||R_\epsilon^{(n)}||^{\omega^{(n)}, U^{(n)}, \delta}$$

Substituting these estimates in (5.4) we obtain

(5.14)
$$||R_\epsilon^{(n+1)}||^{\omega^{(n)}, U^{(n+1)}, \delta - 4\sigma} \leq K\sigma^{-a} ||R_\epsilon^{(n)}||^{\omega^{(n)}, U^{(n)}, \delta} \left(||R_\epsilon^{(n)}||^{\omega^{(n)}, U^{(n)}, \delta} + K e^{-\delta 2^{n+1}} \right)$$

Finally, we note that since

$$\omega^{(n+1)} = \omega^{(n)} + \nabla \int_{\mathbf{T}^d} R^{(n)}$$

we can obtain applying Lemma 4.4 as well as Cauchy bounds – again provided that smallness conditions that guarantee that the compositions can be defined are satisfied –.

(5.15)
$$||R_\epsilon^{(n+1)}||^{\omega^{(n+1)},U^{(n+1)},\delta-4\sigma} \leq K\sigma^{-a}(||R^{(n)}||^{\omega^{(n)},U^{(n)},\delta-\sigma})^2(1+\sigma^{-2}||R^{(n)}||^{\omega^{(n)},U^{(n)},\delta-\sigma})$$

Once we have estimates such as those above, it is quite standard in K. A. M. theory that choosing at each iterative step $\sigma^{(n)} = \sigma_0 2^{-n}$, with σ_0 small enough so that the domain controlled does not reduce to the empty set, if the initial remainder is small enough we obtain a procedure that is converging quadratically. It is also possible to show that the smallness hypothesis that one needs are also satisfied.

Once we have this quadratic convergence, it is also quite straightforward to show that the composition $g_\epsilon^{(0)} \circ g_\epsilon^{(1)} \circ \cdots \circ g_\epsilon^{(n)}$ converges in a non empty domain. ∎

6. Acknowledgements.

I thank S. Tompaidis and R. Paul for several comments on a preliminary version of the manuscript and E. Meletlidou for providing Example 3.5. This research has been supported by NSF and TARP grants as well as an AMS Centennial Fellowship and and a URI from U.T. Austin.

7. References

[Ar] V.I. Arnol'd: Small Denominators I: On the mappings of circle to itself. Translations AMS. **46**, 213–284 (1961).

[Bar] R. Barrar: Convergence of the VonZeipel procedure. Cel. Mech. **2**, 491-504 (1970).

[CEG] P. Collet, H. Epstein, G. Gallavotti: Perturbations of geodesic flows on surfaces of constant negative curvature and their mixing properties. Comm. Math. Phys. **95**, 65–112 (1984).

[CW] W. Craig, P. Worfolk: An integrable normal form for water waves in infinite depth. Physica **84D**, 513–531 (1995).

[DZ] A.I. Dyachenko, V.E. Zakharov: Is free-surface hydrodynamics an integrable system?. Phys. Lett. **A 190**, 144–148 (1994).

[Du] J.J. Duistermaat: On global action variables. Comm. Pure Appl. Math. **XXXIII**, 687-706 (1980).

[Ga] G. Gallavotti: A criterion of integrability for perturbed harmonic oscillators. "Wick ordering" of the perturbations in classical mechanics and invariance of the frequency spectrum. Comm. Math. Phys. **87**, 365–383

[LMM] R. de la Llave, J.M. Marco, R.Moriyón: Canonical perturbation theory for Anosov systems and regularity properties of Livsic's cohomology equation. Ann. Math. **123**, 537–611 (1986).

[Li] A. Livsic: Homology properties of Y-systems. Math. Notes **10**, 754–757 (1971).

[Me] E. Meletlidou: Thesis, Univ. Thesalonika.

[Po] H. Poincaré: *"Les methodes nouvelles de la méchanique céleste"*, Gauthier Villars, Paris (1891–1899).

[Ve] W. A. Veech: Periodic points and invariant pseudomeasures for toral endomorphisms. Ergod. Theo. Dyn. Syst. **6**, 449-473 (1986).

RICARDO MAÑÉ

The Lyapunov exponents of generic area preserving diffeomorphisms

Editor's note. This article contains an outline of a proof of a result announced by the author around 1980. The theorem and its generalization to higher dimensions are reported in the author's address at the International Congress in 1983 [M]. The proofs have not been published. For the 2-dimensional case, the closest to a written draft that is available are some notes handed out by the author at a seminar in Paris. Except for some very minor editing, this article is a direct transcription of those notes. We stress that what is presented here is only the sketch of a proof, and that a substantial amount of work is needed to fill in the details.

Let M be a compact connected boundaryless smooth 2-manifold and let μ be a probability on its Borel σ-algebra generated by a smooth volume form. Let $D_\mu(M)$ be the space of μ-preserving C^1 diffeomorphisms endowed with the C^1 topology.

Theorem. *There exists a residual subset $\mathcal{A} \subset D_\mu(M)$ such that every $f \in \mathcal{A}$ is either Anosov or*

$$\lim_{n \to \pm\infty} \frac{1}{n} \log \|(D_x f^n)\| = 0$$

for a.e. x.

In these notes we shall sketch the proof of this result. This sketch will consist of the statements of all the lemmas conducing to the proof of the theorem. With few exceptions these lemmas will be stated without proof. The sequence of lemmas is divided in several steps. In the first step we introduce an "entropy" function $h : D_\mu(M) \to \mathbb{R}$ that is upper semicontinuous. Then its set $\mathcal{A}_0$ of points of continuity is residual. In the second step we shall prove the existence of a residual set $\mathcal{A}_1 \subset D_\mu(M)$ of diffeomorphisms such that all its hyperbolic sets have measure 0 or 1. The third, fourth and fifth steps are devoted to show that diffeomorphisms in $\mathcal{A}_0$ do not exhibit certain pathological behaviours, because these behaviours would produce discontinuities of h. The perturbations required to show this discontinuity are constructed with the lemmas given in section III. Using this absence of pathological behaviour we shall show that a diffeomorphism $f \in \mathcal{A}_0 \cap \mathcal{A}_1$ satisfies one of the alternatives of the theorem.

§1 First step: the entropy function.

If $f \in D_\mu(M)$ define $\Sigma(f)$ as the set of points $x \in M$ such that the limits:

$$\lim_{n \to \pm\infty} \frac{1}{|n|} \log \|(D_x f^n)\|$$

exist and coincide. By Oseledec's theorem:

$$\mu(\Sigma(f)) = 1 \,.$$

Define $\lambda^+ : \Sigma(f) \to M$ by

$$\lambda^+(x) = \lim_{n \to +\infty} \frac{1}{n} \log \|(D_x f^n)\|$$

and

$$\Sigma_H(f) = \{x \in \Sigma(f) \mid \lambda^+(x) \neq 0\} \,.$$

By Oseledec's theorem:

$$\lambda^+(x) \geq 0$$

for all $x \in \Sigma(f)$ and if $\lambda^+(x) \neq 0$ there exists a splitting $T_x M = E^s(x) \oplus E^u(x)$ such that:

$$\lim_{n \to \pm\infty} \frac{1}{n} \log \|(D_x f^n) \mid E^s(x)\| = -\lambda^+(x)$$

$$\lim_{n \to \pm\infty} \frac{1}{n} \log \|(D_x f^n) \mid E^u(x)\| = \lambda^+(x) \,.$$

Define the entropy function $h : D_\mu(M) \to \mathbb{R}$ by

$$h(f) = \int_{\Sigma(f)} \lambda^+ d\mu \,.$$

Lemma. *h is upper semicontinuous.*

Proof. Define

$$c_n = \int_{\Sigma(f)} \log \|(D_x f^n) \mid E^u(x)\| d\mu(x)$$

then:

$$c_{n+m} \leq c_n + c_m$$

for all $n > 0$, $m > 0$. Hence:

$$h(f) = \lim_{n \to +\infty} \frac{1}{n} \int_{\Sigma(f)} \log \|(D_x f^n) \mid E^u(x)\| d\mu(x) = \lim_{n \to +\infty} \frac{1}{n} c_n$$

$$= \inf_n c_n = \inf_n \frac{1}{n} \int_{\Sigma(f)} \log \|(D_x f^n) \mid E^u(x)\| d\mu(x) \,.$$

111

Definition. $\mathcal{A}_0 = $ *set of points of continuity of h.*

Corollary. $\mathcal{A}_0$ *is residual.*

§2 Second step: avoiding "fat" hyperbolic sets

Proposition. *There exists a residual subset $\mathcal{A}_1 \subset D_\mu(M)$ such that if $f \in \mathcal{A}_1$ all its hyperbolic sets have measure 0 or 1.*

Proof. Let $\mathcal{B}$ be a basis of neighborhoods of M and let $\tilde{\mathcal{B}}$ be the set of all finite subsets of $\mathcal{B}$ such that $\overline{\cup_{B \in F} B} \neq M$. If $f \in D_\mu(M)$ and $F \in \tilde{\mathcal{B}}$ define:

$$S(F, f) = \cap_n f^n \left(\cup_{B \in F} B \right) .$$

If $\varepsilon > 0$ and $F \in \tilde{\mathcal{B}}$ define $D(\varepsilon, F)$ as the set of diffeomorphisms $f \in D_\mu(M)$ such that one of the following properties is satisfied:

(1) There exists a neighborhood $\mathcal{U}$ of f such that $S(F, g)$ is not hyperbolic for all $g \in \mathcal{U}$.

(2) $\mu(S(F, f)) < \varepsilon$.

Obviously $D(\varepsilon, F)$ is open. Moreover it is dense because if $f \in D\mu(M)$ is C^2 (that is a dense property in $D_\mu(M)$ as proved by Zehnder) then every hyperbolic set of f has measure 0 or 1.

§3 The fundamental perturbation lemma

As we explained in the introduction in the next section we shall have to produce perturbations of diffeomorphisms exhibiting certain behaviours. The key tool in the construction of these perturbations will be the next lemma.

Definition. *If $f : M \circlearrowleft$ is a diffeomorphism and $F \in \mathbb{Z}^+$ we say that a sequence of linear maps $L_j : T_{f^j(x)} M \to T_{f^{j+1}(x)} M$, $x \in M$, $j = 0, 1, \ldots m$, is F-adapted if $L_j = D_{f^j(x)} f$ for all j except for at most F values of j or if*

$$(L_n L_{n-1} \cdots L_0) B_1(0) = (D_x f^n) B_1(0)$$

for all $0 \leq n \leq m$.

Fundamental Lemma. *If $f \in D_\mu(M)$, $F \in \mathbb{Z}^+$ and $\mathcal{U}$ is a neighborhood of f there exists $\varepsilon > 0$ such that if $\Lambda_1 \subset M$ is a Borel set, $0 < \varepsilon_1 < \varepsilon$ and $N : \Lambda_1 \to \mathbb{Z}^+$ is a measurable function with the property that for all $x \in \Lambda_1$ and $m > N(x)$ there exists an F-adapted sequence $L_j : T_{f^j(x)} M \to T_{f^{j+1}(x)} M$, $0 \leq j \leq m$, satisfying*

$$\|L_j - (D_{f^j(x)} f)\| \leq \varepsilon$$

for all $0 \leq j \leq m$ and

$$\| \prod_{j=0}^{m} L_j \| \leq \exp m\varepsilon_1$$

then there exists $g \in \mathcal{U}$ such that

$$h(g) \leq h(f) + 2\varepsilon_1 - \int_{\cup_{n \geq 0} f^n(\Lambda_1)} \lambda^+ d\mu.$$

This lemma is obtained from the following sequence of properties:

Lemma. *If $f \in D_\mu(M)$, $0 < k < 1$ and $\mathcal{U}$ is a neighborhood of f, there exist $\varepsilon > 0$ and $r_0 > 0$ such that if $x \in M$, $0 < r < r_0$, and $L : T_x M \to T_{f(x)} M$ is a linear map satisfying:*

$$\|L - (D_x f)\| < \varepsilon$$

then there exists $g \in \mathcal{U}$ such that:

$$g(y) = f(y)$$

when $y \notin B_r(x)$ and

$$D_y g = L$$

if $y \in B_{kr}(x)$.

Lemma. *Given $f \in D_\mu(M)$, $F \in \mathbb{Z}^+$, $0 < k < 1$, $n > 0$ and a neighborhood $\mathcal{U}$ of f there exist $\varepsilon > 0$ and $r_0 > 0$ such that for all $0 < r < r_0$ and every nonperiodic point $x \in M$, if $L_j : T_{f^j(x)} M \to T_{f^{j+1}(x)} M$, $0 \leq j < n$, is an F-adapted sequence satisfying*

$$\|L_j - (D_{f^j(x)} f)\| \leq \varepsilon$$

for all $0 \leq j < n$ then for all $0 < r < r_0$ there exists $g \in \mathcal{U}$ and a compact set $K \subset B_r(x)$ with $\mu(K) > k\mu(B_r(x))$ such that

$$f(y) = g(y)$$

if $y \notin \cup_{m=0}^{n} f^m(B_r(x))$ and

$$D_y g^m = L_{m-1} L_{m-2} \cdots L_0$$

if $y \in K$ and $0 \leq m < n$.

Definition. *Given $f \in D_\mu(M)$ we say that a family of Borel sets $\{U_0, \ldots, U_m\}$ is a tower if they are disjoint and $U_j = f^j(U_0)$ for all $0 \leq j \leq m$.*

Lemma. *Given* $f \in D_\mu(M)$, $0 < k < 1$, $F \in \mathbb{Z}^+$ *and a neighborhood* $\mathcal{U}$ *of* f *there exists* $\varepsilon > 0$ *such that if* $\{U_0^{(n)}, \ldots, U_{j_n}^{(n)}\}$, $n = 1, \ldots, \ell$, *is a family of disjoint towers and for all* $x \in \cup_{n=1}^\ell U_0^{(n)}$ *we have an* F-*adapted sequence* $L_j(x) : T_{f^j(x)}M \to T_{f^{j+1}(x)}M$, $0 \leq j < j_n$, *such that*

$$\|L_j(x) - (D_{f^j(x)}f)\| \leq \varepsilon$$

for all $0 \leq j < j_n$, *then there exists a compact set* $K \subset \cup_{n=1}^\ell U_0^{(n)}$ *and* $g \in \mathcal{U}$ *satisfying:*

(1) $g(x) = f(x)$ *for* $x \in S$

(2) $\mu(K) \geq k\mu(\cup_{n=1}^\ell U_0^{(n)})$

(3) *If* $x \in K$ *there exist* $1 \leq n \leq \ell$ *and* $y \in U_0^{(n)}$ *such that:*
$$D_x g^{j_n} = L_{j_n - 1}(y) \cdots L_0(y).$$

[Editor's remark: S apparently not defined.]

Lemma. *If* $f \in D_\mu(M)$, $F \in \mathbb{Z}^+$ *and* $\mathcal{U}$ *is a neighborhood of* f *there exists* $\varepsilon > 0$ *such that if* $0 < \varepsilon_1 < \varepsilon$ *and* $\{U_0^{(n)}, \ldots, U_{j_n}^{(n)}\}$, $n = 1, \ldots,$ *are disjoint towers whose union is* f-*invariant and for all* $n \geq 1$ *and* $x \in U_0^{(n)}$ *there exists an* F-*adapted sequence* $L_j : T_{f^j(x)}M \to T_{f^{j+1}(x)}M$, $0 \leq j < j_n$, *such that*

(1) $\|L_j - (D_{f^j(x)}f)\| \leq \varepsilon$ *for all* $0 \leq j < j_n$

(2) $\|L_{j_n - 1} \cdots L_0\| \leq \exp n\varepsilon_1$

[Editor's remark: Author probably meant $\exp(j_n \varepsilon_1)$.]

then there exists $g \in \mathcal{U}$ *satisfying*

$$h(g) \leq h(f) + 2\varepsilon_1 - \int_{\cup_{n,j} U_j^{(n)}} \lambda^+ d\mu.$$

The fundamental lemma follows from the last one and the following well known result about decompositions of invariant sets in union of disjoint towers:

Lemma. *If* $f \in D_\mu(M)$ *and* $\Lambda \subset M$ *is a Borel set such that the measure of the set of periodic points of* f *contained in* Λ *is* 0, *then for every* $N > 0$ Λ *can be decomposed in a countable union of disjoint towers with its basis contained in* Λ_1 *and heights larger that* N.

[Editor's remark: The statement of this lemma is probably not intended as is.]

§4 third step: bounding the angles of the Oseledec splitting

If $f \in D_\mu(M)$ define $\delta_1 : \Sigma_H(f) \to \mathbb{R}$ by:

$$\delta_1(x) = \inf_{n \geq 0} \alpha(E^s(f^n(x)), E^u(f^n(x)))$$

where $\alpha(\cdot, \cdot)$ is the angle between the subspaces between brackets. Let $\Sigma_1(f) = \{x \in \Sigma_H(f) \mid \delta_1(x) = 0\}$. Then $f(\Sigma_1(f)) = \Sigma_1(f)$.

Proposition. *If* $f \in D_\mu(M)$ *and* $\mu(\Sigma_1(f)) \neq 0$ *then for every* $\varepsilon_0 > 0$ *and all neighborhood* $\mathcal{U}$ *of* f *there exists* $g \in \mathcal{U}$ *with*

$$h(g) \leq h(f) + \varepsilon_0 - \int_{\Sigma_1(f)} \lambda^+ d\mu \, .$$

Corollary. $f \in \mathcal{A}_0 \Rightarrow \mu(\Sigma_1(f)) = 0$.

Sketch of the proof. Given $f \in D_\mu(M)$ with $\mu(\Sigma_1(f)) > 0$ and a neighborhood $\mathcal{U}$ of f take $\varepsilon > 0$ given by the fundamental lemma. Fix any $c > 0$ and define:

$$\Lambda_c = \{x \in \Sigma_1(f) \mid \lambda^+(x) > c\} \, .$$

Then:

$$\lim_{n \to -\infty} \frac{1}{n} \log \frac{\|(D_x f^n) \mid E^s(x)\|}{\|(D_x f^n) \mid E^\mu(x)\|} \geq -2c$$

for all $x \in \Lambda_c$. [Editor's remark: Author probably meant "$\leq -2c$".] Then there exists $a > 0$ such that if $x \in \Lambda_c$ then $\alpha(E^s(f^m(x)), E^u(f^m(x))) \geq a$ for some $m > 0$. This means that if we define:

$$\Lambda_a = \{x \in \Lambda_c \mid \alpha(E^s(x), E^u(x)) > a\}$$

we have:

$$\Gamma_c = \cup_{n>0} f^n(\Gamma_a) \, .$$

Take $A > 0$ and $0 < \varepsilon_1 < \min(\varepsilon, a)$ and a measurable function $N : \Gamma_a \to \mathbb{Z}^+$ such that if $x \in \Gamma_a$ and $n > N(x)$ then in every interval $[j, j + n\varepsilon_1] \subset [0, n]$ there exists m such that:

$$\alpha(E^s(f^m(x)), E^u(f^m(x))) \leq \varepsilon_1$$
$$\|(D_{f^m(x)} f^{n-m}) \mid E^s(f^m(x))\| \leq \exp(n-m)(-\lambda^+(x) + A\varepsilon_1)$$
$$\|(D_x f^m)\| \leq \exp m(\lambda^+(x) + A\varepsilon_1)$$
$$\|(D_{f^m(x)} f^{n-m})\| \leq \exp(n-m)(\lambda^+(x) + A\varepsilon_1)$$
$$\|(D_x f^m) \mid E^s(x)\| \leq \exp m(-\lambda^+(x) + A\varepsilon_1) \, .$$

Choose $m \in [(n-n\varepsilon_1)/2, (n+n\varepsilon_1)/2]$ with all these properties and define a sequence of linear mappings $L_j : T_{f^j(x)}M \to T_{f^{j+1}(x)}M$ putting $L_j = (D_{f^j(x)} f)$ when $0 \leq j \leq m - 2$ or $m \leq j \leq n - 1$ and

$$L_{m-1} = R(D_{f^{m-1}(x)} f)$$

where R is the rotation that sends $E^u(f^m(x))$ on $E^s(f^m(x))$. Then:

$$\|\prod_{j=0}^{n-1} L_j|E^s(x)\| \le \exp 2(\lambda^+(x) + A\varepsilon_1)\varepsilon_1 n$$

$$\|\prod_{j=0}^{n-1} L_j|E^u(x)\| \le \exp 2(\lambda^+(x) + A\varepsilon_1)\varepsilon_1 n$$

Since $\alpha(E^s(x), E^u(x)) \ge a$ we can choose $A = A(a)$ so small that these inequalities imply:

$$\|\prod_{j=0}^{n-1} L_j\| \le \exp 3C\varepsilon_1 n$$

where:

$$C = \sup_{x\in\Sigma_\mu(f)} \lambda^+(x).$$

If $\varepsilon_1 > 0$ is small enough, the rotation R is so near to the identity that:

$$\|L_j - (D_{f^j(x)}f)\| < \varepsilon$$

for all $0 \le j \le n-1$. Then the hypothesis of the fundamental lemma are satisfied and there exists $g \in \mathcal{U}$ with

$$h(g) \le h(f) + 6C\varepsilon_1 - \int_{\Lambda_c} \lambda^+ d\mu$$

$$= h(f) + 6C\varepsilon_1 - \int_{\Sigma_1(f)} \lambda^+ d\mu + \int_{\Sigma_1(f)-\Lambda_c} \lambda^+ d\mu$$

$$\le h(f) + 6C\varepsilon_1 + 6\mu(\Sigma_1(f) - \Lambda_c) - \int_{\Sigma_1(f)} \lambda^+ d\mu.$$

Since ε_1 is arbitrarily small the proposition is proved.

§5 Fourth step: the domination inequalities

Define $\delta_2 : \Sigma_H(f) \to \mathbb{R}^*$ (i.e. $\mathbb{R} \cup \{+\infty\}$) taking as $\delta_2(x)$ the supremum over all $n > 0$, $m > 0$ of

$$\|(D_{f^n(x)}f^m)\,|\,E^s(f^n(x))\| \cdot \|(D_{f^{n+m}(x)}f^{-m})\,|\,E^u(f^{n+m}(x))\|.$$

Define $\Sigma_2(f)$ as the set of points $x \in \Sigma_H(f)$ where $\delta_2(x) = +\infty$ then $f(\Sigma_2(f)) = \Sigma_2(f)$.

Proposition. *If $f \in D_\mu(M)$ and $\mu(\Sigma_2(f) \cap \Sigma_1(f)^c) \neq 0$ then for every neighborhood $\mathcal{U}$ of f and every $\varepsilon > 0$ there exists $g \in \mathcal{U}$ such that:*

$$h(g) \leq h(f) + \varepsilon - \int_{\Sigma_2(f) \cap \Sigma_1(f)^c} \lambda^+ d\mu .$$

Corollary. $f \in \mathcal{A}_0 \Rightarrow \mu(\Sigma_2(f)) = 0.$

This proposition also follows from the fundamental lemma. To apply it take $K > 0$, $n_0 > 0$, $k > 0$ and define Λ as the set of points $x \in \Sigma_2(f) \cap \Sigma_1(f)^c$ such that

$$\|(D_x f^{n_0}) \mid E^s(x)\| \cdot \|(D_{f^{n_0}(x)} f^{-n_0}) \mid E^u(f^{n_0}(x))\| > K ,$$

$$\delta_1(x) > a .$$

[Editor's question: $k = a$?]

Let

$$\Lambda(K, n_0, k) = \cup_{n>0} f^{-n}(\Lambda) .$$

If K and n_0 are large and k small, the measure of $\Sigma_2(f) \cap \Sigma_1(f)^c - \Lambda(K, n_0, k)$ is small. Now we take $\varepsilon_1 > 0$ and a measurable function $N : \Lambda(K, n_0, k) \to \mathbb{Z}^+$ such that if $x \in \Lambda(K, n_0, k)$ and $n > N(x)$ in every interval $[j, j + n\varepsilon_1]$ contains some m such that $f^m(x) \in \Lambda$. We define the sequence of linear maps L_j as follows: choose $m \in [(n - n\varepsilon_1)/2, (n + n\varepsilon_1)/2]$ with $f^m(x) \in \Lambda$. This property implies that there exists a subspace $S \in T_{f^m(x)} M$ with the properties described in the picture:

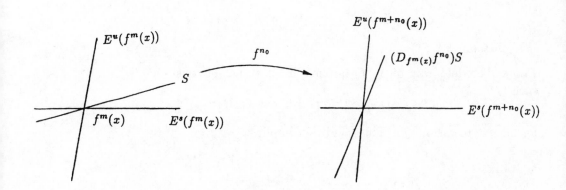

Define L_{m-1} as the composition of $(D_{f^{m-1}(x)} f)$ with a linear map near to I and sending $E^s(f^m(x))$ on S and leaving $E^u(f^m(x))$ invariant. Define L_{m+n_0-1} as $(D_{f^{m+n_0-1}(x)} f)$ composed with a rotation R mapping $(D_{f^m(x)} f^{n_0})(S)$ on

117

$E^u(f^{m+n_0}(x))$. Finally define L_{n-1} as the composition of $D_{f^{n-1}(x)}f$ with a linear map L near to the identity that maps $(D_{f^{m+n_0}(x)}f^{n-m})\mathrm{RE}^u(f^{m+n_0}(x))$ on $E^s(f^n(x))$ and leaves $E^u(f^n(x))$ invariant. The other linear maps L_j are defined as $(D_{f^j(x)}f)$. Then $\prod_{j=0}^{n-1} L_j$ maps $E^s(x)$ on $E^u(f^n(x))$ and $E^u(x)$ on $E^s(f^n(x))$ and the norm can be bounded by $\exp 3Cn\varepsilon_1$ as in the previous section. In fact this estimate as well as the possibility of taking L near to the identity require some more precautions in the choice of m that we didn't explain, hoping that this technical oversimplification may help to put in evidence the underlying idea of the proof.

§6 Fifth step: more domination inequalities

Define $\tilde{\delta}_3 : \Sigma_H(f) \to \mathbb{R}^*$ and $\delta_3 : \Sigma_H(f) \to \mathbb{R}^*$ by

$$\tilde{\delta}_3(x) = \sup\{n > 0 \mid \|(D_xf^m)|E^s(x)\| \cdot \|(D_{f^m(x)}f^{-m})|E^u(f^m(x))\| > 1/2$$
$$\text{for all} \quad 0 \leq m \leq n\}$$

$$\delta_3(x) = \sup_{n>0} \tilde{\delta}_3(f^n(x)).$$

Let $\Sigma_3(f) = \{x \mid \delta_3(x) = +\infty\}$.

Proposition. If $f \in D_\mu(M)$ and $\mu(\Sigma_3(f) \cap \Sigma_2(f)^c \cap \Sigma_1(f)^c) > 0$ [then] *for all $\varepsilon > 0$ and every neighborhood $\mathcal{U}$ of f there exists $g \in \mathcal{U}$ satisfying:*

$$h(g) \leq h(f) + \varepsilon - \int_{\Sigma_3(f) \cap \Sigma_2(f)^c \cap \Sigma_1(f)^c} \lambda^+ d\mu.$$

Corollary. $f \in \mathcal{A}_0 \Rightarrow \mu(\Sigma_3(f)) = 0$.

Lemma. *If $f \in D_\mu(M)$ then for all $c > 0$ the closure of the set*

$$\Lambda_c(f) = \{x \mid \delta_3(x) < c\}$$

is hyperbolic if $\Lambda_c(f) \neq \phi$.

Proof. The definition of $\Lambda_c(f)$ implies the existence of constants $K > 0, 0 < \lambda < 1$ such that:
$$\|(D_xf^n)|E^s(x)\| \cdot \|(D_{f^n(x)}f^{-n})|E^u(f^n(x))\| \leq K\lambda^n$$
for all $n \geq 0$, $x \in \Lambda_c(f)$. This condition implies
$$\inf_{x \in \Lambda_c(f)} \alpha(E^s(x), E^u(x)) > 0.$$

Using the fact that f is area preserving and $\dim M = 2$ it follows that there exist $\tilde{K} > 0, 0 < \tilde{\lambda} < 1$ such that:
$$\|(D_xf^n) \mid E^s(x)\| \leq K\lambda^n$$
$$\|(D_{f^n(x)}f^{-n} \mid E^u(x)\| \leq K\lambda^n$$

for all $x \in \Lambda_c(f)$, $n \geq 0$. From the uniformity of these estimates, the hyperbolicity of $\overline{\Lambda_c(f)}$ follows easily.

Proof of the Theorem. Suppose that $f \in \mathcal{A}_0 \cap \mathcal{A}_1$ and $\mu(\Sigma_H(f)) \neq 0$. We shall prove that f is Anosov. By the corollary in section 1, the hypothesis $f \in \mathcal{A}_0$ implies:

$$\mu(\Sigma_3(f)) = 0 \,.$$

Hence $\mu(\Sigma_H(f) - \Sigma_3(f)) > 0$. But

$$\Sigma_H(f) - \Sigma_3(f) = \cup_{c>0} \Lambda_c(f) \,.$$

Therefore $\mu(\Lambda_c(f)) > 0$ for some $c > 0$. By the lemma the set $\overline{\Lambda_c(f)}$ is hyperbolic and since $f \in \mathcal{A}_1$ its measure must be 1. Then f is Anosov.

REFERENCES

[M] R. Mañé, *Oseledec's theorem from the generic viewpoint*, Proceedings of the International Congress of Mathematicians, August 1983, Warszawa, Volume 2, p. 1259-1276 (Theorem C and its Corollary in particular).

RICARDO MAÑÉ

Lagrangian flows: the dynamics of globally minimizing orbits

Editor's note. There notes were prepared by Bicardo Mañé as the subject of his talk at the conference. A complete version of this paper is written by Gonzalo Contreras, Jorge Delgado and Renato Iturriaga, and will lie published elsewhere.

The objective of this note is to present some results, to be proved in a forth-coming paper, about certain special solutions of Euler-Lagrange equations on closed manifolds. Our main results extend to time dependent periodic Lagrangians, with minor modifications, as we shall explain in Section II.

We have chosen to begin the exposition with the autonomous case because this formally simpler framework allows to reach more easily the core of our concepts and results. Moreover the autonomous case exhibits certain special features involving the energy as a first integral that deserve special attention. They are closely related to the link found by Carneiro [C] between the energy and Mather's action function [Ma].

§1 The autonomous case.

Let L be a Lagrangian on a closed manifold M, i.e. $L : TM \to \mathbb{R}$ is C^α and has positive definite Hessian on the fibers. The Euler-Lagrange equation,

$$\text{(E-L)} \qquad \frac{d}{dt}\left(\frac{\partial L}{\partial v}(x, \dot{x})\right) - \frac{\partial L}{\partial x}(x, \dot{x}) = 0$$

generates a smooth flow $f_t : TM \hookleftarrow$ defined as follows. Given $w \subset (p, v) \in TM$, denote

$$x_w : \mathbb{R} \to M$$

the solution of (E-L) with initial condition :

$$x_w(0) = p \ ,$$
$$\dot{x}_w(0) = v \ .$$

Now define $f_t : TM \hookleftarrow$ by

$$f_t(w) = (x_w(t), \dot{x}_w(t)) .$$

Hence every orbit $\gamma(t)$ of this flow can be uniquely written as $\gamma(t) = (x(t), \dot{x}(t))$, where $x(t)$ is a solution of (E-L).

It is well known that solutions of (E-L), and, through them, orbits of the flow $f_t : TM \hookleftarrow$, are characterized by local variational properties.

Here we shall revisit an old subject : orbits of the flow f_t, selected in the intrincate phase portrait of f_t, by requiring of them to satisfy global variational properties instead of the local ones that every orbit satisfies. Research on these special orbits goes back to Morse ([Mo], 1924) and Hedlund ([H], 1932), and has recently reappeared in the works of Bangaert ([B]) and Mather ([Ma1], [Ma2]). Our approach, while visibly conceptually indebted to those works, will be independent and selfcontained, leading to knew result and also to stronger forms of already known ones, like Mather's Graph Theorem ([Ma]) or the Coboundary Property ([M]).

Recall that the action of the lagrangian L on a absolutely continuous curve $x : [a, b] \to M$ is defined by

$$S_L(x) = \int_a^b L(x(t), \dot{x}(t)) dt .$$

Given two points, $p_i \in M, i = 1, 2$, denote $Ac(p_1, p_2)$ the set of absolutely continuous curves $x : [0, T] \to M$, with $x(0) = p_1, x(T) = p_2$. For each $k \in \mathbb{R}$ we define the **action potential** $\Phi_k : M \times M \to \mathbb{R}$ by

$$\Phi_k(p_1, p_2) = \inf\{S_{L+k}(x) | x \in Ac(p_1, p_2)\} .$$

Theorem I - *There exists $c(L) \in \mathbb{R}$ (called the critical value of L) such that :*

a) $k < c(L) \Rightarrow \Phi_k(p_1, p_2) = -\infty, \forall p_1, p_2$.

b) $k \geq c(L) \Rightarrow \Phi_k(p_1, p_2) >= -\infty, \forall p_1, p_2$, and Φ_k is a Lipschitz function.

c) $\quad k \geq c(L) \Rightarrow \Phi_k(p_1, p_3) \leq \Phi_k(p_1, p_2) + \Phi_k(p_2, p_3), \forall p_1, p_2, p_3$
$\quad\quad\quad \Phi_k(p_1, p_2) + \Phi_k(p_2, p_1) \geq 0, \forall p_1, p_2$

d) $k > c(L) \Rightarrow \Phi_k(p_1, p_2) + \Phi_k(p_2, p_1) > 0, \forall p_1 \neq p_2$.

Defining $d_k : M \times M \to \mathbb{R}$ by $d_k(p_1, p_2) = \Phi_k(p_1, p_2) + \Phi_k(p_2, p_1)$, the properties above say that $d_k(\cdot, \cdot)$ is a metric for $k > c(L)$ and a pseudometric for $k = c(L)$.

Denote $\mathcal{M}(L)$ the set of μ-invariant probabilities of the flow f_t.

Theorem II - *(Ergodic Determination of $c(L)$)*

$$c(L) = -\min\{\int L d\mu | \mu \in \mathcal{M}(L)\} .$$

Definition - We say that $\mu \in \mathcal{M}(L)$ is a minimizing measure if

$$c(L) = -\int L d\mu .$$

Generically, in the sense of [M] the structure of $\mathcal{M}(L)$ is simple.

Theorem III - *For a generic $L, \mathcal{M}(L)$ contains a single measure, and this measure is uniquely ergodic. When this measure is supported by a periodic orbit, this orbit is hyperbolic.*

But we can hope even more : **Conjecture.**

For a **generic** *$L, \mathcal{M}(L)$ consists of a single measure* **supported in a periodic orbit**.

The prerequisite of the next definition is this remark : for every absolutely continuous $x : [a, b] \to M$ and all $k \geq c(L)$:

$(*)$ $\qquad\qquad S_{L+k}(x) \geq \Phi_k(x(a), x(b)) \geq -\Phi_k(x(b), x(a)) .$

Definition - Set $c = c(L)$. We say that $x : [a, b] \to M$ is a **semistatic** curve if it is absolutely continuous and :

$(1) \qquad\qquad S_{L+c}(x/[t_0, t_1]) = \Phi_c(x(t_0), x(t_1))$

for all $a < t_0 \leq t_1 < b$; and that is a **static** curve if

$(2) \qquad\qquad S_{L+c}(x/[t_0, t_1]) = -\Phi_c(x(t_1), x(t_0)) .$

By $(*)$, equality (2) implies (1). Hence static curves are semistatic. Semistatic curves are solutions of (E-L). This follows from classic results that grant that absolutely continuous curves with much weaker variational properties are solutions.

Definition -
$$\Sigma(L) = \{w \in TM | \ x_w \ : \ \mathbb{R} \to M \text{ is semistatic}\}$$
$$\widehat{\Sigma}(L) = \{w \in TM | \ x_w \ : \ \mathbb{R} \to M \text{ is static}\}$$
$$\Sigma^+(L) = \{w \in TM | \ x_w/[0, +\infty) \text{ is semistatic}\} .$$

Remarks :

122

a) Replacing c by any other real number is the definition of semistatic solution, the set $\Sigma^+(L)$ (and then $\widehat{\Sigma}(L) \subset \Sigma(L) \subset \Sigma^+(L)$) becomes empty.

b) In [Ma2], Σ denotes what in our setting would be the closure of the union of the supports of the minimizing measures. This set is in general much smaller than the set we called $\widehat{\Sigma}(L)$.

Theorem IV - *(Characterization of minimizing measures). A measure $\mu \in \mathcal{M}(L)$ is minimizing if and only if*

$$\text{supp}\,(\mu) \subset \widehat{\Sigma}(L) \ .$$

Theorem V - *(Recurrence properties).*
a) $\Sigma(L)$ is chain transitive.

b) $\widehat{\Sigma}(L)$ is chain recurrent.

c) The ω-limit set of a semistatic orbit is contained in $\widehat{\Sigma}(L)$.

Theorem VI - *(Graph properties).*
*a) If $\gamma(t), t \geq 0$ is an orbit in $\Sigma^+(L)$, then, denoting $\pi : TM \to M$ the canonical projection, the map $\pi/\{\gamma(t)/t \geq 0\}$ is **injective with Lipschitz inverse**.*

*b) Denoting $\Sigma_0(L) \subset M$ the projection of $\widehat{\Sigma}(L)$, for every $p \in \Sigma_0(L)$ there exists a **unique** $\xi(p) \in T_pM$ such that*

$$(p, \xi(p)) \in \Sigma^+(L) \ .$$

Moreover

$$(p, \xi(p)) \in \widehat{\Sigma}(L) \ ,$$

*and the vectorfield ξ is **Lipschitz**. Obviously*

$$\widehat{\Sigma}(L) = \text{graph}\,(\xi) \ .$$

The following result will imply the **covering property** :

$$\pi\Sigma^+(L) = M \ ,$$

while also dealing with the **injectivity of π on certain subsets** of $\Sigma^+(L)$.

Define the pseudometric $d_c(\cdot, \cdot)$ on M by

$$d_c(a, b) = \Phi_c(a, b) + \Phi_c(b, a) \ .$$

Denote $\mathcal{S}$ the set of equivalence classes of the equivalence relation $d_c(a,b) = 0$ **in** $\widehat{\Sigma}(L)$. If $\Gamma \in \mathcal{S}$ set

$$\Gamma^+ := \{w \in \Sigma^+(L) | \omega(w) \subset \Gamma\} \ .$$

Obviously Γ^+ is forward invariant. Set :

$$\Gamma_0^+ := \bigcup_{t>0} \pi f_t \Gamma^+ \ .$$

Theorem VII - *(Covering property). If $\Gamma \in \mathcal{S}$.*

a) $\pi \Gamma^+ = M$.

b) *For all $p \in \Gamma_0^+$, there exists a* **unique** *$\xi_\Gamma(p) \in T_p M$ such that*

$$(p, \xi_\Gamma(p)) \in \Gamma^+ \ .$$

Moreover Γ_0^+ is an open and dense subset of M and ξ_Γ is Lipschitz.

Remarks :
a) The solution of

$$\dot{x} = \xi_\Gamma(x)$$

are defined in $[0, +\infty)$ and are semistatic curves.

b) On $\widehat{\Sigma}(L) \cap \Gamma, \xi_\Gamma = \xi$.

The next result is an stronger form of Theorem II.

Theorem VIII - *(Generic structure of $\Sigma(L)$). For a generic Lagrangian L, $\widehat{\Sigma}(L)$ is a uniquely ergodic set and, if it is a periodic orbit, it is a hyperbolic periodic orbit.*

Now we can state the extension to all $\widehat{\Sigma}(L)$ a property proved in [M] for supports of ergodic minimizing measures.

Theorem IX - *(Coboundary property). If $c = c(L)$, then $(L+c)/\widehat{\Sigma}(L)$ is a Lipschitz coboundary. More precisely, taking any $p \in M$ and defining $G : \widehat{\Sigma}(L) \to \mathbb{R}$ by*

$$G(w) = \Phi_c(p, \pi(w)) \ ,$$

then

$$(L+c)/\widehat{\Sigma}(L) = \frac{dG}{df} \ ,$$

where

$$\frac{dG}{df}(w) := \lim_{h \to 0} \frac{1}{h} \left(G(f_h(w)) - G(w) \right) \ .$$

Exploiting that the energy, $E : TM \to \mathbb{R}$, defined as usual by $E(x, v) = \frac{\partial L}{\partial v} v - L$ is a first integral of the flow generated by L, leads to **information on the position of** $\widehat{\Sigma}(L)$. First observe that it is easy to check that a semistatic curve $x : [a, b] \to M$ satisfies :

$$(*) \qquad\qquad E(x(t), \dot{x}(t)) = c .$$

This follows from calculating the derivative at $\lambda = 1$ of the function $F : \mathbb{R} \to \mathbb{R}$ given by :

$$F(\lambda) = \int_a (L + c)(x_\lambda(t), \dot{x}_\lambda(t)) dt ,$$

where $x_\lambda : [a, \lambda b] \to M$ is given by $x_\lambda(t) = x(\lambda t)$. From $(*)$ follows that :

$$\Sigma(L) \subset E^{-1}(c) ,$$

that together with $\pi \Sigma(L) = M$ implies :

$$\pi E^{-1}(c) = M .$$

Hence,

$$c \geq \max_q E(q, 0) .$$

Moreover $\widehat{\Sigma}(L) \subset E^{-1}(c)$ implies :

Corollary - $\mu \in \mathcal{M}(L)$ *is minimizing if and only if*

$$\int \left(\frac{\partial L}{\partial v} \right) v \, d\mu = 0$$

$$supp \, (\mu) \subset E^{-1}(c(L))$$

From the viewpoint of variational calculus, the relevance of the critical value appears in the following results.

Theorem X - *If $k > c(L)$, for all $a, b \in M$, there exists a solution $x(t)$ of (E-L) such that $x(0) = a, x(T) = b$ for some $T \geq 0$, and*

$$S_{L+k}(x/[0, T]) = \min S_{L+k}(y)$$

where the minimum is taken over all the absolutely continuous $y : [0, T_1] \to M, T_1 \geq 0, y(0) = a, y(T_1) = b$. Moreover, the solution $x(t)$ is contained in the energy level $E^{-1}(k)$.

Using that on $E^{-1}(k)$ we have $L + k = (\partial L / \partial v) v$, it follows that :

Corollary -

a) If $k > c(L)$, for all $a, b \in M$, there exists a solution $x(t)$ of (E-L) such that $x(0) = a, x(T) = b$ for some $T > 0$, $E(x(t), \dot{x}(t)) = k$ for all $t \in \mathbb{R}$, and :

$$(*) \qquad \int_0^T \frac{\partial L}{\partial v}(x, \dot{x})\dot{x}dt = \min \int_0^{T_1} \frac{\partial L}{\partial v}(y, \dot{y})\dot{y}dt$$

where the minimum is taken over all the absolutely continuous $y : [0, T_1] \to M, T_1 \geq 0$, with $y(0) = a, y(T_1) = b$ and $E(y(t), \dot{y}(t)) = k$ for a.e. $t \in [0, T_1]$.

b) Conversely, if given $k > c(L)$ and $a, b \in M$, there exists an absolutely continuous $x : [0, T] \to M$ with $x(0) = a, x(T) = b, E(x(t), \dot{x}(t)) = k$ for a.e. $t \in [0, T]$ and satisfying the minimization property $(*)$, then $x(t)$ is a solution of (E-L).

An interesting characterization of the critical value, in terms of an analogous to Tonelli's Theorem ([Ma]) in a prescribed energy level is given by the following result.

Theorem XI - Suppose that $k \in \mathbb{R}$ has the following property : for all $a, b \in \pi(E^{-1}(k))$ there exists an absolutely continuous $x : [0, T] \to M, T \geq 0$, such that :

i) $E(x(t), \dot{x}(t)) = k$ for a.e. $t \in [0, T]$.

ii) $x(0) = a, x(T) = b$.

iii) $\int_0^T \left(\frac{\partial L}{\partial v}\right)(x, \dot{x})\dot{x}dt = \min \int_0^{T_1} \left(\frac{\partial L}{\partial v}\right)(y, \dot{y})\dot{y}dt$ where the minimum is taken over all the absolutely continuous $y = [0, T_1] \to M, y(0) = a, y(T_1) = b, E(y(t), \dot{y}(t)) = k$ for a.e. $t \in [0, T_1]$.

Then $k > c(L)$ and $x(t)$ is a solution of (E-L).

Now let us prove the characterization of minimizing measures given in Theorem IV. Assume $c(L) = 0$. Define $A_n : TM \to \mathbb{R}, F_i : TM \to \mathbb{R}, i = 1, 2$, by

$$A_n(\theta) = S_L(x/[0, n])$$
$$F_1(\theta) = \Phi_0(\pi(\theta), \pi(f_1(\theta)))$$
$$F_2(\theta) = \Phi_0(\pi(f_1(\theta)), \pi(\theta))$$

where $x : \mathbb{R} \to M$ is the solution of (E-L) with $(x(0), \dot{x}(0)) = \theta$.

Then, by Birkhoff's Theorem, for all $\mu \in \mathcal{M}(L)$ we have :

$$\int L d\mu = \int A_n d\mu \quad \forall n$$

and

$$\int F_1 d\mu \geq 0 ,$$

this last property because

$$\int F_1 d\mu = \int \left(\lim_{n \to +\infty} \frac{1}{n} \sum_{j=0}^{n-1} F_1(f_j(\theta)) \right) d\mu$$

$$= \int \lim_{n \to +\infty} \frac{1}{n} \sum_{j=0}^{n-1} \Phi_0(f_j(\theta), f_{j+1}(\theta)) d\mu$$

$$\geq \int \lim_{n \to +\infty} \frac{1}{n} \Phi_0(\theta, f_n(\theta)) = 0 .$$

Moreover, from the definition of Φ_0 :

(1) $$A_1 \geq F_1 .$$

Now suppose that μ is minimizing. Then :

$$c(L) = 0 = \int L d\mu = \int A_1 d\mu \geq \int F_1 d\mu \geq 0 .$$

Hence,

$$\int A_1 d\mu = \int F_1 d\mu ,$$

that, by (1), implies $A_1 = F_1 \mu$ a.e.. By the continuity of both functions, we get :

(2) $$A_1(\theta) = F_1(\theta), \quad \forall \theta \in \text{supp } (\mu) .$$

Then, if we prove :

(3) $$A_1(\theta) = -F_2(\theta), \quad \forall \theta \in \text{supp } (\mu) ,$$

the proof of supp $(\mu) \subset \widetilde{\Sigma}(L)$ will be complete. To prove (3) we use that :

$$\int L d\mu = 0$$

implies that for a.e. θ, there exists a sequence $n_j \to +\infty$ such that :

(4) $$A_{n_j}(\theta) \to 0$$

(5) $$d(\theta, f_{n_j}(\theta)) \to 0 .$$

From (5) follows that

$$F_2(\theta) \leq \lim_{j \to +\infty} A_{n_j-1}(f_{n_j-1}(\theta)) \ .$$

Since $A_{n_j}(\theta) = A_1(\theta) + A_{n_j-1}(f_{n_j-1}(\theta))$ and $A_{n_j}(\theta) \to 0$ we obtain $F_2(\theta) \leq -A_1(\theta)$. Using (2), we get $F_2(\theta) \leq -F_1(\theta)$. Using that $F_2 + F_1 \geq 0$ (because $d_0(\cdot, \cdot)$ is a pseudometric), we obtain (3).

Now suppose supp $(\mu) \subset \widehat{\Sigma}(L)$. On $\widehat{\Sigma}(L)$ we have $A_1 = F_1 = -F_2$ by definition of $\widehat{\Sigma}(L)$. Then

$$\int L d\mu = \int A_1 d\mu = \int F_1 d\mu \geq 0 \ .$$

Moreover the same argument used to prove that the μ-average of F_1 is positive can be used to prove that the μ-average of F_2 is positive. Then :

$$\int L d\mu = \int A_1 d\mu = - \int F_2 d\mu \leq 0 \ .$$

Finally, let us recall a weaker concept of global minimization (taken from Bangaert [B]) that, as the next three results will show, has many interesting connexions with the forms of global minimization introduced above.

Definition - We say that a solution $x(t)$ of (E-L) is a minimizer (resp. forward minimizer) if

$$S_L(x[t_0, t_1]) \leq S_L(y)$$

for every $t_0 \leq t_1$ (resp. $0 < t_0 \leq t_1$) and every absolutely continuous $y : [t_0, t_1] \to M$, with $y(t_i) = x(t_i), i = 0, 1$.

Denote $\wedge(L)$ (resp. $\wedge^+(L)$) the set of $(p, v) \in TM$ such that the solution $x(t)$ of (E-L) with initial condition $(x(0), \dot{x}(0)) = (p, v)$ is a minimizer (resp. a forward minimizer).

Theorem XIII - *The ω-limit set of an orbit in $\wedge^+(L)$ is contained in $\widehat{\Sigma}(L)$.*

Theorem XIV - $f_t / \wedge(L)$ *is chain transitive.*

Theorem XV -

a) *There exists $C > 0$ such that, setting $c = c(L)$,*

$$|S_{L+c}(x|[t_0, t_1])| \leq C$$

for every forward minimizer $x(t)$ and all $0 \leq t_0 \leq t_1$.

b) If $x(t)$ is a forward minimizer and $p \in M$ is such that $p = \lim\limits_{n \to +\infty} x(t_n)$, for some sequence $t_n \to +\infty$, then the limit

$$\lim_{n \to +\infty} S_{L+c}(x|[0, t_n]) ,$$

exists and depends only on p.

Now let us recall a device, exploited at length and in a protagonic rôle in [B], [Ma1], [Ma2], to enlarge the scope of the methods presented above. It consists in observing that the Lagrangians L and $L - \theta$, where θ is a closed 1-form on M, generate the **same** flow. Then the set of minimizing measures of $L - \theta$, to be denoted $\mathcal{M}^\theta(L)$, is contained in $\mathcal{M}(L)$, and the subsets of TM given by $\Sigma(L - \theta), \widehat{\Sigma}(L - \theta), \wedge^+(L - \theta), \wedge(L - \theta)$ are invariant sets of f_t. All these sets, as well as the critical value $c(L - \theta)$ of $L - \theta$, **depend only on the cohomology class** $[\theta] \in H^1(M, \mathbb{R})$ of θ. It is not difficult to check the convexity of the function $\beta^* : H^1(M, \mathbb{R}) \to \mathbb{R}$ defined by :

$$\beta^*([0]) = c(L - \theta) .$$

(This is the dual of Mather's Action Function $\beta : H_1(M, \mathbb{R}) \to \mathbb{R}$). Define the **strict critical value** $c_0(L)$ by :

$$c_0(L) = \min_\theta c(L - \theta) .$$

Using the concept of homology (or asymptotic cycle) of a $\mu \in \mathcal{M}(L)$, it can be proved that :

$$-c_0(L) = \min\{\int L d\mu | \mu \in \mathcal{M}(L), \rho(\mu) = 0\} .$$

This is part of the duality between β^* and Mather's Action Function.

Observe that in an energy level $E^{-1}(c)$ with $c > c_0(L)$, the results in Theorem X and its Corollary can be applied replacing L by $L - \theta$ where θ is a closed 1-form satisfying :

$$c > c(L - \theta) > c_0(L) .$$

Moreover observe that :

$$c_0(L) \geq \max_x \in (x, 0) ,$$

because the energy functions of L and $L - \theta$ coincide. The equality holds when L is a mechanical Lagrangian, i.e. of the form

$$L(x, v) = \frac{1}{2} < v, v >_x - V(x)$$

where $< \cdot, \cdot >_x$ is a Riemannian structure on M. In fact, in this case $c(L) = \max\limits_x E(x, 0)$ and the minimizing measures are linear combinations of the Dirac probabilities concentrated at the maximums of V.

But in general the inequality doesn't hold. An example will be given after the following Corollary of Theorem X.

Corollary - *If $k > c_0(L)$, for every free homotopy class of M, there exists a periodic orbit in $E^{-1}(k)$ such that its projection on M belongs to that free homotopy class.*

Now let us exhibit a Lagrangian on the two dimensional torus $T^2 = \mathbb{R}^2/\mathbb{Z}^2$ having energy levels $E^{-1}(k), k \in (a, b)$ with $b > \max\limits_{x} E(x, 0)$ such that the Corollary above doesn't hold in $E^{-1}(k)$; hence $c_0(L) > b > \max\limits_{x} E(x, 0)$. The example will be a Lagrangian of the form :

$$L_\lambda(x, v) = \frac{1}{2} < v, v >_x + \lambda\psi_1(x_2) - \psi_2(x_2), \ \lambda \in \mathbb{R}$$

where $< \cdot, \cdot >_x$ is the Euclidean inner product and ψ_1, ψ_2 are bump functions as in the picture

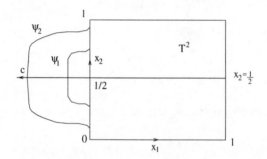

Then the energy is

$$E(x, v) = \frac{1}{2}||v||^2 + \psi_2(x_2)$$

independently of λ. When $\lambda > 0$ the effect of the term $\lambda\psi_1$ is that of magnetic field normal to the plane supported in the band $[0, 1] \times \text{supp}(\psi_1) \cong (\mathbb{R}/\mathbb{Z}) \times \text{supp}(\psi_1)$. Denote $c = \max\psi_2$ and $[k_1, k_2] = \psi_2^{-1}(\{c\})$. Then $0 < k_1 < \frac{1}{2} < k_2 < 1$.

Denote $S_\lambda \subset T^2$ the set of points that can be reached from a point (x_1, x_2), with $x_2 = 1$, through a solution of (E-L) with $L = L_\lambda$, contained in the level $E = c$ (i.e. the initial velocity v satisfies $\frac{1}{2}||v||^2 = c$). For $\lambda = 0, S_0 = T^2 - [k_1, k_2] \times (\mathbb{R}/\mathbb{Z})$. When $\lambda > 0, S_\lambda$ diminishes, say $S_\lambda \subset c[k_1(\lambda), k_2(\lambda)] \times (\mathbb{R}/\mathbb{Z})$ with $0 < k_1(\lambda) < k_1 < k_2 < k_2(\lambda) < 1$. Then the level $E^{-1}(c)$, for $\lambda > 0$, doesn't contain an orbit whose projection is in the homotopy class of $(0, x_2), x_2 \in \mathbb{R}$, because such orbits would intersect $x_2 = 1$.

Since the minimizing measures of $L - \theta$ share the same basic properties, it is convenient, following Mather [Ma1], to extend the term minimizing measures to all the measures in

$$\mathcal{M}^*(L) := \bigcup_\theta \mathcal{M}^\theta(L) \ .$$

Similarly, it is also convenient to extend the term prestatic, static and minimizing orbits to all the orbits having the corresponding property for some $L - \theta$.

The sets

$$\Sigma^*(L) = \bigcup_\theta \Sigma(L - \theta)$$

$$\widehat{\Sigma}^*(L) = \bigcup_\theta \widehat{\Sigma}(L - \theta)$$

$$\wedge(L) = \bigcup_\theta \wedge(L - \theta)$$

are closed, and all dep, $\widehat{\Sigma}(L - \theta)$, and $\wedge(L - \theta)$ are upper semicontinuous functions of $[\theta]$. From this property follows that :

Corollary. $\Sigma^*(L)$ *is chain transitive.*

REFERENCES

[B] V. Bangaert - *Minimal geodesics Erg. Th. and Dyn. Sys..* **10** (1990), 263-287.

[C] M.J. Carneiro - *On the minimizing measures of the action of autonomous Lagrangians.* Preprint IGEX, UFMG (1994).

[H] G.A. Hedlund - *Geodesics on a two dimensional Riemannian manifold with periodic coefficients.* Ann. of Math. **33** (1932), 719-739.

[M] R. Mañé - *Generic properties and problems of minimizing measures.* Preprint, IMPA (1993).

[Ma1] J. Mather - *Action Minimizing Measures for Positive Lagrangian systems.* Math. Z. **207** (1993), 169-207.

[Ma2] J. Mather - *Variational Construction of Connecting orbits.* Ann. Inst. Fourier **43** (1993), 1349-1386.

[Mo] M. Morse - *A fundamental class of geodesics on any closed surface of genus greater than one.* Trans. Amer. Math. Soc. **26** (1924), 25-60.

GABRIEL P PATERNAIN[*] AND MIGUEL PATERNAIN[†]

Anosov geodesic flows and twisted symplectic structures

To Ricardo Mañé, in Memoriam

1 Introduction

Let M^n be a closed n-dimensional manifold endowed with a C^∞ Riemannian metric $< , >$, and let $\pi : TM \to M$ denote the canonical projection. Let ω_0 denote the symplectic form on TM obtained by pulling back the canonical symplectic form of T^*M via the Riemannian metric. Let $H : TM \to \mathbf{R}$ be defined by

$$H(p,v) = \frac{1}{2} < v,v >;$$

the Hamiltonian flow of H with respect to ω_0 gives rise to the geodesic flow of M, $\phi_t : TM \to TM$. Let Ω be a closed 2-form of M and consider the new symplectic form ω_λ defined as:

$$\omega_\lambda \overset{\text{def}}{=} \omega_0 + \lambda \pi^* \Omega \quad \lambda \in \mathbf{R}.$$

Such a form is called a *twisted symplectic structure* [1] and the Hamiltonian flow of H with respect to ω_λ gives rise to a flow $\phi_t^\lambda : TM \to TM$. This flow models the motion of a particle of unit mass and charge λ under the effect of a magnetic field, whose Lorentz force $Y : TM \to TM$ is the bundle map determined uniquely by:

$$\Omega_p(u,v) = < Y_p(u), v >,$$

for all u and v in TM. Observe that ϕ_t^λ preserves all the energy levels $H = const$, in particular $SM \overset{\text{def}}{=} H^{-1}(1/2)$.

Suppose now that $\phi_t = \phi_t^0 : SM \to SM$ is an Anosov flow, then by the Structural Stability Theorem, $\phi_t^\lambda : SM \to SM$ is also an Anosov flow for $\lambda \in (-\varepsilon, \varepsilon)$. Let $h(\lambda)$ denote the topological entropy of $\phi_t^\lambda : SM \to SM$; by the results in [2, 7], the map $\lambda \to h(\lambda)$ is C^∞. One of the aims of the present paper is to use Pollicott's formulas in [12] to show:

---

[*]partially supported by grants from CSIC and CONICYT # 301
[†]partially supported by grants from CSIC and CONICYT # 301

Theorem 1.1 $h'(0) = 0$ and if $\Omega \neq 0$,

$$h''(0) \leq -h(0)\frac{(\int_{SM} \| Y(v) \|^2 \, dm)^2}{\int_{SM}\{\| (\nabla_v Y)(v) \|^2 - < R(v, Y(v))v, Y(v) >\} \, dm} < 0,$$

where ∇ is the Riemannian connection, R is the curvature tensor of M and m is the measure of maximal entropy of ϕ_t.

Corollary 1.2 If $\Omega \neq 0$, the function $\lambda \to h(\lambda)$ presents a strict local maximum at $\lambda = 0$.

Corollary 1.2 shows that if we start with an Anosov geodesic flow and we consider a small perturbation of it by twisting the symplectic structure of TM, the topological entropy strictly decreases. It should be noted that if we drop the hypothesis of ϕ_t being Anosov, nothing can be said in general about the topological entropy: a magnetic field could split an homoclinic connection and increase the topological entropy or it could keep it constant as in the case of the round two sphere and the magnetic field given by the area form.

Our next subject of study is motivated by the following related question: is it possible for ϕ_t^λ to be Anosov for all $\lambda \in \mathbf{R}$? Let us consider an example. Let M be a closed oriented surface of genus ≥ 2 endowed with a Riemannian metric of constant curvature -1 and let Ω be the associated area form. We study this example in Section 5 and we find that ϕ_t^λ is Anosov if and only if $\lambda^2 < 1$, and that the map $\lambda \to h(\lambda)$ is given by:

$$h(\lambda) = \begin{cases} \sqrt{1 - \lambda^2} & \text{if } \lambda^2 < 1 \\ 0 & \text{if } \lambda^2 \geq 1 \end{cases}$$

A complete description of the orbits of the flow ϕ_t^λ is given in [5]. Incidentally, this example shows that Dinaburg's result [3] relating the positivity of the topological entropy of the geodesic flow with the fundamental group, can fail when the symplectic structure is twisted.

In general we have:

Theorem 1.3 If for $\lambda \in \mathbf{R}$, $\phi_t^\lambda : SM \to SM$ is Anosov and $\Omega \neq 0$ then,

$$\lambda^2 < \frac{-(n-1)\int_{SM} Ric(v) \, d\mu}{\int_{SM} \| Y(v) \|^2 \, d\mu},$$

where $Ric : SM \to \mathbf{R}$ is the Ricci curvature function, and μ is the normalized Lebesgue measure of SM.

Let us consider the map $\varphi : TM \to TM$ given by $\varphi(p, v) = (p, \lambda v)$. It is easy to check that $\varphi^* \omega_\lambda = \lambda(\omega_0 + \pi^* \Omega)$ and $\varphi^* H = \lambda^2 H$, thus Theorem 1.3 implies:

Corollary 1.4 *If the Hamiltonian flow of H with respect to the symplectic structure $\omega_0 + \pi^*\Omega$ is Anosov on the energy level $H^{-1}(E/2)$, then E must verify:*

$$E > \frac{\int_{SM} \| Y(v) \|^2 \, d\mu}{-(n-1) \int_{SM} Ric(v) \, d\mu} > 0,$$

where $Ric : SM \to \mathbf{R}$ is the Ricci curvature function, and μ is the normalized Lebesgue measure of SM.

Therefore for a geodesic flow to be Anosov on a twisted tangent bundle, the energy has to be sufficiently high.

It is interesting to note that the example described before shows that all the inequalities are sharp.

2 Preliminaries

We describe in this section our setting. Let us begin by fixing on M a smooth Riemannian metric $< \, , \, >$, with Riemann curvature tensor R. Let $\pi : TM \to M$ denote the canonical projection and let $K : TTM \to TM$ denote the connection map. The latter is defined giving its value on each fibre setting

$$K_v(\xi) = \frac{D\,Z}{dt}\Big|_{t=0},$$

where $Z : (-\epsilon, \epsilon) \to TM$ verifies $Z(0) = v$, $Z'(0) = \xi$, and $\frac{D}{dt}$ denotes covariant derivative along $\pi \circ Z$.

It is well known that TTM splits as the direct sum of the vertical and the horizontal subbundles . The vertical fibre on v is given by

$$V(v) = Ker d_v \pi,$$

and the horizontal fibre on v is defined as

$$H(v) = Ker K_v.$$

Observe that $H(v)$ and $V(v)$ are orthogonal with respect to the Sasaki metric of TM given by

$$<< \xi, \eta >> \stackrel{\text{def}}{=} < d_v\pi(\xi), d_v\pi(\eta) > + < K_v(\xi), K_v(\eta) > .$$

Thus $T_v TM$ can be identified with $T_{\pi(v)}M \oplus T_{\pi(v)}M$, and hence we write in the sequel

$$\xi = (\xi_1, \xi_2),$$

where $\xi_1 = d_v\pi(\xi)$ and $\xi_2 = K_v(\xi)$ for every ξ in T_vTM. The symplectic structure ω_0 described in the Introduction can be written as

$$\omega_0(\xi,\eta) = < d\pi(\xi), K(\eta) > - < K(\xi), d\pi(\eta) > .$$

Fix a closed two-form Ω in M. For any $\lambda \in \mathbf{R}$ define ω_λ by

$$\omega_\lambda = \omega_0 + \lambda\pi^*\Omega.$$

For the rest of this section consider the Hamiltonian $H : TM \to \mathbf{R}$ given by $H(v) = \frac{<v,v>}{2}$. Let $Y : TM \to TM$ be the bundle map such that

$$\Omega_p(u,v) = < Y_p(u), v >$$

for all u and v in T_pM and all p in M. Denote by

$$X^\lambda : TM \to TTM$$

the symplectic gradient of H with respect to ω_λ. Since the identity

$$d_vH(\xi) = \omega_0(X^\lambda(v),\xi) + \lambda < Y(d_v\pi(X^\lambda(v))), d_v\pi(\xi) >$$

holds for every ξ in T_vTM, the identity

$$< \xi_2, v > = < X_1^\lambda(v), \xi_2 > - < X_2^\lambda(v), \xi_1 > + \lambda < Y(X_1^\lambda(v)), \xi_1 >$$

is valid for all ξ_1 and $\xi_2 \in T_{\pi(v)}M$ (obviously we made use of the identification $\xi = (\xi_1, \xi_2)$ as it was explained before and $(X_1^\lambda, X_2^\lambda)$ are the horizontal and vertical components of X^λ). Therefore

$$X^\lambda(v) = (v, \lambda Y(v))$$

for every v in TM. It is easily seen from this equation that a curve of the form $t \to (\gamma(t), \dot\gamma(t)) \in TM$ is an integral curve of X^λ if and only if

$$\frac{D}{dt}\dot\gamma = \lambda Y(\dot\gamma), \tag{1}$$

which is nothing but Newton's law of motion.

Let us derive the Jacobi equation. Denote by $\phi_t^\lambda : TM \to TM$ the flow generated by the symplectic gradient X^λ. Take a curve $Z : (-\epsilon, \epsilon) \to TM$ with $Z(0) = v$, $Z'(0) = \xi$ and the variation $f(s,t) = \pi(\phi_t^\lambda(Z(s)))$. Set $J_\xi(t) \stackrel{\text{def}}{=} \frac{\partial f}{\partial t}(0,t)$, $\gamma_s \stackrel{\text{def}}{=} f(s,t)$ and $\gamma_0 \stackrel{\text{def}}{=} \gamma$.

¿From the well known identity:

$$\frac{D}{ds}\frac{D}{dt}\frac{\partial f}{\partial t} = \frac{D}{dt}\frac{D}{dt}\frac{\partial f}{\partial s} + R(\frac{\partial f}{\partial t}, \frac{\partial f}{\partial s})\frac{\partial f}{\partial t}$$

135

and

$$\frac{D}{dt}\dot{\gamma}_s = \lambda Y(\dot{\gamma}_s) \tag{2}$$

(which obviously follows from (1)) we obtain:

$$\ddot{J}_\xi + R(\dot{\gamma}, J_\xi)\dot{\gamma} = \frac{D}{ds}(\lambda Y(\dot{\gamma}_s)).$$

The right hand side of the last equation can be simplified observing that the map $(p, v) \to Y(p, v)$ is linear in v. Thus

$$\frac{D}{ds}Y(\dot{\gamma}_s) = \nabla_{J_\xi}Y(\dot{\gamma}_s) + Y(\dot{J}_\xi)$$

and we deduce the Jacobi equation:

$$\ddot{J}_\xi + R(\dot{\gamma}, J_\xi)\dot{\gamma} - \lambda Y(\dot{J}_\xi) - \lambda\nabla_{J_\xi}Y(\dot{\gamma}) = 0. \tag{3}$$

We now derive the Riccati equation. Let E be a ω_λ- Lagrangian subspace of T_vTM, i.e. a maximal suspace on which ω_λ identically vanishes. Assume that

$$d\phi_t^\lambda(E) \cap V(\phi_t^\lambda(v)) = 0,$$

for every t in an interval I. Let

$$S(t) : H(\phi_t^\lambda(v)) \to V(\phi_t^\lambda(v))$$

be the unique linear map whose graph is $d\phi_t^\lambda(E)$ for $t \in I$. We shall see that S verifies a Riccati equation. Observe that

$$d\phi_t^\lambda(\xi) = (J_\xi(t), \dot{J}_\xi(t)).$$

If $\xi \in E$, then $\dot{J}_\xi(t) = S(t)J_\xi(t)$ for $t \in I$ and hence

$$\ddot{J}_\xi = \dot{S}J_\xi + S\dot{J}_\xi.$$

Using the Jacobi equation (3) we obtain:

$$\lambda Y S J_\xi - R_{\dot{\gamma}}J_\xi + \lambda\nabla_{J_\xi}Y(\dot{\gamma}) = \dot{S}J_\xi + S^2 J_\xi,$$

where $R_v : T_{\pi(v)}M \to T_{\pi(v)}M$ is the curvature operator given by $R_v(w) = R(v, w)v$ for every v in $T_{\pi(v)}M$. Since $\xi \in E$ is arbitrary we obtain:

$$\dot{S} + S^2 - \lambda Y S + R_{\dot{\gamma}} - \lambda Q_{\dot{\gamma}} = 0, \tag{4}$$

where $Q_v : T_{\pi(v)}M \to T_{\pi(v)}M$ is the linear operator given by $Q_v(w) = \nabla_w Y(v)$ for every w in $T_{\pi(v)}M$. Equation (4) can be simplified observing that

$$S^* - S + \lambda Y = 0$$

where S^* is the adjoint operator of S. This last identity follows directly from the condition $\omega_\lambda(\xi, \eta) = 0$ for every ξ and η in the graph of S. Then, the desired Riccati equation is:

$$\dot{S} + S^*S + R_{\dot{\gamma}} - \lambda Q_{\dot{\gamma}} = 0. \tag{5}$$

136

3 Derivatives of topological entropy

We describe in this Section the results obtained by M. Pollicott in [12].

Consider a C^∞ family of Anosov flows $\lambda \to \phi_t^\lambda$, $\lambda \in (-\varepsilon, \varepsilon)$, on a compact manifold X. Recall from the Structural Stability Theorem, that there exist functions $\alpha^\lambda \in C^s(X)$, $\Theta^\lambda \in C^s(X, X)$ ($s > 0$ depends on the stable and unstable foliations) such that:

(i) $\alpha^0 \equiv 1$; $\Theta^0 \equiv I_X$,

(ii) Θ^λ carries ϕ_t^0 orbits to ϕ_t^λ orbits,

(iii) α^λ is a change in speed in ϕ_t^0 to make Θ^λ a conjugacy, and furthermore, the maps $\lambda \to \alpha^\lambda$, Θ^λ are C^∞.

Let us expand α^λ as follows:

$$\lambda \to \alpha^\lambda = 1 + \lambda(D_0\alpha) + (\lambda^2/2)(D_0^2\alpha) + \ldots$$

We denote by $h(\lambda)$ the topological entropy of the flow ϕ_t^λ. Pollicott's main result in [12] is :

Theorem 3.1 *The first derivative of $h(\lambda)$ at $\lambda = 0$ is*

$$h'(0) = h(0) \int_X D_0\alpha \, dm,$$

and the second derivative of $h(\lambda)$ at $\lambda = 0$ is

$$h''(0) = h(0)\{Var(D_0\alpha) + \int_X D_0^2\alpha \, dm +$$

$$2(\int_X D_0\alpha \, dm)^2 - 2\int_X (D_0\alpha)^2 \, dm\},$$

where m is the measure of maximal entropy, and Var is the variance for ϕ_t^0.

We remark that the formula for the first derivative of $h(\lambda)$ was first obtained in [6].

4 Proof of Theorem 1.1

Let $\Theta^\lambda : SM \to SM$ be the map from Section 3 that carry orbits of ϕ_t to orbits of ϕ_t^λ. Consider a vector $v \in SM$ so that $c_0(t) \overset{\text{def}}{=} \pi(\phi_t v)$ is a closed geodesic with period T. Now consider the variation of c_0 defined as

$$c_\lambda(t) \overset{\text{def}}{=} \pi(\phi_t^\lambda \Theta^\lambda v).$$

This variation is C^∞ since the dependence of Θ^λ on λ is C^∞. The curves c_λ are all closed with periods, let us say, T_λ.

For a smooth curve $c : [0, a] \to M$, let $l(c)$ denote the length of c and let $E(c)$ denote its energy defined as,

$$E(c) = \int_0^a < \dot{c}(t), \dot{c}(t) > dt.$$

Let ψ_t^λ denote the reparametrization of the flow ϕ_t by a change of speed α^λ. We know that Θ^λ is a conjugacy between ψ_t^λ and ϕ_t^λ, therefore the orbit $t \to \psi_t^\lambda v$ is closed with period T_λ. On the other hand since ψ_t^λ is a reparametrization of the flow ϕ_t by a change of speed α^λ, it follows easily that T_λ and T are related by:

$$T_\lambda = \int_0^T \frac{1}{\alpha^\lambda(\dot{c}_0(t))} dt.$$

If we set $\beta^\lambda \stackrel{\text{def}}{=} 1/\alpha^\lambda$, and we observe that the curves c_λ have unit speed we deduce:

$$l(c_\lambda) = T_\lambda = \int_0^T \beta^\lambda(\dot{c}_0(t)) \, dt. \tag{6}$$

For each $\lambda \in (-\varepsilon, \varepsilon)$, consider a new family of curves $\bar{c}_\lambda : [0, T] \to M$ defined by

$$\bar{c}_\lambda(t) \stackrel{\text{def}}{=} c_\lambda(tT_\lambda/T).$$

Then, $\lambda \to \bar{c}_\lambda$ is a C^∞ variation by closed curves of period T of the closed geodesic c_0 (they may not have unit speed though).

¿From the definitions,

$$E(\lambda) \stackrel{\text{def}}{=} E(\bar{c}_\lambda) = T_\lambda^2/T.$$

Differentiating this equality with respect to λ we obtain:

$$\frac{2T_\lambda}{T} \frac{dT_\lambda}{d\lambda} = \frac{dE}{d\lambda},$$

$$\frac{1}{T} \{ (\frac{dT_\lambda}{d\lambda})^2 + T_\lambda \frac{d^2 T_\lambda}{d\lambda^2} \} = \frac{1}{2} \frac{d^2 E}{d\lambda^2}.$$

Evaluating the last two equalities at $\lambda = 0$ and using equation (6) we obtain:

$$\int_0^T D_0 \beta(\dot{c}_0(t)) \, dt = \frac{dT_\lambda}{d\lambda} |_{\lambda=0} = \frac{1}{2} \frac{dE}{d\lambda} |_{\lambda=0},$$

$$\int_0^T D_0^2 \beta(\dot{c}_0(t)) \, dt = \frac{d^2 T_\lambda}{d\lambda^2} |_{\lambda=0} = \frac{1}{2} \frac{d^2 E}{d\lambda^2} |_{\lambda=0} - \frac{1}{T} (\frac{dT_\lambda}{d\lambda})^2 |_{\lambda=0}.$$

On the other hand, since c_0 is a closed geodesic, it is a critical point of the energy functional, i.e., $\frac{dE}{d\lambda} |_{\lambda=0} = 0$, and thus we have proved:

138

Lemma 4.1 *For every closed geodesic c_0 with period T we have:*

$$\int_0^T D_0\beta(\dot{c}_0(t))\,dt = 0$$

$$\int_0^T D_0^2\beta(\dot{c}_0(t))\,dt = \frac{1}{2}\frac{d^2 E}{d\lambda^2}\big|_{\lambda=0}.$$

Observe now that the first equality in Lemma 4.1 and the Livsic Theorem [9], imply that $v \to D_0\beta(v)$ is zero up to a coboundary. Therefore (cf. [12] for properties of the variance) we have:

$$\int_{SM} D_0\beta\,dm = 0,$$

$$Var(D_0\beta) \equiv 0.$$

Using that $\beta^\lambda = 1/\alpha^\lambda$ and Theorem 3.1 we deduce the following:

Lemma 4.2 $h'(0) = 0$ *and*

$$h''(0) = -h(0)\int_{SM} D_0^2\beta\,dm.$$

Thus, on account of Lemma 4.2, to prove Theorem 1.1 it suffices to show:

$$\int_{SM} D_0^2\beta\,dm \geq \frac{(\int_{SM} \|\,Y(v)\,\|^2\,dm)^2}{\int_{SM}\{\|\,(\nabla_v Y)(v)\,\|^2 - <R(v,Y(v))v,Y(v)>\}\,dm} > 0.$$

We shall make use of the following lemma, whose proof will be given only after completing the proof of Theorem 1.1.

Lemma 4.3 *The variational field W associated with the variation $\lambda \to \bar{c}_\lambda$ of the closed geodesic c_0 verifies:*

$$\ddot{W}(t) + R(\dot{c}_0(t), W(t))\dot{c}_0(t) = Y(\dot{c}_0(t)),$$

$$W(0) = W(T),$$

$$\dot{W}(0) = \dot{W}(T),$$

where $\dot{W}$ indicates the covariant derivative of W along c_0, R is the curvature tensor of M and Y is the Lorentz force.

Let us complete now the proof of Theorem 1.1. Let Λ denote the set of piecewise smooth vector fields $V : [0, T] \to TM$ along c_0, so that $V(0) = V(T)$. Then the *index form* $I : \Lambda \times \Lambda \to \mathbf{R}$ is defined by:

$$I(V, U) = \int_0^T \{< \dot{V}, \dot{U} > - < R(\dot{c}_0, V)\dot{c}_0, U >\} dt.$$

It is well known that

$$\frac{1}{2} \frac{d^2 E}{d\lambda^2} \mid_{\lambda=0} = I(W, W), \tag{7}$$

where W is the variational field of the variation $\lambda \to \bar{c}_\lambda$. On the other hand, since ϕ_t is Anosov, every closed geodesic has Morse index zero [8], and thus if $V \in \Lambda$,

$$I(V, V) \geq 0. \tag{8}$$

Take $\mu \in \mathbf{R}$, clearly the field $W(t) + \mu Y(\dot{c}_0(t))$ belongs to Λ (cf. Lemma 4.3), and thus by (8)

$$I(W + \mu Y, W + \mu Y) \geq 0,$$

then

$$I(W, W) \geq -2\mu I(W, Y) - \mu^2 I(Y, Y) \tag{9}$$

Next, we use the definition of the index form and Lemma 4.3 to deduce:

$$I(W, Y) = \int_0^T \{- < \ddot{W}, Y > - < R(\dot{c}_0, W)\dot{c}_0, Y >\} dt =$$

$$- \int_0^T < Y(\dot{c}_0(t)), Y(\dot{c}_0(t)) > dt.$$

Therefore from (9):

$$I(W, W) \geq 2\mu \int_0^T \| Y(\dot{c}_0(t)) \|^2 dt - \mu^2 I(Y, Y). \tag{10}$$

But $\dot{Y}(t) = (\nabla_{\dot{c}_0(t)} Y)(\dot{c}_0(t))$ since c_0 is a geodesic, then

$$I(Y, Y) = \int_0^T \{\| (\nabla_{\dot{c}_0} Y)(\dot{c}_0) \|^2 - < R(\dot{c}_0, Y(\dot{c}_0))\dot{c}_0, Y(\dot{c}_0) >\} dt,$$

Combining (7) and (10) and using Lemma 4.1 we obtain:

$$\int_0^T D_0^2 \beta(\dot{c}_0(t)) dt \geq 2\mu \int_0^T \| Y(\dot{c}_0(t)) \|^2 dt -$$

$$\mu^2 \int_0^T \{\| (\nabla_{\dot{c}_0} Y)(\dot{c}_0) \|^2 - < R(\dot{c}_0, Y(\dot{c}_0))\dot{c}_0, Y(\dot{c}_0) >\} dt, \tag{11}$$

for all $\mu \in \mathbf{R}$ and every closed geodesic c_0.

Recall that for a volume preserving Anosov flow, the invariant probability measures supported on closed orbits are dense (in the weak* topology) in the set of all invariant probability measures, thus inequality (11) implies:

$$\int_{SM} D_0^2 \beta \, dm \geq 2\mu \int_{SM} \| Y(v) \|^2 \, dm-$$

$$\mu^2 \int_{SM} \{\| (\nabla_v Y)(v) \|^2 - < R(v, Y(v))v, Y(v) > \} \, dm \stackrel{\text{def}}{=} 2\mu A - \mu^2 B \qquad (12)$$

for all $\mu \in \mathbf{R}$. Observe that if Y is not identically zero (i.e., $A \neq 0$) then inequality (12) implies that $B > 0$, since it has to hold for all μ. Moreover, the maximum of the expression $2\mu A - \mu^2 B$ is A^2/B, thus concluding the proof of Theorem 1.1.

$\diamond$

Finally, we prove Lemma 4.3.

Proof: Recall the definition of $\bar{c}_\lambda$:

$$\bar{c}_\lambda(t) = c_\lambda(tT_\lambda/T),$$

where

$$c_\lambda(t) = \pi(\phi_t^\lambda \Theta^\lambda v).$$

Therefore, $\bar{c}_\lambda(0) = \bar{c}_\lambda(T)$ and $\dot{\bar{c}}_\lambda(0) = \dot{\bar{c}}_\lambda(T)$. If $W(t) \stackrel{\text{def}}{=} \frac{\partial \bar{c}_\lambda}{\partial \lambda}|_{\lambda=0}$, then we obtain:

$$W(0) = W(T),$$

$$\dot{W}(0) = \dot{W}(T),$$

where for the last equality we have used the symmetry of the Riemannian connection.

Recall form basic Riemannian Geometry, that if $(t, \lambda) \to f(t, \lambda)$ is a parametrized surface, then

$$\frac{D}{dt}\frac{D}{dt}\frac{\partial f}{\partial \lambda} + R(\frac{\partial f}{\partial t}, \frac{\partial f}{\partial \lambda})\frac{\partial f}{\partial t} = \frac{D}{d\lambda}(\frac{D}{dt}\frac{\partial f}{\partial t}).$$

If we use this equality for $f(t, \lambda) = \bar{c}_\lambda$ and we evaluate at $\lambda = 0$, it follows:

$$\ddot{W} + R(\dot{c}_0, W)\dot{c}_0 = \frac{D}{d\lambda}(\frac{D}{dt}\dot{\bar{c}}_\lambda)|_{\lambda=0} . \qquad (13)$$

On account of Newton's law (cf. equation (1)):

$$\frac{D}{dt}\dot{c}_\lambda(t) = \lambda Y(\dot{c}_\lambda(t)),$$

hence

$$\frac{D}{dt}\dot{\bar{c}}_\lambda(t) = \frac{\lambda T_\lambda}{T}Y(\dot{\bar{c}}_\lambda(t)),$$

and therefore

$$\frac{D}{d\lambda}\left(\frac{D}{dt}\dot{c}_\lambda\right)|_{\lambda=0} = Y(\dot{c}_0(t)).$$

The last equality together with equation (13) give the desire ODE for W.

$\diamond$

Remark 4.4 Let us observe that in fact our proof, yields a more general result than the one stated in Theorem 1.1. Note that a perturbation of an Anosov geodesic flow by a magnetic field amounts to adding a "small vertical part"; i.e.

$$X^\lambda(v) = (v, 0) + \lambda(0, Y(v)).$$

However, the Y induced by a magnetic field is special, namely, it is linear on the fibres and antisymmetric. Our proof works in fact for an **arbitrary** fibre preserving map Y, i.e., an **arbitrary** first order vertical perturbation.

5   An example

Let M be a closed oriented surface endowed with a Riemannian metric with sectional curvature $K : M \to \mathbf{R}$. Let Ω be the area form; then $Y : TM \to TM$ is an almost complex structure such that $\nabla Y = 0$. Note that in this case Theorem 1.1 tells us that

$$h''(0) \leq \frac{h(0)}{\int_{SM} K(\pi v)\, dm} < 0.$$

Observe that since $\nabla Y = 0$ the Jacobi equation (3) reduces to

$$\ddot{J} + R(\dot{c}, J)\dot{c} = \lambda Y(\dot{J}).$$

Since $\{\dot{c}, Y(\dot{c})\}$ is an orthonormal basis of $T_{\dot{c}}M$, we can write:

$$J = x\dot{c} + yY(\dot{c}).$$

Then

$$\dot{J} = (\dot{x} - \lambda y)\dot{c} + (\dot{y} + \lambda x)Y(\dot{c}),$$

and

$$\ddot{J} = (\ddot{x} - 2\lambda\dot{y} - \lambda^2 x)\dot{c} + (\ddot{y} + 2\lambda\dot{x} - \lambda^2 y)Y(\dot{c}).$$

On the other hand

$$\lambda Y(\dot{J}) = -\lambda(\dot{y} + \lambda x)\dot{c} + \lambda(\dot{x} - \lambda y)Y(\dot{c}).$$

Thus the Jacobi equation yields the following ODE's:

$$\ddot{x} - \lambda\dot{y} = 0, \tag{14}$$

$$\ddot{y} + \lambda\dot{x} + Ky = 0. \tag{15}$$

If we consider a variation contained in SM, i.e., if the initial conditions verify:

$$(J(0), \dot{J}(0)) \in T_{\dot{c}(0)}SM,$$

then

$$< \dot{J}, \dot{c} >= 0,$$

that is, $\dot{x} - \lambda y = 0$, which implies equation (14). If we substitute $\dot{x} = \lambda y$ in equation (15) we obtain:

$$\ddot{y} + (\lambda^2 + K)y = 0. \tag{16}$$

Suppose now $K \equiv -1$, then equation (16) is easily solved and it follows right away that ϕ_t^λ is Anosov if and only if $\lambda^2 - 1 < 0$ with constant Lyapunov exponent $\sqrt{1 - \lambda^2} = h(\lambda)$. In addition, if $\lambda^2 - 1 \geq 0$, the Lyapunov exponent is identically zero and thus $h(\lambda) = 0$, for $\lambda^2 \geq 1$.

6   Proof of Theorem 1.3

Suppose the flow $\phi_t^\lambda : SM \to SM$ is Anosov and let E denote the stable or unstable subbundle. It is well known that E is continuous, ϕ_t^λ-invariant and Lagrangian and from the results in [10, 11] we know that:

$$E(v) \cap V(v) = \{0\},$$

for every $v \in SM$. Therefore for each $v \in SM$, there exists a map $S_v : T_{\pi(v)}M \to T_{\pi(v)}M$ so that its graph is $E(v)$; moreover the correspondence $v \to S_v$ is continuous.

Next observe that since E is Lagrangian and ϕ_t^λ-invariant, $X^\lambda \in E(v)$ for every v in SM and thus (recall the formula for X^λ from Section 2):

$$S_v v = \lambda Y(v). \tag{17}$$

Since $< Sv, Sv >\leq tr(S^*S)$ ("tr" denotes the trace of an operator) we deduce from equation (17):

$$\lambda^2 < Y(v), Y(v) >\leq tr(S^*S).$$

If we take traces in the Riccati equation (5) we obtain:

$$tr\dot{S} + tr(R_{\dot{\gamma}} - \lambda Q_{\dot{\gamma}}) \leq -\lambda^2 ||Y(\dot{\gamma})||^2,$$

for an integral curve $(\gamma, \dot{\gamma})$ of X^λ in SM. If $(\gamma(0), \dot{\gamma}(0))$ is a Birkhoff generic point with respect to the normalized Lebesgue measure μ of SM (which is known to be ϕ_t^λ-invariant), we deduce

$$\lim_{T \to \infty} \left(\frac{1}{T}\int_0^T tr\dot{S}\, dt + \frac{1}{T}\int_0^T tr(R_{\dot{\gamma}} - \lambda Q_{\dot{\gamma}})\, dt\right) \leq -\lim_{T \to \infty}\frac{1}{T}\int_0^T \lambda^2 ||Y(\dot{\gamma})||^2\, dt$$

which yields

$$\int_{SM} tr(R_v - \lambda Q_v)\, d\mu \leq -\lambda^2 \int_{SM} ||Y(v)||^2\, d\mu.$$

Observe that since $Q(v) = -Q(-v)$ one has

$$\int_{SM} tr Q_v\, d\mu = 0,$$

which implies

$$\int_{SM} tr R_v\, d\mu = (n-1) \int_{SM} Ric(v)\, d\mu \leq -\lambda^2 \int_{SM} ||Y(v)||^2\, d\mu.$$

The last inequality combined with the fact that Anosov flows form an open family, concludes the proof of the theorem.

$\diamond$

Remark 6.1 Note that the arguments used in the proof of Theorem 1.3 are very similar to those used by Green in [4]. In fact these arguments show a bit more. Recall that the flow $\phi_t^\lambda : SM \to SM$ does not have conjugate points if for all $v \in SM$,

$$d\phi_t^\lambda(V(v)) \cap V(\phi_t^\lambda(v)) = 0,$$

for all $t \neq 0$. If there are no conjugate points, the arguments in [4] applied to our Riccati equation (5) also imply the existence of bounded and measurable solutions S_v of the Riccati equation (5) and thus the proof of Theorem 1.3 can be used to obtain the following:

"If the flow $\phi_t^\lambda : SM \to SM$ does not have conjugate points, then

$$(n-1) \int_{SM} Ric(v)\, d\mu \leq -\lambda^2 \int_{SM} ||Y(v)||^2\, d\mu".$$

As an amusing corollary we deduce that any non-zero magnetic field on a flat manifold creates conjugate points.

References

[1] V.I. Arnold, A.B. Givental, *Symplectic Geometry*, Dynamical Systems IV, Encyclopaedia of Mathematical Sciences, Springer Verlag: Berlin 1990

[2] G. Contreras, *Regularity of topological and metric entropy of hyperbolic flows*, Math. Z. **210** (1992) 97-111.

[3] E. I. Dinaburg, *On the relations among various entropy characteristics of dynamical systems*, Math USSR Izv. **5** (1971) 337-37.

[4] L. W. Green, *A theorem of E. Hopf*, Michigan Math. J. **5** (1958) 31-34.

[5] V. Guillemin, A. Uribe, *Circular symmetry and the trace formula*, Invent. Math. **96** (1989) 385-423.

[6] A. Katok, G. Knieper, H. Weiss, *Formulas for the Derivative and Critical Points of Topological Entropy for Anosov and Geodesic Flows*, Comm. Math. Phys. **138** (1991) 19-31.

[7] A. Katok, G. Knieper, M. Pollicott, H. Weiss, *Differentiability and Analyticity of Topological Entropy for Anosov and Geodesic Flows*, Invent. Math. **98** (1989) 581-597.

[8] W. Klingenberg, *Riemannian manifolds with geodesic flows of Anosov type*, Ann. of Math., **99** (1974) 1-13.

[9] A. Livsic, *Cohomology of dynamical systems*, Math. USSR-Izv. **6** (1972) 1278-1301.

[10] G. P. Paternain, M. Paternain, *On Anosov Energy Levels of Convex Hamiltonian Systems*, Math. Z. **217** (1994) 367-376.

[11] G. P. Paternain, *On Anosov energy levels of Hamiltonians on twisted cotangent bundles*, Bulletin of the Brazilian Math. Society Vol 25 **2** (1994) 207-211.

[12] M. Pollicott, *Derivatives of topological entropy for Anosov and geodesic flows*, J. Diff. Geom. **39** (1994) 457-489.

Gabriel P. Paternain
Centro de Matemática
Facultad de Ciencias
Eduardo Acevedo 1139
Montevideo CP 11200
Uruguay
E-mail: gabriel@cmat.edu.uy

Miguel Paternain
IMERL-Facultad de Ingeniería
Julio Herrera y Reissig 565, C.C. 30
Montevideo, Uruguay
E-mail: miguel@cmat.edu.uy

MARK POLLICOTT
Entropy and geodesic arcs on surfaces

Ricardo Mañé, in memorium

0. INTRODUCTION

It is a well established principle that entropy quantifies the growth of many important quantities in Riemannian geometry (cf.[2] [10], [11],[13], [15]). In this paper we shall consider another interesting instance of this phenomenon.

Let V denote a compact C^∞ Riemannian surface then for any two points $x, y \in V$ we use $N_{xy}(t)$ to denote the number of distinct geodesic arcs from x to y of length at most $t > 0$. (This number maybe zero, finite, or infinity). We want to consider the behaviour of this function as t tends to infinity, for typical values of x and y. For the special case of metrics of *negative curvature* this problem was studied by Margulis [11].

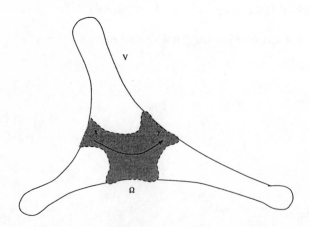

FIGURE 0. Counting geodesic arcs between $x, y \in \Omega$.

The author is supported by a Royal Society 1983 University Research Fellowship. I would like to thank Keith Burns and Gabriel Paternain for informing us of their results prior to publication. I am particularly grateful to Luísa Magalhães for her help with the geometric aspects of this paper.

R. Mañé observed that for almost all points $x, y \in V$ (with respect to the volume)

$$\limsup_{t \to +\infty} \frac{1}{t} \log^+ N_{xy}(t) \leq h \tag{0.1}$$

where $h = h(\phi)$ is the topological entropy of the associated geodesic flow $\phi_t : SV \to SV$ on the universal cover [9, Appendix]. (This estimate makes essential use of the work of G.Paternain and M. Paternain [16]). Mañé also posed the question of whether the inequality in (0.1) is actually an equality [9, Problem 1]. However, Burns and G. Paternain answered this question in the negative by constructing an example of a compact C^∞ surface V_0 such that the inequality (0.1) is strict for $x, y \in U$, for an open set $U \subset V_0$.

Let $\pi : SV \to V$ denote the canonical projection of vectors onto their foot points (i.e. $\pi(x, v) = x$). Given an ergodic ϕ-invariant probability measure μ on SV we call $\Omega \subset V$ a μ-shaded region if there exists a set $\Lambda \subset SV$ with $\mu(\Lambda) = 1$ and $\Omega = \pi(\Lambda)$. We formulate our main result as follows.

Theorem. *Given any ergodic ϕ-invariant probability measure μ there exists a μ-shaded region $\Omega \subset SV$ such that*

$$h_\mu(\phi) \leq \limsup_{t \to +\infty} \frac{1}{t} \log^+ N_{xy}(t), \quad \forall x, y \in \Omega \tag{0.2}$$

By considering different choices of measure μ we get the three corollaries below. We begin by taking μ to be a measure of maximal entropy.

Corollary 1. *There exists $x_0 \in M$ such that*

$$\limsup_{t \to +\infty} \frac{1}{t} \log^+ N_{x_0 x_0}(t) \geq h \tag{0.3}$$

For the next corollary we take μ to be any fully supported measure on SV.

Corollary 2. *If $h_\mu(\phi)$ is the entropy of a fully supported ergodic measure μ on SV then there exists a dense set of points $D \subset V$ such that*

$$\limsup_{t \to +\infty} \frac{1}{t} \log^+ N_{x_0 x_0}(t) \geq h_\mu(\phi), \quad \forall x, y \in D$$

Finally, of particular interest in light of the results of Mañé and Burns-Paternain mentioned above is where the invariant measure is taken to be an (ergodic) Liouville measure ν (i.e. $d\nu = d(\text{Volume}) \times d(\text{Haar})$).

Corollary 3. *If ν is ergodic then for almost all $x, y \in V$ (relative to the volume) we have that*

$$h(\nu) \leq \limsup_{t \to +\infty} \frac{1}{t} \log^+ N_{xy}(t) \leq h \qquad (0.4)$$

Clearly, these Corollaries only become interesting when the associated entropies giving lower bounds are nonzero. Below we consider some examples.

Example 1 (Burns-Gerber-Donnay Examples). These are examples of C^∞ (and even real analytic) Riemannian metrics on the 2 sphere for which the associated geodesic flow $\phi_t : SV \to SV$ is Bernoulli with respect to the Liouville measure μ [3], [5] These examples are constructed from surfaces of constant curvature -1 with boundary with the addition of "caps" at the boundary (cf. Figure 0).

In particular, the Bernoulli property means that $h_\mu(\phi) > 0$ and μ is ergodic. Corollary 1 applies to give the interesting lower bound

$$0 < h_\nu(\phi) \leq \limsup_{t \to +\infty} \frac{1}{t} \log^+ N_{xy}(t)$$

for almost all $x, y \in V$.

Example 2 (Burns-Paternain Example). This example is of a C^∞ metric on the 2-sphere which has non-zero topological entropy but for which there exists an open set $U \subset V$ with an upper bound

$$\limsup_{t \to +\infty} \frac{1}{t} \log^+ N_{xy}(t) < h, \text{ and } x, y \in U.$$

In light of the Theorem this "failure" can be interpreted by saying that for any measure μ of maximal entropy (i.e. $h_\nu(\phi) = h$) the bundle $S_U V$ above the open set U has zero measure with respect to the measure μ.

Example 3 (No conjugate points). We now consider a class of metrics for which the result is already well-known.

We say that V has *no conjugate points* if for the universal cover $\tilde{V}$ (with the lifted Riemannain metric) there is a unique geodesic joining every pair of points $x', y' \in \tilde{V}$. Under this special assumption we have

$$\lim_{t \to +\infty} \frac{1}{t} \log^+ N_{xy}(t) = h$$

for all points $x, y \in V$ [9]. Notice that no Riemannian metric on the 2 sphere can satisfy the no conjugate points condition and so this does not apply to Examples 1 and 2 above.

In the particular case that V has a metric of negative curvature an asymptotic estimate for $N_{xy}(t)$ can be given [11], [18]. However, unless the surface has constant negative curvature then by a result in [8] we have $h_\nu(\phi) < h$. This allows us to see that the inequalities in (0.4) may be far from optimal.

In section 1 we recall some basic properties of entropy. In section 2 we recall some results for geodesic flows and show how the three corollaries follow from the Theorem. In section 3 we consider the theory of non-uniformly hyperbolic diffeomorphisms for surfaces and non-uniformly hyperbolic flows for three dimensional manifolds. This section introduces most of the machinary we shall require. In section 4 we prove a preliminary result for diffeomorphisms and in section 5 the corresponding result for flows. In section 6, we specialise to the case of geodesic flows and establish the Thereom. Finally, in section 7 we comment on generalisations of these results.

Our approach is motivated by a combination of Katok's result on periodic orbits [7] and Newhouse's result [12] on volume growth. One would conjecture that the "limsup" in the statements could be replaced by "liminf". The difficulty probably corresponds to that in the analogous question in [7].

1. ENTROPY

Let $f : M \to M$ be a diffeomorphism on a compact manifold M. Let μ be an f-invariant probability measure then we write $h_\mu(f)$ for the measure theoretic entropy of μ.

For our purposes it is convenient to work with a characterisation of $h_\mu(f)$ due to A. Katok. Let d denote the metric on M then for $n \geq 1$ we define a new metric d_n by

$$d_n(x, y) = \max_{0 \leq i \leq n-1} d(f^i x, f^i y).$$

Given $n \geq 1$ the (n, ϵ)-ball around $x \in M$ is defined to be

$$B_{d_n}(x, \epsilon) = \{y \in M : d_n(x, y) < \epsilon\}.$$

Definition. Given $0 < \beta \leq 1$ we call a finite set $F \subset M$ a (n, ϵ, β)-spanning set if $\mu\left(\cup_{x \in F} B_{d_n}(x, \epsilon)\right) \geq \beta$. We let $\mathcal{N}(n, \epsilon, \beta)$ be the least cardinality of an (n, ϵ, β)-spanning set.

The following result will be useful later.

Proposition 1. [7]

$$h_\mu(f) = \lim_{\epsilon \to 0} \limsup_{n \to +\infty} \frac{1}{n} \log^+ \mathcal{N}(n, \epsilon, \beta)$$

(The limit exists and is independent of β)

Next we come to the definition of the entropy for a flow.

Definition. For a flow $\phi_t : M \to M$ the entropy of a ϕ-invariant measure μ is defined to be the entropy of the time-one homeomorphism i.e. $h_\mu(\phi) := h_\mu(\phi_{t=1})$

Two well known properties of the time t_0 flow viewed as a homeomorphism are given by the following.

Lemma 1. *Let $\phi_t : M^3 \to M^3$ be a flow with an ergodic measure μ.*
  (i) *For any $t \in \mathbb{R}$ we have that $h_\mu(\phi_t) = |t| h_\mu(\phi)$; and*
  (ii) *For almost all $t_0 \in \mathbb{R}$ the homeomorphism $\phi_{t_0} : M^3 \to M^3$ is ergodic with respect to μ.*

Proof. Part (i) is due to Abramov [1]. Part (ii) is a well-known result result using representation theory [19].

2. Geodesic flows

In this section we shall recall some basic properties of the geodesic flows and show how the three corollaries follow from the Theorem.

Let V denote a compact C^∞ Riemannian surface. The unit tangent bundle SV denote the space of all tangent vectors of unit length with respect to the Riemannian metric. This is again a C^∞ manifold. Given $x \in V$ we let $S_x V$ denote the fibre in the unit tangent bundle above x.

We define the *geodesic flow* $\phi_t : SV \to SV$ as follows. Given a unit tangent vector $v \in S_x V$ we let $\gamma_v : \mathbb{R} \to V$ denote the unique unit speed geodesic with $\dot\gamma_v(0) = v$. We then set $\phi_t(v) = \dot\gamma_v(t)$.

The following result of Newhouse shows the existance of measures μ of maximal entropy μ.

Proposition 2 [13]. *For any C^∞ flow $\phi_t : M \to M$ on a compact manifold M there exists a ψ-invariant probability measure μ such that $h_\nu(\psi) = h(\psi)$*

We now move on to deducing the corollaries from the Theorem.

Proof of Corollary 1. We can apply Proposition 2 to the geodesic flow $\phi_t : SV \to SV$ to deduce that there exists a ϕ-invariant probability measure μ such that $h_\nu(\phi) = h(\phi)$. Moreover, we can assume without loss of generality that μ is ergodic (since otherwise we can replace μ by an extremal ergodic measure in its choquet decomposition).

We apply the Theorem with this ϕ-invariant probabiliy measure and we let $\Lambda \subset SV$ be the associated set. For any point $x_0 \in \Omega := \pi(\Lambda)$ we see from the Main Theorem that

$$\limsup_{t \to +\infty} \frac{1}{t} \log^+ N_{x_0 x_0}(t) \geq h_\mu(\phi) = h$$

This completes the proof.

Proof of Corollary 2. If μ is fully supported then a set Λ of full measure must be dense in SV. In particular, $D := \pi(\Lambda)$ must be dense in V. This completes the proof.

Proof of Corollary 3. Since Λ has full measure with respect to the Liouville measure ν the image $\Omega := \pi(\Lambda) \subset V$ has full measure with respect to the volume on V and if we choose $x, y \in \Omega$ then

$$\limsup_{t \to +\infty} \frac{1}{t} \log^+ N_{xy}(t) \geq h_\nu(\phi).$$

This completes the proof.

3. NON-UNIFORMLY HYPERBOLIC SYSTEMS

In this section we want to recall some of the essential features of the theory of non-uniformly hyperbolic diffeomorphisms and flows and to derive some preliminary results.

Let $f : M^2 \to M^2$ be a C^2 diffeomorphism on a compact surface M^2 and let μ be an ergodic f-invariant probability measure. Let $h_\mu(f)$ be the entropy of the measure μ.

Definition. We call a f-invariant Borel set $\Lambda \subset M$ a *Pesin set* for μ if there exists $\lambda_1, \lambda_2 > 0$ and measurable functions $\epsilon_1 : \Lambda \to \mathbb{R}$ and $\eta : \Lambda \to \mathbb{R}$ such that

(i) $\mu(\Lambda) = 1$;

(ii) for $x \in \Lambda$, $\begin{cases} \lim_{n \to \pm\infty} \frac{\log \epsilon_1(f^n x)}{n} = 0, \\ \lim_{n \to \pm\infty} \frac{\log \eta(f^n x)}{n} = 0, \end{cases}$

(iii) for $x \in \Lambda$, we can write $T_x M^2 = E_x^s \oplus E_x^u$ as a sum of one-dimensional bundles where

$$\begin{cases} \|Df^n|E_x^s\| \leq e^{\epsilon_1(f^n x)} e^{-\lambda_1 n} \text{ for } n \geq 0 \text{ and} \\ \|Df^{-n}|E_x^u\| \leq e^{\epsilon_1(f^{-n} x)} e^{-\lambda_2 n} \text{ for } n \geq 0 \end{cases}$$

and angle $\angle(E_x^s, E_x^u) \geq \eta(x)$.

151

The existence of a set Λ with $\mu(\Lambda) = 1$ and associated *Lyapunov exponents* $\lambda_1, -\lambda_2 \in \mathbb{R}$ is due to Oseledec [14]. The Liapunov exponents quantify the asymptotic behaviour of the tangent map $D_x f : T_\Lambda M \to T_\Lambda M$. It follows from the Pesin-Ruelle entropy inequality (cf. [20], [17]) that these Liapunov exponents are related to the entropy $h(\mu)$ by

$$h_\mu(f) \leq \lambda_1 \text{ and } h_\mu(f^{-1})(= h_\mu(f)) \leq \lambda_2$$

and so since $h_\mu(f) > 0$ we have that $\lambda_1 > 0$ and $\lambda_2 > 0$. This leads to the following conclusion.

Proposition 3. *Assume that $h_\mu(f) > 0$ then there exists a Pesin set Λ.*

To formulate the results we require in the following result.

Proposition 4. *There exists a set $\Lambda' \subset \Lambda$ with $\mu(\Lambda') > 0$ with the following properties. Given any $\alpha > 0$ there exists a family of neighbourhoods $B(x)$ of $x \in \Lambda$ which are differentiable squares (i.e. there exists a neighbourhood $[-2, 2] \times [-2, 2] \subset U_x \subset \mathbb{R}^2$ and a diffeomorphism $\Phi_x : U_x \to \Phi_x(U) \subset M^2$ with $\Phi_x ([-1, 1] \times [-1, 1]) = B(x))$ such that if $x, f^n(x) \in \Lambda'$ then*

(a) *the diameter of $B(f^i x)$ is less than α for $0 \leq i \leq n$;*
(b) *the image $f\left(B(f^i x)\right)$ intersects $B(f^{i+1}(x))$ transversely (i.e. there is a diffeomorphism $\Phi_{f(x)}$ from a neighbourhood of $(0, 0) \in \mathbb{R}^2$ to a neighbourhood of $f(x) \in V$ such that $\Phi_{f(x)} ([-1, 1] \times [-1, 1]) = B(f(x))$ and $\Phi_{f(x)} ([-2, 2] \times [-\frac{1}{2}, \frac{1}{2}]) = f(B(x)))$*
(c) *The neighbourhoods $B(x)$ vary continuously on Λ'*

Moreover the splitting $\Lambda' \ni x \mapsto E_x^s \oplus E_x^u$ varies continuously.

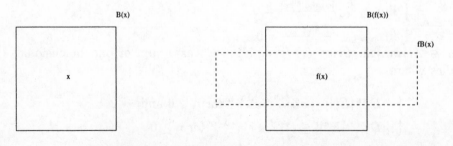

FIGURE 1. Transverse intersection property

Proof. Following Fathi-Herman-Yoccoz [6] we define a new (measurable) norm for $v = v_u \oplus v_s \in T_x M = E_x^s \oplus E_x^u$ (where $x \in \Lambda$) by

$$|||v|||_x = \max \left(\sum_{n=0}^{\infty} e^{-\lambda_1 n} ||D_x f^n(v_s)||, \ \sum_{n=-\infty}^{0} e^{\lambda_2 n} ||D_x f^n(v_s)|| \right).$$

This has the desirable property that

$$\begin{cases} |||D_x f(v)||| \leq e^{-\lambda_1} |||v|||, \text{ for } v \in E^s \\ |||D_x f^{-1}(v)||| \leq e^{-\lambda_2} |||v|||, \text{ for } v \in E^u \end{cases} \tag{3.1}$$

(cf. [6]) and

$$\frac{1}{2}||v||_x \leq |||v|||_x \leq A(x)||v||_x, \quad \forall v \in V$$

where $\lim_{n \to \pm\infty} \frac{1}{n} \log A(f^n(x)) = 0$. (cf. [6, Proposition 4]).

Fix $\epsilon > 0$. It is easy to see that $\Lambda = \cup_{T>0} \Lambda_T$ where we denote

$$\Lambda_T = \{x \in \Lambda : \eta(f^n x) \geq \frac{1}{T} e^{-\epsilon|n|}, \epsilon_1(f^n x) \leq T e^{\epsilon|n|} \text{ and } A(f^n x) \leq T e^{\epsilon|n|}\}.$$

Observe that $\Lambda_{T_1} \supset \Lambda_{T_2}$ for $T_1 \geq T_2$. In particular, if we fix $T > 0$ sufficiently large then $\mu(\Lambda_T) > 0$. Moreover, by its definition this set is closed.

Finally, we want to consider the set

$$\Lambda' = \{x \in \Lambda_T : \forall \delta > 0, \mu(B(x, \delta)) > 0\}$$

and we observe that $\mu(\Lambda') = \mu(\Lambda_T) > 0$.

We can use the C^2 hypotheses on f to extend (3.1) to a neighbourhood of $x \in \Lambda$. Given an arbitrary point $x \in \Lambda$ we want to fix local charts $\Phi : U_x \to \Phi(U_x)$ (with $(0,0) \in U \subset \mathbb{R}^2$ and $x \in \Phi(U_x) \subset V$) and $\Phi : U_{f(x)} \to \Phi(U_{f(x)})$ (with $(0,0) \in U_{f(x)} \subset \mathbb{R}^2$ and $f(x) \in \Phi(U_{f(x)}) \subset V$). The charts allow us to treat $\Phi(U_x)$ and $\Phi(U_{f(x)})$ as though they were the corresponding neighbourhoods U_x and $U_{f(x)}$ of euclidean space. We begin by using $|| \cdot ||$ as the euclidean norm. Since f is C^2 there is a constant $L > 0$ such that locally:

(1) $||D_{z_1} f - D_{z_2} f|| \leq L||z_1 - z_2||$ for $z_1, z_2 \in \Phi(U_x)$;
(2) $||f(z_1) - f(z_1)|| \leq L||z_1 - z_2||$ for $z_1, z_2 \in \Phi(U_x)$.

Following Newhouse [12, proof of Theorem 1] we can extend the splitting $T_x M = E_x^s \oplus E_x^u$ to the neighbourhood $\Phi(U_x)$ by defining a splitting $T_z M = \tilde{E}_z^s \oplus \tilde{E}_z^u$ ($z \in \Phi(U_x)$) by *translating* the splitting above x using the euclidean structure associated to U_x. We can also extend the norm $||| \cdot |||_x$ to a norm $||| \cdot |||_z$ on $T_z M$ by translating vectors from $T_z M$ to $T_x M$ using the euclidean structure.

The euclidean structure on U_x also gives rise to a natural system of local co-ordinates, as follows. If let $e_x^s \in E_x^s$ and $e_x^u \in E_x^u$ satisfy $|||e_x^s|||_x = 1 = |||e_x^u|||_x$, then we can (locally) map

$$(u_1, u_2) \mapsto \Phi_x \left(x + u_1 D\Phi_x^{-1}(e_1) + u_2 D\Phi_x^{-1}(e_2) \right).$$

With these conventions we have that for

$$\delta(x) := \min\left(1, \frac{e^{\lambda_1 + 2\epsilon} - e^{\lambda_1 + \epsilon}}{2LA(f(x))}, \frac{e^{\lambda_1 + 2\epsilon} - e^{\lambda_1 + \epsilon}}{2LA(f(x))} \right)$$

we know

(a) $|||D_z f(v)|||_{f(z)} \le e^{-\lambda_1 + 2\epsilon} |||v|||_z$ for $v \in \tilde{E}_z^s$ where $||z - x|| \le \delta(x)$;
(b) $|||D_z f(v)|||_{f(z)} \ge e^{\lambda_2 + 2\epsilon} |||v|||_z$ for $v \in \tilde{E}_z^u$ where $||z - x|| \le \delta(x)$;
(c) $\delta(f^n x) \ge e^{-\epsilon|n|} \delta(x)$

(cf. [12]). For example, to show (a) we observe that for $v \in \tilde{E}_z^s$:

$$\begin{aligned}
|||D_z f(v)|||_{f(z)} &\le |||D_x f(v)||| + |||\left(D_x f - D_z f \right)(v)||| \\
&\le e^{\lambda_1} |||v||| + A(x)||(D_x f - D_z f)(v)|| \\
&\le e^{\lambda_1} |||v||| + A(x)L||x - z||.||v|| \\
&\le \left(e^{\lambda_1} + 2A(x)L\delta(x) \right) ||v|| \\
&\le \left(e^{\lambda_1 + 2\epsilon} \right) |||v|||
\end{aligned}$$

The proof of part (b) is similar and (c) follows from the definitions.

When $x, f^n(x) \in \Lambda'$ then we see that

$$\begin{cases}
A(f^i x) \le T e^{\epsilon i} & \text{for } 0 \le i \le \frac{[n]}{2} \\
A(f^i x) \le T e^{\epsilon(n-i)} & \text{for } \frac{[n]}{2} < i \le n
\end{cases}$$

and so, in particular,

$$\delta(f^i(x)) \ge D \max\left(e^{-\epsilon i}, e^{-\epsilon(n-i)} \right).$$

for some constant $D > 0$.

We want to define a "box" about x by

$$B(x) = \Phi_x \left([-\alpha\epsilon(x), \alpha\epsilon(x)] e_x^s \times [-\alpha\epsilon(x), \alpha\epsilon(x)] e_x^u \right)$$

where $0 < \alpha < 1$ is some (small) constant whose value we shall specify later.

Since $D_x f(E_x^s) = E_{f(x)}^s$ and $D_x f(E_x^u) = E_{f(x)}^u$ we see that f preserves the axes of the local co-ordinates in $\Phi(U_x)$ and $\Phi(U_{f(x)})$. Properties (a) and (b) above allow

154

us to see that f contracts $||| \cdot |||$-distances in the e_x^s co-ordinate by at least $e^{-\lambda_1} < 1$ and f expands $||| \cdot |||$-distances in the e_x^u co-ordinate by at least $e^{\lambda_2} > 1$. This allows us to conclude that $f(B(x))$ meets $B(x)$ transversely.

The following notation will be convenient in the sequel.

Notation. Using the notation of the Proposition 4 we call the two curves

$$\partial_1^u = \Phi_x([-1,1] \times \{-1\}) \text{ and } \partial_2^u = \Phi_x([-1,1] \times \{1\}) \subset B(x)$$

the *expanding sides* of $B(x)$.
We call the two curves

$$\partial_1^s = \Phi_x(\{-1\} \times [-1,1]) \text{ and } \partial_2^s = \Phi_x(\{1\} \times [-1,1]) \subset B(x)$$

the *contracting sides* of $B(y)$.

The following simple geometric lemma for these diffeomorphisms will prove very useful later.

Lemma 2. *Assume that:*

   (a) $x, f(x), f^2(x), \ldots, f^{n-1}(x)$ *be a finite orbit in Λ;*
   (b) C_1 *is a curve connecting the two contracting sides of $B(x)$; and*
   (c) C_2 *is curve connecting the two expanding sides of $B(f^{n-1}x)$.*

Then $f^n(C_1) \cap C_2 \neq \emptyset$.
   Moreover, there exists $z \in f^{n-1}(C_1) \cap C_2$ such that $f^{-i}(z) \in B(f^{n-1-i}(x))$ for $0 \leq i \leq n-1$.

Proof. By Proposition 3 we know that $f(B(x))$ meets $B(f(x))$ transversely. It is therefore clear that some subcurve $C_1^{(1)} \subset f(C_1)$ must connect the two contracting sides of $B(f(x))$. Proceeding inductively, suppose there exists a subcurve $C_1^{(i)}$ of $f^i(C_1)$ that connects the two contracting sides in $B(f^i x)$ for $1 \leq i \leq n-2$. We easily see that there is subcurve $C_1^{(i+1)}$ of $f(C_1^{(i)})$ that connects the two contracting sides of $B(f^{i+1}x)$.

It is now clear that $C_1^{(n-1)} \cap C_2 \neq \emptyset$ and we choose $z \in C_1^{(n-1)} \cap C_2$. Since

$$C_1^{(n-1)} \subset f C_1^{(n-2)} \subset \ldots \subset f^i C_1^{(n-1-i)} \subset \ldots \subset f^{n-1} C_1$$

it follows that $z \in f^{n-1} C_1 \cap C_2$ and so we conclude that $f^{n-1} C_1 \cap C_2 \neq \emptyset$.

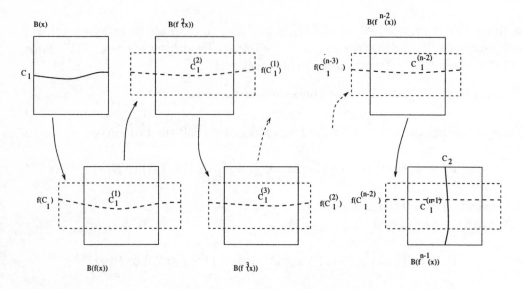

FIGURE 2. The proof of Lemma 2

Finally, by construction we have that

$$\begin{cases} f^{-i}(z) \in C_1^{(n-1-i)} \subset B(f^{n-1-i}(x)), & \text{for } 1 \le i \le n-1, \\ f^{-(n-1)}z \in C_1 \subset B(x) \end{cases}$$

This completes the proof of the lemma.

We can now give the corresponding results for C^2 flows $\phi_t : M^3 \to M^3$ on a compact 3-dimensional manifold. Let μ be an ergodic ϕ-invariant probability measure.

Definition. We call a ϕ-invariant Borel set $\Lambda \subset M$ a *Pesin set* for μ if $\exists \lambda_1, \lambda_2 > 0$ and measurable functions $\epsilon_1 : \Lambda \to \mathbb{R}$, $\eta : \Lambda \to \mathbb{R}$ such that

(i) $\mu(\Lambda) = 1$;
(ii) for $x \in \Lambda$,

$$\begin{cases} \lim_{t \to \pm\infty} \frac{\log \epsilon_1(\phi_t(x))}{t} = 0, \\ \lim_{t \to \pm\infty} \frac{\log \eta(\phi_t(x))}{t} = 0, \end{cases}$$

(iii) for $x \in \Lambda_k$, $k \ge 1$, we can write $T_x M^2 = E_x^0 \oplus E_x^s \oplus E_x^u$ as a sum of one-dimensional bundles, where E_x^0 is tangent to the flow direction

$$\begin{cases} ||D\phi_t|E_x^s|| \le e^{\epsilon_1(\phi_t x)} e^{-\lambda_1 t} \text{ for } t \ge 0 \text{ and} \\ ||D\phi_{-t}|E_x^u|| \le e^{\epsilon_1(\phi_t x)} e^{-\lambda_2 t} \text{ for } t \ge 0 \end{cases}$$

and $\angle(E_x^s, E_x^u) \ge \eta(x)$.

Many of the properties from the diffeomorphism case have analogues in the flow case. Corresponding to Proposition 3 we have the following.

Proposition 5. [7] *Assume that $h_\mu(\phi) > 0$ then there exists a Pesin set Λ.*

Corresponding to Proposition 4 we have the following.

Proposition 6. *Fix $t_0 > 0$, then there exists $\Lambda' \subset M^3$ with $\mu(\Lambda') > 0$ and the following properties. For any sufficiently small $\alpha > 0$ and $\eta > 0$ there exists a family of neighbourhoods $C(x)$ of $x \in \Lambda'$ which are differentiable cubes (i.e. there exists a neighbourhood $[-1, 1] \times [-1, 1] \times [-\eta, \eta] \subset U \subset \mathbb{R}^3$ and a diffeomorphism $\Phi_x : U \to \Phi_x(U) \subset M^3$ such that $\Phi_x([-1, 1] \times [-1, 1] \times [-\eta, \eta]) = C(x)$ and $\Phi_x((u, v, t + t')) = \phi_t \Phi_x((u, v, t'))$, for $-\eta \leq t', t + t' \leq \eta$) and if $x, \phi_{nt_0}(x) \in \Lambda'$ then*

(a) *the diameter of each "slice" $\Phi_{\phi_{it_0}(x)}([-1, 1] \times [-1, 1] \times \{t\})$ ($x \in \Lambda_k$, $-\eta \leq t \leq \eta$) is less than α for $0 \leq i \leq n$;*

(b) *the image $\phi_{t_0}(C(\phi_{it_0}x))$ intersects $C(\phi_{(i+1)t_0}(x))$ transversely (i.e. there is a diffeomorphism $\Phi_{\phi_{t_0}x}$ from a neighbourhood of $(0, 0) \in \mathbb{R}^3$ to a neighbourhood of $\phi_{t_0}(x) \in V$ such that $\Phi([-1, 1] \times [-1, 1] \times [-\eta, \eta]) = C(\phi_{t_0}x)$ and $\Phi([-2, 2] \times [-\frac{1}{2}, \frac{1}{2}] \times [-\eta, \eta]) = \phi_{t_0}(C(x))$).*

Moreover, the splitting $\Lambda' \ni x \mapsto E_x^0 \oplus E_x^s \oplus E_x^u$

The following notation will prove convenient

Notation. Using the notation of Proposition 6 we call the two sets

$$\partial_1^{wu} = \Phi_x([-1, 1] \times \{-1\} \times [-\eta, \eta]) \subset C(x) \text{ and}$$
$$\partial_2^{wu} = \Phi_x([-1, 1] \times \{1\} \times [-\eta, \eta]) \subset C(x)$$

the *weak expanding sides* of $C(x)$. We call the two sets

$$\partial_1^{ws} = \Phi_x(\{-1\} \times [-1, 1] \times [-\eta, \eta]) \subset C(x) \text{ and}$$
$$\partial_2^{ws} = \Phi_x(\{1\} \times [-1, 1] \times [-\eta, \eta]) \subset C(x)$$

the *weak contracting sides* of $C(x)$.

The proof of Lemma 2 easily adapts to show the following analogous result for flows.

Lemma 3. *Assume that:*

(a) *We have points* $x, \phi_{t_0}(x), \phi_{2t_0}(x), \ldots, \phi_{(n-1)t_0}(x) \in \Lambda$ *on a* ϕ-*orbit;*

(b) C_1 *is a curve connecting the two weak contracting sides of* $C(x)$; *and*

(c) C_2 *is a curve connecting the two weak expanding sides of* $C(\phi_{(n-1)t_0}(x))$.

Then $\exists - \eta \leq t_1 \leq \eta$ *such that* $\phi_{(n-1)t_0+t_1}(C_1) \cap C_2$.

Moreover, there exists $z \in \phi_{(n-1)t_0+t_1}(C_1) \cap C_2$ *such that*

$$\phi_{-it_0}(z) \in C(\phi_{(n-1-i)t_0+t_1} x)$$

for $0 \leq i \leq n-1$.

4. Results for diffeomorphisms

In this section we shall present a result for a C^2 diffeomorphism $f : M^2 \to M^2$ on a compact surface M^2. This should be viewed as a prelude to the corresponding result for a C^2 flow $\psi_t : M^3 \to M^3$ on a three dimensional compact manifold M^3 in the next section. It is this result for flows which when applied to the geodesic flow $\phi_t : S_1 V \to S_1 V$ will imply the Theorem.

Proposition 7. *Let* $f : M^2 \to M^2$ *be a* C^2 *diffeomorphism on a compact surface and let* μ *be a* f-*invariant ergodic probability measure.*

Then there exists a f-*invariant set* $\Lambda \subset M$ *with* $\mu(\Lambda) = 1$ *such that for any pair of* C^1 *curves* $C_1, C_2 \subset M^2$ *such that*

(i) $C_1 \cap \Lambda \neq \emptyset$ *and* $\exists x_0 \in C_1 \cap \Lambda$ *such that the tangent to the curve* C_1 *at* x_0 *is* **not** *in the contracting direction (i.e.* $T_{x_0} C_1 \neq E^s_{x_0}$); *and*

(ii) $C_2 \cap \Lambda \neq \emptyset$ *and* $\exists y_0 \in C_2 \cap \Lambda$ *such that the tangent to the curve* C_2 *at* y_0 *is* **not** *in the expanding direction (i.e.* $T_{y_0} C_2 \neq E^u_{y_0}$)

we have that

$$\limsup_{n \to +\infty} \frac{1}{n} \log^+ Card(f^n C_1 \cap C_2) \geq h(\mu)$$

Remark. It is easy to see that the conditions (i) and (ii) are required by considering a linear Anosov diffeomorphism of $\mathbb{T}^2$ with C_1 and C_2 being (small) pieces of stable and unstable manifolds, respectively

We take Λ' as introduced in section 3. Let us suppose that

(a) there exists $x_0 \in C_1 \cap \Lambda'$ such that $T_{x_0} C_1$ does not lie in the stable direction $E^s_{x_0}$

(b) there exists $y_0 \in C_2 \cap \Lambda'$ such that $T_{y_0} C_2$ does not lie in the unstable direction $E^u_{y_0}$

158

We choose small neighbourhoods U_1 and U_2 of x_0 and y_0, respectively, so that

   (i) the tangents to C_1 and C_2 are in neither the expanding nor contracting direction at all points $x \in U_1 \cap \Lambda'$ and all points $y \in U_2 \cap \Lambda'$;

   (ii) $\mu(U_1 \cap \Lambda') > 0$ and $\mu(U_2 \cap \Lambda') > 0$

(using that the tangent to C_1 and C_2 and the stable and unstable directions for Λ' vary continuously). We fix (small) values $\epsilon > 0$ and $\beta > 0$. Let

$$\Gamma_N = \{x \in U_1 \cap \Lambda' : \exists N \le n \le [N(1+\beta)], f^n x \in U_2 \cap \Lambda'\}$$

where $[N(1+\beta)]$ is the integar part of $N(1+\beta)$.

Lemma 4. *Let μ be ergodic and choose $\mu(U_1 \cap \Lambda') > \beta_0 > 0$ then $\mu(\Gamma_N) \ge \beta_0 > 0$ for N sufficiently large.*

Proof. Using the Birkhoff ergodic theorem we know that

$$\frac{1}{\beta N}\text{Card}\{N \le i \le N(1+\beta) : f^i(x) \in U_2 \cap \Lambda'\}$$

$$= \frac{1}{\beta N}\left(\sum_{i=0}^{N(1+\beta)} \chi_{U_2 \cap \Lambda'}(f^i x) - \sum_{i=0}^{N} \chi_{U_2 \cap \Lambda'}(f^i x)\right)$$

$$\rightarrow \left(\frac{1+\beta}{\beta} - \frac{1}{\beta}\right)\mu(U_2 \cap \Lambda')$$

as $N \to +\infty$, for almost all $x \in U_1 \cap \Lambda'$. The result easily follows.

We let $K_N(\epsilon)$ denote a maximal set of points in Γ_N such that $d_N(x,y) \ge \epsilon$ ($x \ne y$) where $x, y \in K_N(\epsilon)$. It is easy to see that the union of the $(N, 2\epsilon)$-balls around points in $K_N(\epsilon)$ contains Γ_N. Therefore, we see that $\text{Card}(K_N(\epsilon)) \ge \mathcal{N}(N, 2\epsilon, \beta_0)$. In particular, we have that

$$\limsup_{N \to +\infty} \frac{1}{N} \log^+ \text{Card}(K_N(\epsilon))$$

$$\ge \limsup_{N \to +\infty} \frac{1}{N} \log^+ \mathcal{N}(N, 2\epsilon, \beta_0) \tag{4.1}$$

$$\ge h_\mu(f)(1 - \eta(\epsilon))$$

where $\eta(\epsilon) \to 0$ as $\epsilon \to 0$.

Lemma 5. *Every point* $x \in K_N(\epsilon)$ *gives rise to a distinct point in the union* $\cup_{n=N}^{N(1+\beta)} f^n(C_1) \cap C_2$.

Proof. We can assume without loss of generality that

(a) the boxes $B(x)$ $(x \in U_1 \cap \Lambda')$ have the additional property that the intersection of the curve C_1 with each $B(x)$ connects the two contracting sides

(b) the boxes $B(y)$ $(y \in U_2 \cap \Lambda')$ have the additional property that the intersection of the curve C_2 with each $B(y)$ connects the two expanding sides.

If this were not the case then it can be achieved achieve by choosing the neighbourhoods U_1 and U_2 sufficiently small and the following simple device

(i) replacing the rectangle $B(x)$ by a subrectangle

$$B'(x) = \Phi_x \left([-\eta, \eta] \times [-1, 1]\right) \subset \Phi_x \left([-1, 1] \times [-1, 1]\right) = B(x)$$

and replacing the rectangle $B(f^n x)$ by the subrectangle

$$B'(f^n x) := \Phi_x \left([-1, 1] \times [-\gamma, \gamma]\right) \subset \Phi_x \left([-1, 1] \times [-1, 1]\right) = B(f^n x),$$

for some sufficiently small $\eta, \gamma > 0$;

(ii) choosing $\eta = \eta_0 < \eta_1 < \eta_2 < \ldots < \eta_m = 1$ such that replacing $B(f^i x)$ by

$$B'(f^i x) := \Phi_{f^i x} \left([-1, 1] \times [-\eta_i, \eta_i]\right) \subset \Phi_{f^i x} \left([-1, 1] \times [-1, 1]\right) = B(f^i x),$$

(for $i = 0, \ldots, m$) we have that for $n \geq m$ the boxes associated to $x, f(x),$ $\ldots, f^n(x)$ still satisfy the transversality property; and

(iii) choosing $\gamma = \gamma_0 < \gamma_1 < \gamma_2 < \ldots < \gamma_{m'} = 1$ such that replacing $B(f^i x)$ by

$$B'(f^i x) := \Phi_{f^i x} \left([-1, 1] \times [-\gamma_i, \gamma_i]\right) \subset \Phi_{f^i x} \left([-1, 1] \times [-1, 1]\right) = B(f^i x),$$

(for $i = 0, \ldots, m'$) we have for $n \geq m+m'$ the boxes associated to $f^{(n-m')}(x)$, $f^{n-m'+1}(x), \ldots, f^n(x)$ still satisfy the transversality property.

For each $x \in K_N(\epsilon)$ we can apply Lemma 1 with $N \leq n \leq [N(1+\beta)]$ satisfying $f^n(x) \in U_2 \cap \Lambda_k$ to find a point $z(x) \in f^n(C_1) \cap C_2$.

Finally, if $x' \in K_N(\epsilon)$ is different from x then we need to show that $z(x) \neq z(x')$. To see this we can assume that the boxes $B(x)$, $x \in \Lambda$, have diameter less than $\frac{\epsilon}{4}$ (assuming that α is sufficiently small). With this assumption $B(f^i(x))) \cap B(f^i(x')) = \emptyset$ for some $0 \leq i \leq N$ such that $d(f^i(x), f^i(x')) = d_N(x, x') \geq \epsilon$. Therefore, $z \neq z'$ because by construction $f^{n-i}(z) \in B(f^i(x))$ and $f^{n-i}(z') \in B(f^i(x'))$.

This completes the proof of the lemma.

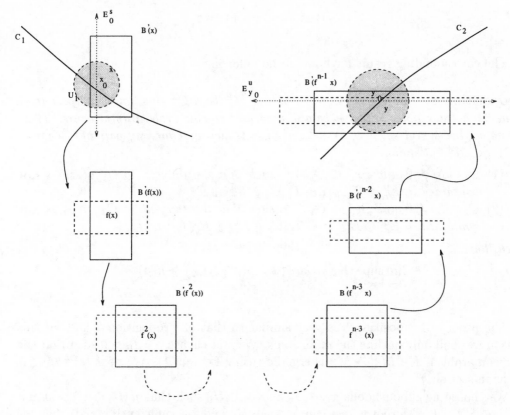

FIGURE 3. Creating points in $\cup_{n=N}^{N(1+\beta)} f^n(C_1) \cap C_2$

Proof of Proposition 6. By the Lemma 5 we know that

$$\text{Card}\,(K_N(\epsilon)) \leq \text{Card}\left(\cup_{n=N}^{N(1+\beta)} f^n(C_1) \cap C_2\right)$$

and therefore

$$\limsup_{n\to+\infty} \frac{1}{n} \log^+ \text{Card}\,(f^n(C_1) \cap C_2)$$

$$\geq \frac{1}{1+\beta} \limsup_{N\to+\infty} \frac{1}{N} \log^+ \text{Card}\left(\cup_{n=N}^{N(1+\beta)} f^n(C_1) \cap C_2\right)$$

$$\geq \frac{1}{1+\beta} \limsup_{N\to+\infty} \frac{1}{N} \log^+ \text{Card}\,(K_N(\epsilon))$$

$$\geq h_\mu(f) \frac{(1-\eta(\epsilon))}{1+\beta}$$

where we have used (4.1) to deduce the last line. Finally, since we can choose ϵ (and thus η) and β arbitrarily small the result follows.

5. Results for Flows

The corresponding result for flows is the following.

Proposition 8. *Fix $\rho > 0$. Let $\psi : M^3 \to M^3$ be a C^2 flow on a compact three dimensional manifold and let μ be a ψ-invariant ergodic probability measure. There exists a ψ-invariant set $\Lambda \subset M$ with $\mu(\Lambda) = 1$ such that for any pair of C^1 curves $C_1, C_2 \subset M^3$ satisfying*

(1) $C_1 \cap \Lambda \neq \emptyset$ and $\exists x_0 \in C_1 \cap \Lambda$ such that the tangent to C_1 at x_0 is **not** contained in $E_{x_0}^0 \oplus E_{x_0}^s$ (i.e. $T_{x_0}C_1 \not\subset E_{x_0}^0 \oplus E_{x_0}^s$)

(2) $C_2 \cap \Lambda \neq \emptyset$ and $\exists y_0 \in C_1 \cap \Lambda$ such that the tangent to C_2 at y_0 is **not** contained in $E_{y_0}^0 \oplus E_{y_0}^u$ (i.e. $T_{y_0}C_2 \not\subset E_{y_0}^0 \oplus E_{y_0}^u$)

such that

$$\limsup_{t \to +\infty} \frac{1}{t} \log^+ Card \left(\psi_{[t-\rho, t]} C_1 \cap C_2 \right) \geq h(\mu)$$

The proof of Proposition 8 is very similar to that of Proposition 7 and for this reason we shall only outline the main steps. We first choose $\rho > t_0 > 0$ such that the homeomorphism $\phi_t : M^3 \to M^3$ is ergodic (using Lemma 1). We next take $\Lambda' \subset \Lambda$ be as in section 4.

We choose neighbourhoods $x_0 \in U_1$ and $y_0 \in U_1$ such that $\mu(U_1 \cap \Lambda') > 0$ and $\mu(U_2 \cap \Lambda') > 0$. Moreover, we can arrange U_1 and U_2 sufficiently small that for $x \in U_1 \cap \Lambda'$ and $y \in U_2 \cap \Lambda'$ the tangents $T_y C_0$ and $T_y C_1$ are still not contained in $E_x^s \oplus E_x^0$ and $E_x^u \oplus E_x^0$, respectively. Finally, we also arrange that $U_1 \cap \Lambda'$ and $U_2 \cap \Lambda'$ are both contained in flow boxes of height smaller than $\gamma < \frac{\rho}{2}$ in the flow direction.

Fix $\beta > 0$ and let

$$\Gamma_N' = \{ x \in U_1 \cap \Lambda' : \exists N \leq n \leq [N(1 + \beta)], \phi_{(n-1)t_0} \in U_2 \cap \Lambda' \}.$$

By analogy with the proof of Lemma 4, we can apply the Birkhoff ergodic theorem to ϕ_{t_0} to deduce the following.

Lemma 6. $\mu(\Gamma_N') \to \mu(U_1 \cap \Lambda')$ *as* $N \to +\infty$.

We henceforth assume that N is sufficiently large that $\mu(\Gamma_N') > 0$. We let $K_N(\epsilon)$ denote a maximal set of points in Γ_N' such that for $x, y \in K_N(\epsilon)$ with $x \neq y$ we have

$$\max_{0 \leq i \leq n-1} d\left(\phi_{it_0}x, \phi_{it_0}y \right) \geq \epsilon$$

We have the following analogue of Lemma 3.

Lemma 7. *Assume that*

(a) $x, \phi_{t_0}(x), \ldots, \phi_{(n-1)t_0}(x)$ *is a finite orbit with* $x, \phi_{(n-1)t_0}(x) \in \Lambda'$;

(b) C_1 *is a curve connecting the two weak expanding sides of* $C(x)$

(c) C_2 *is a curve connecting the two weak expanding sides of* $C(\phi_{(n-1)t_0}(x))$.

Then $\phi_{(n-1)t_0}(C_1) \cap \phi_{t_1}(C_2) \neq \emptyset$ *for some* $-\gamma \leq t_1 \leq \gamma$. *Moreover, there exists* $z \in \phi_{(n-1)t_0}(C_1) \cap \phi_{t_1}(C_2)$ *such that*

$$\phi_{-it_0} z \in C(\phi_{(n-1-i)t_0} x)$$

for $0 \leq i \leq n-1$.

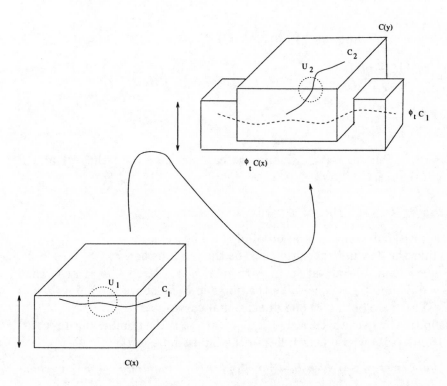

FIGURE 3. Creating intersections under the flow

Using this we can derive the following analogue of Lemma 5 by an essentially similar proof.

Lemma 8. *Every point* $x \in K_N(\epsilon)$ *gives rise to a distinct point in*

$$\phi_{[Nt_0-\gamma, [N(1+\beta)]t_0+\gamma]}(C_1) \cap C_2$$

The final part of the proof of Proposition 7 now proceeds as did the final part of the proof of Proposition 6.

Proof of Proposition 7. By Lemma 8 we know that

$$\text{Card}\,(K_N(\epsilon)) \leq \text{Card}\,(\phi_{[Nt_0-\gamma,[N(1+\beta)]t_0+\gamma]}(C_1) \cap C_2)$$

and therefore

$$\limsup_{t\to+\infty} \frac{1}{t} \log^+ \text{Card}\,(\phi_{[t,t-\rho]}(C_1) \cap C_2)$$

$$\geq \frac{1}{(1+\beta)t_0} \limsup_{N\to+\infty} \frac{1}{N} \log^+ \text{Card}\,(\phi_{[Nt_0-\gamma,[N(1+\beta)]t_0+\gamma]}(C_1) \cap C_2)$$

$$\geq \frac{1}{(1+\beta)t_0} \limsup_{N\to+\infty} \frac{1}{N} \log^+ \text{Card}\,(K_N(\epsilon))$$

$$\geq \frac{h_\mu(\phi_{t_0})}{t_0} \frac{(1-\eta(\epsilon))}{1+\beta}$$

$$= h_\mu(\phi) \frac{(1-\eta(\epsilon))}{1+\beta}$$

where we have used Lemma 1 (i). Finally, we can choose ϵ (and thus η) and β sufficiently small the result follows.

6. Proof of the Theorem

In this section we shall conclude the proof of the Theorem.

We want to take the flow in Proposition 8 to be the geodesic flow $\phi_t : S_1V \to S_1V$. Proposition 8 gives us a ϕ-invariant set Λ with $\mu(\Lambda) = 1$. If $x, y \in \Omega = \pi(\Lambda)$ then we choose $C_1 = S_x V$ and $C_2 = S_y V$ to be the fibres in S_1M above the points x and y, respectively. These are both C^∞ one dimensional curves.

By these definitions we see that $\text{Card}\,(\phi_{[t-\rho,t]}C_1 \cap C_2)$ is the number of (directed) geodesic arcs from x to y whose length lies in the interval $[t-\rho, t]$.

Lemma 9. *There exists $\epsilon > 0$ such that for any $t \in (0, \epsilon)$ and any $v \in SV$ we have*

$$T_{\phi_t(v)}\left(S_{\pi(\phi_t(v))}V\right) \neq D_v\phi_t\left(T_{\pi(v)}\left(S_{\pi(v)}V\right)\right)$$

This geometric lemma is easily seen for constant curvature surfaces, and so for variable curvature surfaces by taking sufficiently small neighbourhoods.

Example $V = \mathbb{T}^2$. As an illustration of Lemma 9, consider the special case where M is the flat torus $\mathbb{T}^2$. Using the co-ordinates of $\mathbb{R}^2$ we can assume without loss of generality that v is above $(0,0) \in \mathbb{T}^2$ in the direction $(1,0)$ (i.e. $v =$

$((0,0),(1,0)) \in S_{(0,0)}\mathbb{T}^2 \subset T\mathbb{T}^2 = \mathbb{T}^2 \times \mathbb{R}^2)$. Flowing for a small time $t > 0$ under the geodesic flow moves v to $\phi_t(v) = ((t,0),(1,0))$ and so $\pi(\phi_t(v)) = (t,0)$. We can parameterise $S_{\pi(\phi_t v)}\mathbb{T}^2$ by $\alpha(\theta) = ((t,0),(\cos\theta,\sin\theta))$ $(0 \leq \theta < 2\pi)$ and $\phi_t S_{\pi(v)}\mathbb{T}^2$ by $\beta(\theta) = ((t\cos\theta, t\sin\theta),(\cos\theta,\sin\theta))$ $(0 \leq \theta < 2\pi)$. Taking derivatives we have that $\alpha'(0) = ((t,0),(0,1)) \in T_{\phi_t(v)}\left(S_{\pi(\phi_t(v))}\mathbb{T}^2\right)$ and $\beta'(0) = ((0,t),(0,1)) \in D_v\phi_t\left(T_{\pi(v)}\left(S_{\pi(v)}\mathbb{T}^2\right)\right)$ Each is a non-zero vector which spans the respective one-dimensional space. In particular, since they are independent we see that
$$T_{\phi_t(v)}\left(S_{\pi(\phi_t(v))}\mathbb{T}^2\right) \neq D_v\phi_t\left(T_{\pi(v)}\left(S_{\pi(v)}\mathbb{T}^2\right)\right)$$

A consequence of this is the following.

Proposition 9. *Let*

$$\mathcal{E}^s = \{v \in SV : T_vS_{\pi(v)}V = E_v^s\} \text{ and } \mathcal{E}^u = \{v \in SV : T_vS_{\pi(v)}V = E_v^u\}$$

then $\mu(\mathcal{E}^s) = 0$ and $\mu(\mathcal{E}^u) = 0$ for any ϕ-invariant measure μ.

Proof. We observe that if $v \in \mathcal{E}^s$ then by $D\phi_t$ invariance of the bundle E^s we have that $T_{\phi_t(v)}\phi_t\left(S_{\pi(v)}V\right) = D\phi_t E_v^u = E_{\phi_t(v)}^u$. However, by Lemma 9 for $t \in (0,\epsilon)$ we have $T_{\phi_t(v)}S_{\pi(\phi_t(v))}V \neq D_v\phi_t\left(T_{\pi(v)}\left(S_{\pi(v)}V\right)\right)$ and so $\phi_t(v) \notin \mathcal{E}^s$. However, this implies that $\mu(\mathcal{E}^s) = 0$. A similar argument applies to show that $\mu(\mathcal{E}^u) = 0$.

If we denote $\Lambda'' = \Lambda' - (\mathcal{E}^s \cup \mathcal{E}^u)$ then $\mu(\Lambda'') = \mu(\Lambda') > 0$. We finally define $\Omega = \pi(\Lambda'')$. Since $x,y \in \Omega$ we see that $C_1 \cap \Lambda'' \neq \emptyset$ and $C_2 \cap \Lambda'' \neq \emptyset$. We can choose $v \in C_1 \cap \Lambda''$ and $w \in C_2 \cap \Lambda''$, say. Thus all of the hypotheses of Proposition 8 are satisfied and we can deduce that

$$\limsup_{t \to +\infty} \frac{1}{t}\log^+ N_{x,y}(t)$$

$$\geq \limsup_{t \to +\infty} \frac{1}{t}\log^+ \text{Card}\left(\phi_{[t-\rho,t]}C_1 \cap C_2\right)$$

$$\geq h(\mu)$$

This completes the proof of the Theorem.

7. FINAL COMMENTS

We conclude with a few comments and problems

(1) We have, in fact, established a slightly stronger result. Given an ergodic measure μ let Ω be the set constructed above. For any $\rho > 0$ and $x,y \in \Omega$ we denote by $N_{xy}(t,t-\rho)$ the number of distinct geodesic arcs from x to

y whose length lies in the interval $[t - \rho, t]$. The argument in the prceeding section gives that that

$$\limsup_{T \to +\infty} \frac{1}{T} \log^+ N_{xy}(t, t - \rho) \geq h(\mu)$$

(2) If μ is *not* ergodic then Lemma 4 may fail, and our proof does not apply and we can say nothing about this case.

(3) The proof of the Theorem is closely tied to the assumption that V is a surface. In higher dimensions our proof does not apply and we can again say nothing.

References

1. L. Abramov, *On the entropy of a flow*, Dokl. Akad. Nauk. SSSR **128** (1959), 873-875.
2. G. Besson, G. Courtois and S. Gallot, *Les variétés hyperboliques sont des minmal locaux de l'entropie topologique*, Invent. Math. **117** (1994), 403 - 446.
3. K. Burns and M. Gerber, *Real analytic Bernoulli geodesic flows on S^2*, Ergod. Th. and Dynam. Sys. **8** (1989), 27-45.
4. K. Burns and G.Paternain, Preprint.
5. V. Donnay, *Geodesic flow on the two sphere Part II: Ergodicity*, Dynamical Systems (J. Alexander, ed.), Lecture Notes in mathematics, 1342, 1988.
6. A. Fathi, M. Herman and J.-C. Yoccoz, *A proof of Pesin's stable manifold theorem*, Dynamical Systems (J. Palis, ed.), Lecture Notes in mathematics, 1007, 1983.
7. A. Katok, *Lyapunov exponents, entropy and periodic orbits for diffeomorphisms*, Publ. Math. (IHES) **51** (1980), 137-173.
8. A. Katok, *Entropy and closed geodesics*, Ergod. Th. and Dynam. Sys. **2** (1982), 339-365.
9. R. Mañé, *On the Topological Entropy of Geodesic Flows*, Preprint.
10. A. Manning, *Topological entropy for geodesic flows*, Annals of Math. **110** (1979), 567-573.
11. G. Margulis, *On some applications of ergodic theory to the study of manifolds of negative curvature*, Fun. Anal. Appll **3** (1969), 89 - 90.
12. S. Newhouse, *Entropy and volume*, Ergod. Th. and Dynam. Sys. **8** (1989), 283 - 299.
13. S. Newhouse, *Continuity properties of entropy*, Annals of Math. **129** (1989), 215 - 237.
14. V. Oseldec, *A multiplicative ergodic theorem*, Trans. Moscow Math. Soc. **19** (1968), 197 - 231.
15. W. Parry and M. Pollicott, *An analogue of the prime number theorem for closed orbits of Axiom A flows*, Annals of Math. **118** (1983), 573 - 591.
16. G. Paternain and M. Paternain, *Topological entropy versus geodesic entropy*, Internat. J. Math. **5** (1994), 213 - 218.
17. Y.Pesin, *Lyapunov characteristic exponents and ergodic properties of smooth dynamical systems with an invariant measure*, Soviet Math. Dok. **17** (1976), 196 - 199.
18. M. Pollicott, *A new proof of a theorem of Margulis on geodesic ars on negatively curved manifolds*, Amer. J. Math. **117** (1995), 289 - 305.
19. C. Pugh and M.Shub, *Elements of ergodic actions*, Comp. Math. **23** (1971), 115-122.
20. D.Ruelle, *An inequality for the entropy of differentiable maps*, Bol. Soc. Bras. Mat. **9** (1978), 83-87.

MARK POLLICOTT, DEPARTMENT OF MATHEMATICS, UNIVERSITY OF MANCHESTER, OXFORD ROAD, MANCHESTER, CV4 7AL, ENGLAND

FELIKS PRZYTYCKI[*]

On measure and Hausdorff dimension of Julia sets for holomorphic Collet–Eckmann maps

Dedicated to the memory of Ricardo Mañé

Introduction.

It is well-known that if $f : \bar{\mathbb{C}} \to \bar{\mathbb{C}}$ a rational map of the Riemann sphere is hyperbolic, i.e. expanding on its Julia set $J = J(f)$ namely $|(f^n)'| > 1$ for an integer $n > 0$, then its Hausdorff dimension $\mathrm{HD}(J) < 2$.

The same holds for a more general class of "subexpanding" maps, namely those maps, whose all critical points in $J(f)$ are non-reccurrent, supposed $J(f) \neq \bar{\mathbb{C}}$, see [U] (periodic parabolic points are allowed).

On the other hand there is an abundance of rational maps with $J \neq \bar{\mathbb{C}}$ and $\mathrm{HD}(J) = 2$, [Shi].

Recently Chris Bishop and Peter Jones proved that for every finitely generated not geometrically finite Kleinian groups for the Poincaré limit set Λ one has $\mathrm{HD}(\Lambda) = 2$. As geometrically finite exhibits some analogy to subexpanding in the "Kleinian Groups – Rational Maps" dictionary, the question arised, expressed by Ch. Bishop and M. Lyubich at the MSRI Berkeley conference in January 1995, isn't it true for every non-subexpanding rational map with connected Julia set, that $HD(J) = 2$?

Here we give a negative answer. For a large class of "non-uniformly" hyperbolic so called Collet-Eckmann maps, studied in [P1], satisfying an additional Tsujii condition, $\mathrm{HD}(J) < 2$.

---

* The author acknowledges support by Polish KBN Grant 2 P301 01307 "Iteracje i Fraktale, II". He expresses also his gratitude to the MSRI at Berkeley (partial support by NSF grant DMS-9022140) and ICTP at Trieste, where parts of this paper were written.

167

Notation. For a rational map $f : \bar{\mathbb{C}} \to \bar{\mathbb{C}}$ denote by $\mathrm{Crit}(f)$ the set of all critical points of f, i.e. points where $f' = 0$. Let $\nu := \sup\{$ multiplicity of f^n at $c : c \in \mathrm{Crit}(f) \cap J\}$. Finally denote by $\mathrm{Crit}'(f)$ the set of all critical points of f in $J(f)$ whose forward trajectories do not contain other critical points.

We prove in this paper the following:

Theorem A. Let f be a rational map on the Riemann sphere $f : \bar{\mathbb{C}} \to \bar{\mathbb{C}}$, and suppose there exist $\lambda > 1, C > 0$ such that for every f-critical point $c \in \mathrm{Crit}'(f)$

$$|(f^n)'(f(c))| \geq C\lambda^n, \tag{0.1}$$

there are no parabolic periodic points, and $J(f) \neq \bar{\mathbb{C}}$.

Then $\mathrm{Vol}(J(f) = 0$, where Vol denotes Riemann measure on $\bar{\mathbb{C}}$.

Theorem B. Under the conditions of Theorem A, assume additionally that

$$\lim_{t \to \infty} \limsup_{n \to \infty} \frac{1}{n} \sum_{j=1}^{n} \max(0, -\log(\mathrm{dist}(f^j(c), \mathrm{Crit}(f))) - t) = 0. \tag{0.2}$$

Then $\mathrm{HD}(J(f)) < 2$.

For $f(z) = z^2 + c$, $c \in [-2, 0]$ real, it is proved in [T] that (0.1) and (0.2) are satisfied for a positive measure set of parameters c for which there is no sink in the interval $[c, c^2 + c]$. Tsujii's condition in [T], called there *weak regularity*, is in fact apparently stronger than (0.2). The set of subexpanding maps satisfying (0.1) and weak regularity has measure 0, [T]. Thus Theorem B answers Bishop-Lyubich's question.

Remark. In [DPU] it is proved that for every rational map $f : \bar{\mathbb{C}} \to \bar{\mathbb{C}}$, $c \in \mathrm{Crit}'$

$$\limsup_{n \to \infty} \frac{1}{n} \sum_{j=1}^{n} -\log \mathrm{dist}(f^j(c), \mathrm{Crit}(f)) \leq C_f$$

where C_f depends only on f. Here in the condition (0.2) it is sufficient, for Theorem B to hold, to have a positive constant instead of 0 on the right hand side, unfortunately apparently much smaller than C_f.

Crucial in proving Theorems A and B is the following

Theorem 0.1 (on the existence of pacim), see [P1]. Let $f : \overline{\mathbb{C}} \to \overline{\mathbb{C}}$ satisfy the assumptions of Theorem A. Let μ be an α-conformal measure on the Julia set $J = J(f)$ for an arbitrary $\alpha > 0$. Assume[1]

$$\mu \text{ has no atoms at critical points of } f \tag{0.3}$$

Assume also that there exists $1 < \lambda' < \lambda$ such that for every $n \geq 1$ and every $c \in \mathrm{Crit}'(f)$

$$\int \frac{d\mu}{\mathrm{dist}(x, f^n(c))^{(1-1/\nu)\alpha}} < C^{-1}(\lambda')^{\alpha n/\nu}. \tag{0.4}$$

Then there exists an f-invariant probability measure m on J absolutely continuous with respect to μ (pacim).

Recall that a probability measure μ on J is called α-conformal if for every Borel $B \subset J$ on which f is injective $\mu(f(B)) = \int_B |f'|^\alpha d\mu$. In particular $|f'|^\alpha$ is Jacobian for f and μ. (A function φ such that $\mu(f(B)) = \int_B \varphi d\mu$ for every B as above is called Jacobian.) The number α is called the exponent of the conformal measure.

If $\mathrm{Vol}(J) > 0$ then the restriction of Vol to J, normalized, is 2-conformal and obviously satisfies (0.3) and (0.4). If $\mathrm{HD}(J) = 2$ then by [P1] we know there exists a 2-conformal measure μ on J but we do not know whether it is not too singular, namely whether it satisfies (0.3) and (0.4). Fortunately for every f satisfying the assumptions of Theorem A and additionally the assumption (0.2) we can prove that (0.4) holds indeed for every α-conformal measure and we can construct a $\mathrm{HD}(J)$-conformal measure satisfying (0.3) repeating the construction from [DU].

Gathering together the results which we prove along the paper and refering to [P1] we obtain the following extension of Theorem B:

---

[1] In the Dijon preprint version of [P1] this assumption is missing. I thank J. Graczyk for pointing me out this error.

Theorem C. For every rational map $f : \overline{\mathbb{C}} \to \overline{\mathbb{C}}$ satisfying Collet-Eckmann condition (0.1), Tsujii condition (0.2), having no parabolic periodic points and with Julia set J not the whole sphere the following holds: $\mathrm{HD}(J) = \mathrm{Cap}(J) < 2$, there exists a $\mathrm{HD}(J)$-conformal probability measure on J not having atoms at critical points and there exists a probability f-invariant measure m absolutely continuous with respect to μ and such that $dm/d\mu > \mathrm{Const} > 0$. The measure m is ergodic, of positive entropy, and has positive Lyapunov exponent.

(We write $\mathrm{Cap}(J)$ for Minkowski dimension. Other names: box dimension, limit capacity.)

Notation. Const will denote various positive constants which may change from one formula to another, even in one string of estimates.

Section 1. More on pacim. Proof of Theorem A.

Proposition 1.1. In the situation of Theorem 0.1 there exists $K > 0$ such that μ-a.e. $\frac{dm}{d\mu} \geq K$.

Proof. In Proof of Theorem 0.1 [P1] one obtains m as a weak* limit of a subsequence of the sequence of measures $\frac{1}{n} \sum_{j=0}^{n-1} f_*^j(\mu)$.

It is sufficient to prove that there exists $K > 0$ and $n_0 > 0$ such that for μ-a.e. $y \in J(f)$

$$\frac{df_*^n(\mu)}{d\mu}(y) = \mathcal{L}^n(\mathbb{1}) \geq K. \tag{1.1}$$

Here $\mathcal{L}$ denotes the transfer operator, which can be defined for example by $\mathcal{L}(\varphi)(y) = \sum_{f(z)=y} |f'(z)|^{-\alpha} \varphi(z)$. $\mathbb{1}$ is the constant function of value 1. We can assume $y \notin \bigcup_{n>0} f^n(\mathrm{Crit}(f))$ because

$$\mu\left(\bigcup_{n>0} f^n(\mathrm{Crit}(f))\right) = 0 \tag{1.2}$$

If a critical value for f^n were an atom then a critical point would have μ measure equal to ∞.

170

(The equality in (1.1) follows from the definition of $\mathcal{L}$. However pay attention that it makes use of the assumption (0.4) if one considers $\mathcal{L}\mathbb{1}$ as a classical function.)

It is sufficient to prove the inequality (1.1) for $y \in B(x,\delta) \cap J(f)$ for an *a priori* chosen x and an arbitrarily small δ and next to use the fact that there exists $m \geq 0$ such that $f^m(B(x,\delta)) \supset J(f)$ (called *topological exactness*). Indeed

$$\mathcal{L}^n(\mathbb{1})(w) = \sum_{f^m(y)=w} \mathcal{L}^{n-m}(\mathbb{1})(y)|(f^m)'|^{-\alpha} \geq (\sup|(f^m)'|)^{-\alpha}\mathcal{L}^{n-m}(y_0)$$

where $y_0 \in f^{-m}(\{w\}) \cap B(x,\delta)$.

Recall the estimate from [P1]. For an arbitrary $\gamma > 1$ there exists $C > 0$ such that for every $x \in J(f)$

$$\mathcal{L}^n(\mathbb{1})(x) \leq C + C \sum_{c\in\mathrm{Crit}(f)\cap J} \sum_{j=0}^{\infty} \frac{\gamma^j \lambda^{-j\alpha/\nu}}{\mathrm{dist}(x, f^j(f(c)))^{(1-1/\nu)\alpha}}. \tag{1.3}$$

By the assumptions (0.1) and (0.3) the above function is μ-integrable if γ is small enough.

Pay attention to the assumption (0.3). It concerns only $c \in \mathrm{Crit}'$. Fortunately there is only a finite number of summands in (1.3) for which $f^{j_0}(c) \in \mathrm{Crit}, j_0 \geq j$. Each summand is integrable because up to a constant it is bounded by $\mathcal{L}^j(\mathbb{1})$.

So

$$\sum_{c\in\mathrm{Crit}(f)\cap J} \sum_{j=s}^{\infty} \frac{\gamma^j \lambda^{-j\alpha/\nu}}{\mathrm{dist}(x, f^j(f(c)))^{(1-1/\nu)\alpha}} \to 0 \quad \mu - \text{a.e. as } s \to \infty. \tag{1.4}$$

Fix from now on an arbitrary $x \in J(f)$ for which (1.4) holds, $(dm/d\mu)(x) \geq 1$ and $x \notin \bigcup_{n>0} \varphi^n(\mathrm{Crit}(f))$ (possible by (1.2) and by $\int (dm/d\mu)d\mu = 1$).

We need now to repeat from [P1] a part of the Proof of Theorem 0.1:
For every $y \in B(x,\delta)$ and $n > 0$

$$\mathcal{L}^n(\mathbb{1})(y) = \sum_{y'\in f^{-n}(y),\text{regular}} |(f^n)'(y')|^{-\alpha} + \sum_{(y',s)\text{singular}} \mathcal{L}^{n-s}(\mathbb{1})(y')|(f^s)'(y')|^{-\alpha}$$

$$= \sum_{\text{reg},y} + \sum_{\text{sing},y}. \tag{1.5}$$

171

We shall recall the definitions of *regular* and *singular*: Take an arbitrary subexponentially decreasing sequence of positive numbers $b_j, j = 1, ...$ with $\sum b_j = 1/100$. Denote by $B_{[k}$ the disc $B(x, (\prod_{j=1}^{k}(1-b_j))2\delta)$. We call s the *essentially critical time* for a sequence of compatible components $W_j = \text{Comp} f^{-j}(B_{[j})$, where compatible means $f(W_j) \subset W_{j-1}$, if there exists a critical point $c \in W_s$ such that $f^s(c) \in B_{[s}$.

We call y' *regular* in (1.5) if for the sequence of compatible components $W_s, s = 0, 1, ..., n, W_n \ni y'$ no $s < n$ is essentially critical.

We call a pair (y', s) *singular* if $f^s(y') = y$ and for the sequence of compatible components $W_j, j = 0, 1, ..., s, W_s \ni y'$ the integer s is the first (i.e. the only) essentially critical time.

If δ is small enough then all s in $\sum_{\text{sing},x}$ are sufficiently large that $\sum_{\text{sing},x} \leq 1/2$. This follows from the estimates in [P1, Sec.4]; here is the idea of the proof: Transforming $\sum_{\text{sing},x}$ in (1.5) using the induction hypothesis (1.3) we obtain the summands

$$C \frac{\gamma^j \lambda^{-j\alpha/\nu}}{\text{dist}(x, f^{s+j-1}(f(c)))^{(1-1/\nu)\alpha}}, \quad j = 0, ..., n-s$$

multiplied by

$$\text{Const}|(f^{s-1})'(x')|^{-\alpha/\nu} a_s < \gamma^{s-1} \lambda^{-(s-1)\alpha/\nu}.$$

The numbers a_s are constants arising from distortion estimates, related to b_s. The numbers γ^s swallow them and other constants.

(There is a minor inaccuracy here: (s, x') is a singular pair where the summand appears, provided the captured critical point c is not in the forward trajectory of another critical point, otherwise one moves back to it, see [P1] for details.)

Now $\sum_{\text{sing},x} \leq 1/2$ follows from (1.4).

The result is that $\sum_{\text{reg},x} \geq 1/2$. So by the uniformly bounded distortion along regular branches of f^{-n} on $B(x, \delta)$ we obtain

$$\mathcal{L}^n(\mathbb{1})(y) \geq \sum_{\text{reg},y} \geq \text{Const} \sum_{\text{reg},x} \geq \text{Const} > 0$$

The name *regular* concerned formally $y' \in f^{-n}(y)$ but in fact it concerns the branch of f^{-n} mapping y to y' not depending on $y \in B(x, \delta)$.

172

By distortion of any branch g of f^{-n} on a set U we mean
$\sup_{z \in B} |g'(z)| / \inf_{z \in B} |g'(z)|$.
Proposition 1.1 has been proved. ♣

There exists a decomposition of J in a finite number of ergodic components $E_1, ..., E_k$, see [P1, Theorem B]. Denote the measure m restricted to E_i and normalized, by m_i. Taking this into account we obtain

Corollary 1.2 In the situation of Theorem 0.1 for measure-theoretic entropy $h_{m_i}(f) > 0$, for every $i = 1, ..., k$.

Proof. Denote $dm/d\mu$ by u.

Consider an open set $U \subset \bar{\mathbb{C}}$ intersecting $J(f)$ such that there exist two branches g_1 and g_2 of f^{-1} on it. Then by the f-invariance of m we have $\mathrm{Jac}_m(g_1) + \mathrm{Jac}_m(g_2) \leq 1$ ($= 1$ if we considered all branches of f^{-1}). $\mathrm{Jac}_m(g_i)$ means Jacobian with respect to m for g_i.

We have $m(U) > 0$ because μ does not vanish on open sets in J (by the topological exactness of f on J) and by Proposition 1.2. At m-a.e. $x \in U$

$$\mathrm{Jac}_m(g_i)(x) = u(g_i(x))|g_i'(x)|u(x)^{-1} > 0,$$

(here we also used (1.4)).

Hence $\mathrm{Jac}_m(g_i) < 1$, so $\mathrm{Jac}_m(f) > 1$ on the set $g_i(U)$, $i = 1, 2$. Now we use Rochlin's formula and obtain

$$h_{m_i}(f) = \int_{E_i} \log \mathrm{Jac}_m(f) dm_i > 0$$

♣

Let $\chi_{m_i} = \int \log |f'| dm_i$ denote the Lyapunov characteristic exponent on E_i.

Corollary 1.3 In the situation of Theorem 0.1, $\chi_{m_i} > 0$ for every $i = 1, ..., k$.

Proof. This Corollary follows from Ruelle's inequality $h_{m_i}(f) < 2\chi_{m_i}$, see [R].

Proof of Theorem A. Suppose $\mathrm{Vol}(J(f)) > 0$. After normalization we obtain a 2-conformal measure μ on $J(f)$ and by Theorem 0.1 and Corollary 1.3 a pacim m with $\chi_m > 0$. By Pesin's Theory [Pesin] in the iteration in the dimension 1 case [Le] ([Le] is on the real case, but the complex one is similar), for m-a.e. x, there exists a sequence of integers $n_j \to \infty$ and $r > 0$ such that for every j there exists a univalent branch g_j of f^{-n_j} on $B_j := B(f^{n_j}(x), r)$ mapping $f^{n_j}(x)$ to x and g_j has distortion bounded by a uniform constant. By $\chi_m > 0$ $\mathrm{diam} g_j(B(f^{n_j}(x), r)) \to 0$. (This follows also automatically from the previous assertions by the definition of Julia set [GPS].) Now we can forget about the invariant measure m and go back to Vol. Because $J(f)$ is nowhere dense in $\overline{\mathbb{C}}$, there exists $\varepsilon > 0$ such that for every $z \in J(f)$

$$\frac{\mathrm{Vol}(B(z,r) \setminus J(f))}{\mathrm{Vol}(B(z,r))} > \varepsilon.$$

Bounded distortion for g_j on $B(z,r)$, $z = f^{n_j}(x)$ allows to deduce that the same part of each small disc$\approx g_j(B_j)$ around x is outside $J(f)$, up to multiplication by a constant. This is so because we can write for every $X \subset B(z,r), y \in B(z,r)$

$$\mathrm{Vol}(g_j(X)) \approx |g_j'(y)|^2 \mathrm{Vol}(X) \tag{1.6}$$

where $\approx$ means up to the multiplication by a uniformly bounded factor. So x is not a density point of $J(f)$. On the other hand a.e. point is a density point. So $\mathrm{Vol} J(f) = 0$ and we arrived at a contradiction. ♣

Section 2. Proof of Theorem B.

Definition. We call a probability measure μ on J α-subconformal if the equality in the definition of α-conformal measure (see Introduction) is replaced by the inequality: $\mu(f(B)) \geq \int_B |f'|^\alpha d\mu$.

Lemma 2.1. Suppose f satisfies the assumptions of Theorem B. Then for every $, \sigma > 0$ there exists $C_1 > 0$ such that for every $c \in \mathrm{Crit}'$ and $n_0 > 0$ there exists a sequence $r_j, j = 1, 2, \ldots$ satisfying

$$r_1 > C_1 \exp -n_0 \tag{2.1}$$

174

$$r_{j+1} > r_j^{1+\sigma} \tag{2.2}$$

$$r_{j+1} < r_j/2. \tag{2.3}$$

and

$$\mu(B(f^{n_0}(c), r_j)) \leq C_1 r_j^\alpha \tag{2.4}$$

for every $\alpha \leq 2$ and every α-subconformal measure μ.

Proof. Step 1. Denote the expression from (0.2)

$$\max\left(0, -\log \inf_{c \in \mathrm{Crit}'(f)} \mathrm{dist}(f^n(c), \mathrm{Crit}(f)) - t\right)$$

by $\varphi_t(n)$. Consider the following union of open-closed intervals

$$A'_t := \bigcup_n (n, n + \varphi_t(n) \cdot K_f] \quad \text{and write} \quad A_t := \mathbb{Z}_+ \setminus A'_t ,$$

for an arbitrary constant $K_f > \nu/\log\lambda$ (in the convention that if $\varphi_t(n) = 0$, then the interval in the union is empty).

By (0.2) for every $a > 0$ there exist $t > 0$ and $n(a, t)$ such that for every $n \geq n(a, t)$

$$A_t \cap [n, n(1 + a)] \neq \emptyset \tag{2.5}$$

Moreover, fixing an arbitrary integer $M > 0$, we can guarantee for every $n' \geq n(1 + a)$, $n \geq n(a, t)$

$$\sharp(A_t \cap \{j \in [n, n'] : j \text{ divisible by } M\}) \geq \frac{1}{2M}(n' - n). \tag{2.6}$$

Observe that for every a, n_0, n

$$[n_0 + n, n_0 + n + a(n_0 + n)] = [n_0 + n, n_0 + n + a(\frac{n_0}{n} + 1)n].$$

So if $n \geq bn_0$ for an arbitrary $b > 0$ and $n_0 \geq n(a, t)$, then (2.5) yields

$$A_t \cap [n_0 + n, n_0 + n + a(b^{-1} + 1)n] \neq \emptyset. \tag{2.7}$$

Denote in the sequel $a(b^{-1} + 1)$ by a'.

Step 2. Observe now that if $n \in A_t$ then for every $c \in \text{Crit}'(f)$ there exist branches $g_s, s = 1, 2, \dots n - 1$ of f^{-s} on $B_n := B(f^n(c), \delta)$ such that $g_s(f^n(c)) = f^{n-s}(c)$, distortions bounded by a uniform constant C_2 (i.e. $\sup |g'_s| / \inf |g'_s| \leq C_2$), where $\delta = \varepsilon \exp -t\nu$, for a constant ε small enough. Sometimes to exhibit the dependence on n we shall write $g_{s,n}$.

Indeed, define g_s on $B_{[s} = B(f^n(c), \prod_{j=1}^{s}(1 - b_j)2\delta)$ for $s = 1, 2, \dots n-1$ according to the procedure described in the Proof of Proposition 1.1. If there is an obstruction, namely s an essential critical time, then for every $z \in B_{[s}$

$$|g'_{s-1}(z)| \leq \lambda^{-s}\vartheta^s \leq \exp(-s\nu/K_f) \tag{2.8}$$

for $\vartheta > 1$ arbitrarily close to 1 (in particular such that $K_f > \frac{\nu}{\log \lambda - \log \vartheta}$) and for s large enough. The constant ϑ takes care of distortion. (2.8) holds for $z = f^s(q)$, where q is the critical point making s a critical time, without ϑ by (0.1) (with the constant C instead). The small number ε takes care of s small, which cannot then be essential critical.

The inequality (2.4) and rooting ($1/\nu$ to pass from $s - 1$ to s) imply $\varphi_t(f^{n-s}(c)) \geq s/K_f$, so $n \notin A_t$, a contradiction.

Step 3. We find r_j satisfying the assertions of the Lemma by taking

$$r_j := \frac{1}{2C_2} \text{diam } g_{n_j, n_0 + n_j}(B(f^{n_0 + n_j}(c), \delta))$$

where n_j are taken consecutively so that $n_0 + n_j \in A_t$ and

$$n_{j+1} \in [(1 + \vartheta)n_j, (1 + \vartheta)n_j(1 + a')] \quad \text{for} \quad j \geq 2 \quad \text{and}$$

$$n_1 \in [bn_0, bn_0 + a'bn_0],$$

where $\vartheta > 0$ is an arbitrary constant close to 0. This is possible by (2.7).

This gives for say $\vartheta < a' < 1$

$$r_{j+1}/r_j \geq C_2^{-1} \exp(-3(\log L)a'n_j), \tag{2.9}$$

where $L := \sup |f'|$. One obtains this in 2 steps: first by the branch $g_{n_{j+1} - n_j, n_0 + n_{j+1}}$, next by $g_{n_j, n_0 + n_j}$ which shrinks the ratio by at most C_2^{-1}. In the same way by acting only by $g_{n_1, n_0 + n_1}$ one obtains (2.1).

To conclude we need to know that r_j shrink exponentially fast with $n_j \to \infty$, uniformly on n_0. For that we need the following fact (see for example [GPS], find the analogous fact in the Proof of Theorem A):

(*) For every $r > 0$ small enough and $\xi, C > 0$ there exists m_0 such that for every $m \geq m_0, x \in J(f)$ and a branch g of f^{-m} on $B(x,r)$ having distortion less than C, we have diam $g(B(x,r)) < \xi r$.

Apply now (2.6) to $n = n_0, n' = n_j + n_0$. We obtain a "telescope": For all consecutive $\tau_1, \tau_2, ... \tau_{k(j)} \in A_t \cap [n_0, n_j + n_0]$ divisible by M

$$g_{\tau_{i+1}-\tau_i, \tau_{i+1}}(B(f^{\tau_{i+1}}(c), \delta)) \subset B(f^{\tau_i}(c), \delta/2C_2)$$

for $M \geq m_0$ from (*).

Hence using (2.6)

$$r_j \leq 2^{-n_j/2M} \ . \tag{2.10}$$

The property (2.3) follows from the fact that for n_0 large enough, for every $j > 0$, $n_{j+1} - n_j \geq M$ and the argument the same as for $\tau_{i+1} - \tau_i$ above is valid.

Denote $2a' \log L$ by γ and $(\log 2)/2M$ by γ'. (2.9) and (2.10) give

$$r_{j+1} \geq C_2^{-1} r_j \exp -\gamma n_j \geq C_2^{-1} r_j (\exp -\gamma' n_j)^{\gamma/\gamma'} \geq C_2^{-1} r_j^{1+\gamma/\gamma'}.$$

As γ' is a constant and γ can be made arbitrarily small if a is small enough, we obtain (2.2). C_2^{-1} disappears when we double γ/γ' for δ small enough.

Finally we obtain $\mu(B(f^{n_0}(c), r_j)) \leq \mu g_{n_j, n_0+n_j}(B(f^{n_0+n_j}(c), \delta)) \leq C_2^\alpha \delta^{-\alpha} r_j^\alpha$ what proves (2.4).

We have proved the Lemma for every n_0 large enough. Now by pulling back one easily provees it for every $n_0 > 0$. ♣

Remark 2.2. The only result at our disposal on the abundance of non-subexpanding maps satisfying (0.1) and (0.2) is Tsujii's one concerning $z^2 + c$, c real (see the Introduction). For this class however the exponential convergence of diamComp$f^{-n_j}(B(f^{n_j+n_0}(0), \delta)$ to 0 follows from [N] (the component containing $f^{n_0}(c)$). So restricting our interests to this class we could skip (2.6) and the considerations leading to (2.10) above.

By [N] $\operatorname{diam}\left(\operatorname{Comp}(f^{-n}(B(x,\delta)))\cap I\!\!R\right) < C\tilde{\lambda}^{-n}$ for some constants $C > 0, \tilde{\lambda} > 1, \delta$ small enough and every component Comp. Just the uniform convergence of the diameters to 0 as $n \to \infty$ follows from [P1], but I do not know how fast it is.

Lemma 2.3 Under the assumptions of Theorem B, for every $\lambda' > 1$ there exists $C > 0$ such that for every $\alpha \leq 2$ and α-conformal measure μ the estimate (0.4) holds.

Proof. By Lemma 2.1 we obtain

$$\int \frac{d\mu}{\operatorname{dist}(x, f^{n_0}(c))^{(1-1/\nu)\alpha}}$$

$$\leq \mu(\bar{\mathbb{C}} \setminus B(f^{n_0}(c), r_1)) \frac{1}{r_1^{(1-1/\nu)\alpha}} + \sum_{j\geq 2} \mu(B(f^{n_0}(c), r_{j-1}) \setminus B(f^{n_0}(c), r_j)) \frac{1}{r_j^{(1-1/\nu)\alpha}}$$

$$\leq \operatorname{Const} \exp(n_0(1-1/\nu)\alpha) + \operatorname{Const} \sum_{j\geq 2} \frac{r_{j-1}^\alpha}{r_j^{(1-1/\nu)\alpha}}$$

$$\leq (\exp((1-1/\nu)\alpha))^{n_0} + \operatorname{Const} \sum_{j\geq 2} r_{j-1}^\alpha r_{j-1}^{-(1-1/\nu)\alpha(1+\sigma)}.$$

The latter series has summands decreasing exponentially fast for σ small enough so it sums up to a constant, hence the first summand dominates. We obtain the bound by $(\lambda')^{n_0}$ with $\lambda' > 1$ arbitrarily close to 1. Thus (0.4) has been proved. ♣

For an arbitrary rational map f restricted to a forward invariant set $K \subset J$ we write $\operatorname{HD}_{\mathrm{ess}}(K)$ for the essential Hausdorff dimension, which can be defined for example as the supremum of the Hausdorff dimension of all expanding isolated Cantor sets in K. (We say that an f-invariant set X is isolated if every forward f-trajectory which starts in a sufficiently small neighbourhood U of X either is contained in X or escapes from U.) There always exists an α-conformal measure with the exponent $\alpha = \operatorname{HD}_{\mathrm{ess}}(J)$, this is the minimal possible exponent for conformal measures, see [DU] [P2] and [PUbook]. If f satisfies (0.1) then $\operatorname{HD}_{\mathrm{ess}}(J) = \operatorname{HD}(J)$, see [P1].

In our situation we can say more:

Lemma 2.4. If f satisfies the assumptions of Theorem B, then there exists an α-conformal measure with $\alpha = \operatorname{HD}_{\mathrm{ess}}(J) = \operatorname{HD}(J)$ which does not have atoms at f-critical points.

178

Proof. We repeat the construction from [DU]. Consider for every $n = 1, 2, \dots$ the set $V_n = B(\text{Crit}', \frac{1}{n})$ and construct μ_n a subconformal measure on $K(V_n) = J \setminus \bigcup_{k \geq 0} f^{-k}(V_n)$ as in [DU, Lemma 5.1].

Here the situation is easier than in [DU] because f on $K(V_n)$ is expanding, [P1, Sec.3]. So each μ_n is α_n-subconformal (α_n-conformal on sets disjoint with $\text{cl} V_n$), with $\alpha_n = \text{HD}_{\text{ess}}(K(V_n))$, $\alpha_n \nearrow \alpha$ and $\mu_n \to \mu$ which is an α-conformal measure.

(In [DU] one obtains each μ_n with $\frac{df_* \mu_n}{d\mu_n} \geq e^{c_n} |f'|^{\alpha_n}$ with $c_n \searrow 0$. Here $c_n = 0$. Also μ in [DU] can have an atom at a critical value. Here, due to (0.1) and the subconformality, this is automatically excluded, otherwise the measure of the forward trajectory of the critical value would be infinite.)

By Lemma 2.1 for every $c \in \text{Crit}'$ we have $\mu_n(B(f(c), r_j) \leq C_1 r_j^{\alpha_n}))$. So

$$\mu_n(\text{Comp} f^{-1}(B(f(c), r_j) \setminus B(f(c), r_{j+3}))$$

$$\leq \text{Const } r_{j+3}^{(1/\nu(c)-1)\alpha_n} \mu_n(B(f(c), r_j) \setminus B(f(c), r_{j+3}))$$

$$\leq \text{Const} C_1 r_{j+3}^{(1/\nu(c)-1)\alpha_n} r_j^{\alpha_n} = \text{Const} C_1 r_{j+3}^{(1/\nu(c)-3\sigma)\alpha_n}.$$

again using Lemma 2.1. $\nu(c)$ is the multiplicity of f at the critical point c. Comp means the component close to c. $\sigma \approx 0$. It is crucial that the estimate is uniform on n.

Thus one obtains

$$\mu(\text{Comp} f^{-1}(B(f(c), r_{j+1}) \setminus B(f(c), r_{j+2}))) \leq \text{Const } r_{j+3}^{(1/\nu(c)-3\sigma)\alpha} \to 0 \quad (2.11)$$

as $j \to \infty$. (We passed from $j, j+3$ to $j+1, j+2$ to cope with the case $\mu \partial(\text{Comp} f^{-1}(B(f(c), r_{j+1}) \setminus B(f(c), r_{j+2}))) > 0$. Remember that to conclude $\lim \mu_n B = \mu B$ one assumes $\mu(\partial B) = 0$.)

Similarly by further pulling back one obtains (2.11) around critical points in $J \setminus \text{Crit}'$. Finally by the construction μ_n have no atoms at critical points, because the topological supports of μ_n's do not contain critical points. The Lemma has been proved. ♣

Proof of Theorems B and C. By Lemma 2.4 there exist a $\text{HD}(J)$-conformal measure μ on J satisfying (0.3). By Lemma 2.3 μ satisfies also (0.4). Hence by Theorem 0.1 there exists a pacim $m \ll \mu$. Moreover $\chi_m > 0$ by Corollary 1.3. As in the Proof of Theorem A, by Pesin Theory there exists $X \subset J$, $m(X) = \mu(X) = 1$, such that for every $x \in X$ there exists a sequence of integers $n_j(x) \to \infty$, $r > 0$ and univalent branches g_j of f^{-n_j} on $B(f^{n_j}(x), r)$ mapping f^{n_j} to x with uniformly bounded distortion. Write $B_{x,j} := g_j(B(f^{n_j}(x), r))$.

Suppose now that $\mathrm{HD}(J) = 2$. We obtain for every $x \in X$ by applying (1.6) to Vol and μ (similarly as in the Proof of Theorem A)

$$\mu(B_{x,j}) \leq \mathrm{Const}\,\mathrm{Vol}(B_{x,j}) \leq \mathrm{Const}\ \mathrm{Vol}(B(x, \mathrm{diam}\,B_{x,j})).$$

If $\mathrm{Vol}X = 0$ then there exists a covering of X by discs $B(x_t, \mathrm{diam}\,B_{x_t,j_t})$, $t = 1, 2, \ldots$ whose union has $\mathrm{Vol} < \varepsilon$ for ε arbitrarily close to 0, of multiplicity less than a universal constant (Besicovitch's theorem). Hence

$$\varepsilon \geq \mathrm{Const} \sum_t \mathrm{Vol}B(x_t, \mathrm{diam}\,B_{x_t,j_t}) \geq \mathrm{Const}\mu\Big(\sum_t B_{x_t,j_t}\Big) \geq 1,$$

a contradiction. Hence $\mathrm{Vol}J \geq \mathrm{Vol}X > 0$.

This contradicts Theorem A that $\mathrm{Vol}J = 0$ and the proof of Theorem B is over.

Remark that we could end the proof directly: As in the Proof of Theorem A we show that no point of X is a point of density of the Vol measure. Hence $\mathrm{Vol}X = 0$. (I owe this remark to M. Urbański.)

To finish the proof of Theorem C it remains only to check the ergodicity. However the ergodicity follows easily from [P1, Sec.3], passing (acting by iterates of f) from a neighbourhood of a.e. point x to a neighbourhood of a critical point, and from the Proof of Lemma 2.1. Briefly: the existence of the branches $g_{n-1,n}$ for a growing sequence of n's yields for every invariant set A with $m(A) > 0$ and every $c \in \mathrm{Crit}$, the existence of $r_n \to 0$ such that $\frac{m(B(c,r_n) \cap A)}{m(B(c,r_n))} \geq \mathrm{Const} > 0$. So x cannot be a point of density of $J \setminus A$. If $m(J \setminus A) > 0$ then similarly x cannot be a point of density of A. This can happen only for a set of x's of measure 0. A contradiction. Theorem C is has been proved. ♣

References.

[BJ] Ch. Bishop, P. Jones: Hausdorff dimension and Kleinian groups. Preprint SUNY at Stony Brook, IMS 1994/5.

[DPU] M. Denker, F. Przytycki, M. Urbański: On the transfer operator for rational functions on the Riemann sphere. Preprint SFB 170 Göttingen, 4 (1994). To appear in Ergodic Th. and Dyn. Sys..

[DU] M. Denker, M. Urbański: On Sullivan's conformal measures for rational maps of the Riemann sphere. Nonlinearity 4 (1991), 365-384.

[GPS] P. Grzegorczyk, F. Przytycki, W. Szlenk: On iterations of Misiurewicz's rational maps on the Riemann sphere. Ann. Inst. H. Poincaré, Phys. Théor. 53 (1990), 431-444.

[Le] F. Ledrappier: Some properties of absolutely continuous invariant measures on an interval. Ergod. Th. & Dynam. Sys. 1 (1981), 77-93.

[N] T. Nowicki: A positive Liapunov exponent for the critical value of an S-unimodal mapping implies niform hyperbolicity. Ergodic Th. & Dynamic. Sys. 8 (1988), 425-435.

[Pesin] Ya. B. Pesin: Characteristic Lyapunov exponents and smooth ergodic theory. Russ. Math. Surv. 32 (1977), 45-114.

[P1] F. Przytycki: Iterations of holomorphic Collet-Eckmann maps: conformal and invariant measures. Preprint n° 57, Lab. Top. Université de Bourgogne, Février 1995.

[P2] F. Przytycki: Lyapunov characteristic exponents are non-negative. Proc. Amer. Math. Soc. 119(1) (1993), 309-317.

[PUbook] F. Przytycki, M. Urbański: To appear.

[R] D. Ruelle: An inequality for the entropy of differentiable maps. Bol. Soc. Bras. Mat. 9 (1978), 83-87.

[Shi] M. Shishikura: The Hausdorff dimension of the boundary of the Mandelbrot set and Julia set. Preprint SUNY at Stony Brook, IMS 1991/7.

[T] M. Tsujii: Positive Lyapunov exponents in families of one dimensional dynamical systems. Invent. Math. 111 (1993), 113-137.

[U] M. Urbański: Rational functions with no recurrent critical points. Ergodic Th. and Dyn. Sys. 14.2 (1994), 391-414.

Permanent address: Institute of Mathematics, Polish Academy of Sciences, ul. Śniadeckich 8, 00 950 Warszawa, Poland. e-mail: feliksp@impan.impan.gov.pl

CHARLES PUGH AND MICHAEL SHUB[*]
Stable ergodicity and partial hyperbolicity

Dedicated to the memory of Ricardo Mañé.

1. Introduction

By the results of Anosov in 1967 volume preserving uniformly hyperbolic systems are ergodic and are open. Thus they exhibit robust statistical behavior. Averages are the same for almost all points, not only for the system in question but also for any small perturbation which preserves the same volume. If the perturbation only preserves a close by volume, then the averages of continuous functions are still close by. On the other hand, in 1954, Kolmogorov announced that there are no ergodic Hamiltonian systems in a neighborhood of a completely integrable one. Completely integrable systems have no hyperbolic behavior at all.

In this paper we will review the results of [Grayson, Pugh and Shub, 1994], [Pugh and Shub, 1996], [Pugh, Shub and Wilkinson, 1996], and [Brezin and Shub, 1995] which study the mixed situation in which the system is only partially hyperbolic.

Our themes are:

1) A little hyperbolicity goes a long way toward guaranteeing ergodic behavior.
2) Stably ergodic systems are considerably more general than one might have feared from Kolmogorov's theorem.
3) Some hyperbolicity may be necessary for stable ergodicity.

We consider C^2 diffeomorphisms f of closed manifolds M which preserve a fixed smooth volume on M. We say that f is stably ergodic if there is a neighborhood U of f in the C^2 volume preserving diffeomorphisms of M such that every $g \in U$ is ergodic.

---

[*] Partially supported by an NSF grant.

Each of our main themes is developed in a section below. Finally in section 5, we suggest some generalizations to dissipative systems.

2. Partial Hyperbolicity and Ergodicity

The Main Theorem of this section gives sufficient conditions for a diffeomorphism to be ergodic. We find stably ergodic diffeomorphisms by finding open sets of diffeomorphisms satisfying these conditions. The theorem may be interpreted to say that even for systems which are not uniformly hyperbolic, the same phenomenon which produces chaotic behavior i.e. some hyperbolicity may also guarantee ergodicity.

Main Theorem: [Pugh and Shub, 1996] Let $f : M \to M$ be partially hyperbolic and dynamically coherent. Suppose that the stable and unstable bundles have the accessibility property and that the invariant bundles are sufficiently Hölder. Then f is ergodic.

The accessibility property is a concept from control theory which we apply to the strong unstable and stable foliations of a partially hyperbolic diffeomorphism. We will soon explain these concepts. Partially hyperbolic diffeomorphisms which are dynamically coherent have some more properties which we will eventually come to. All three distributions E^s, E^c and E^u, the strong stable, center and strong unstable sub-bundles of the tangent bundle, of C^2 diffeomorphisms are Hölder. That they are sufficiently Hölder is expressed in terms of relationships of the Hölder exponents and ultimately in terms of the contraction and expansion constants of the various natural invariant bundles for the derivative. We leave these details to be consulted in [Pugh and Shub, 1996] and [Pugh, Shub and Wilkinson, 1996], but note that foliations with C^1 tangent bundles are sufficiently Hölder. Partially hyperbolic systems and the accessibility property were to our knowledge first considered in [Brin and Pesin, 1974].

We say that a C^r diffeomorphism $f : M \to M$ is *partially hyperbolic* iff $r \geq 1$ and there is a continuous Tf-invariant direct sum decomposition

$$TM = E^s \oplus E^c \oplus E^u$$

where E^s and E^u are non-trivial, some Finsler $\| \ \|$ on TM and some real constants $0 < a < b < c < 1 < d < e < g$ such that

$$a\|v\| < \|Tf(v)\| < b\|v\| \text{ for } v \in E^s - \{0\}$$
$$c\|v\| < \|Tf(v)\| < d\|v\| \text{ for } v \in E^c - \{0\}$$
$$e\|v\| < \|Tf(v)\| < g\|v\| \text{ for } v \in E^u - \{0\}.$$

Since $Tf : E^c \to E^c$ may have some contraction and expansion E^s and E^u are strong contracting and strong expanding Tf invariant subbundles. Tangent to E^s and E^u are the strong contracting and strong expanding f invariant foliations which we will denote by W^s and W^u.

Given continuous sub-bundles $F, H \subset TM$ and points $m_0, m_1 \in M$ we say that m_1 is *accessible* from m_0 iff there is a continuous piecewise C^1 path $\phi[0,1]$ joining

m_0 and m_1 i.e. ϕ is continuous, $\phi(0) = m_0, \phi(1) = m_1$ and there are a finite number of reals $0 = t_0 < t_1 < \cdots < t_j = 1$ such that $\phi|[t_i, t_{i+1}]$ is C^1 embedding, and is tangent either to F or H, $i = 0, \ldots, j - 1$. We say the pair F, H has the *accessibility property* iff for any $m_0, m_1 \in M$ m_1, is accessible from m_0.

Only connected manifolds can have the accessibility property. We are assuming throughout that M is connected.

A *partially hyperbolic diffeomorphism* is said to have the *accessibility property* iff E^s, E^u has the accessibility property.

Accessibility is an equivalence relation on points in M. We say that the pair F, H has the *essential accessibility property* iff the only measurable subsets of M saturated by the equivalence relation have measure 0 or 1.

The *partially hyperbolic diffeomorphism* f is said to have the *essential accessibility property* iff E^s, E^u does. We could replace accessibility by the more general essential accessibility in the statement of Main Theorem, but accessibility is an easier property to verify.

Finally we define dynamically coherent. Let $f : M \to M$ be partially hyperbolic with splitting $E^s \oplus E^c \oplus E^u$. Then we say that f is *dynamically coherent* iff

1) there is a foliation $\mathfrak{L}$ (with C^1 leaves) tangent to E^c.
2) For any leaf L of $\mathfrak{L}$, $W^u(L)$ and $W^s(L)$ are unions of leaves of $\mathfrak{L}$.

We call $\mathfrak{L}$ the central foliation. The leaves of $\mathfrak{L}$ are C^1 and their tangent bundles vary continuously, in fact in a Hölder fashion. We call them central leaves, they are normally hyperbolic and hence have stable and unstable manifolds. A priori $\mathfrak{L}$ is not a C^1 foliation. For a C^1 foliation not only has C^1 leaves and a C^0 tangent bundle, but it also has C^1 foliation charts.

We now give a condition which is simple to verify for a diffeomorphism to be partially hyperbolic and dynamically coherent.

Theorem 2.5. [P-S] Let $f : M \to M$ be a C^1 partially hyperbolic diffeomorphism as above. If there is a C^1 foliation $\mathfrak{L}$ tangent to E^c, then f is partially hyperbolic and dynamically coherent, and so are all sufficiently small C^1 perturbations of f.

3. Genericity of Stable Ergodicity?

We do not know if Hölder smoothness without any further hypothesis is sufficient in the Main Theorem of the last section. For C^2 diffeomorphisms some Hölder smoothness follows from partial hyperbolicity. Nor do we know if dynamically coherent is essential. Partially hyperbolic may suffice. Also it is quite likely that the accessibility property is generic. So we conjecture

Conjecture 1: [Pugh and Shub, 1996] Among the partially hyperbolic C^2 volume preserving diffeomorphisms of M the stably ergodic are open and dense.

In particular, Conjecture 1 would imply that the generic C^2 volume preserving perturbation of an ergodic automorphism of the torus is ergodic (see [Grayson, Pugh and Shub, 1994] for some discussion of this). It also would imply that if $A : M \to M$ is a volume preserving C^2 Anosov diffeomorphism then the generic C^2 volume preserving perturbation of $A \times id : M \times N \to M \times N$ is stably ergodic for any compact manifold N. See [Bonatti and Diaz, 1994] for the rather striking topological transitivity version of this second case of Conjecture 1.

We do know a large class of examples of partially hyperbolic diffeomorphisms which are stably ergodic. The time one map of the geodesic flow on a compact surface of constant negative curvature is the most classically studied partially hyperbolic diffeomorphism and it has the accessibility property. In [Grayson, Pugh and Shub, 1994] we proved that it is stably ergodic. Amie Wilkinson [Wilkinson, 1995] removed the hypothesis that the negative curvature be constant. In n-dimensions we have:

Theorem 3. [Pugh and Shub, 1996] The time one map of the geodesic flow on the unit tangent bundle of a compact n-manifold of constant curvature $k, k < 0$ is stably ergodic - it is ergodic and so are all C^2 small volume preserving perturbations of it.

We have also a class of examples which come from the theory of homogeneous spaces of Lie groups. We will assume that our spaces are of the form G/B where G is a connected Lie group and B is a closed subgroup which, in addition, is *admissible* in a certain technical sense (see [Brezin and Shub, 1995]) which we will not make precise here. If G is nilpotent, solvable or semi-simple or if B is discrete then the admissibility condition is satisfied.

For $a \in G$ let L_a denote left translation by a i.e. $L_a(h) = ah$ for all $h \in G$. Then L_a induces a map on G/B which we call L_a as well. Given an automorphism A of G and $a \in G$ we call $L_a A : G \to G$ an affine diffeomorphism of G, we also denote this map by aA. If $A(B) = B$ then we continue to denote the induced map on G/B by $L_a A$ or aA and call it an affine diffeomorphism of G/B. We will assume that the Haar measure on G induces a finite measure on G/B which is invariant under left translation and that $A : G/B \to G/B$ is measure preserving.

Given an affine diffeomorphism $aA : G \to G$, aA induces an automorphism of the Lie Algebra $\mathfrak{g}$ of G by $ad(a)DA(e)$ where e is the identity of G. In particular, $ad(a)DA(e)$ is a linear map. Let $\mathfrak{g}^s$ and $\mathfrak{g}^u$ be the generalized eigenspaces of $\mathfrak{g}$ corresponding to the contracting and expanding eigenvalues of $ad(a)DA(e)$. Let $\mathcal{L} \subset \mathfrak{g}$ be the Lie subalgebra of $\mathfrak{g}$ generated by $\mathfrak{g}^s$ and $\mathfrak{g}^u$. Then it is not hard to see [Pugh and Shub, 1996] that $\mathcal{L}$ is an ideal in $\mathfrak{g}$ which is $ad(a)DA(e)$ invariant. As an ideal $\mathcal{L}$ is tangent to the connected normal subgroup which we denote by H and call the hyperbolically generated subgroup of G.

Theorem 1: Let G/B be a compact manifold and aA be an affine diffeomorphism of G/B. Let r be a positive real. If the eigenvalues of $ad(a)DA(e)$ are sufficiently bunched near the three numbers 1, r or $\frac{1}{r}$ and $HB = G$, then aA is stably ergodic on G/B.

The theorem has examples for semi-simple groups. We specialize to SL(n,R).

Let Γ be a uniform discrete subgroup of $SL(n, R)$.

For $A \in SL(n, R)$ let $L_A : SL(n, R)/\Gamma \to SL(n, R)/\Gamma$ be given by left translation by A.

Theorem 2: [Pugh and Shub, 1996] The following four conditions are equivalent.

a) A has an eigenvalue with modulus different from 1.

b) L_A is partially hyperbolic and the stable and unstable bundles have the accessibility property.

c) The Lie Algebra generated by the contracting and expanding subspace of Adg is the whole Lie Algebra $SL(n, R)$.

d) L_A is stably ergodic among left translations of $SL(n, R)/\Gamma$.

4. Necessity

We say that aA is is stably ergodic under perturbations by left translations if there is a neighborhood U of a in G such that $a'A$ is ergodic for every $a' \in U$. We continue to assume that the pair G,B is admissible.

It is now easy to state our main theorem of this section.

Main Theorem: [Brezin and Shub, 1995] If an affine diffeomorphism is stably ergodic under perturbations by left translations then $\overline{HB} = G$ where H is the hyperbolically generated subgroup of G.

Remark (1). If $\overline{HB} = G$ then the affine diffeomorphism is ergodic, this is essentially Hopf's proof of the ergodicity of the geodesic flow. See [Pugh and Shub, 1996] where generalizations are proven in the C^2 category.

Remark (2). That $\overline{HB} = G$ is the same as the action of H on G/B being ergodic, which in this setting is the same as the essential accessibility property of [Pugh and Shub, 1996].

Remark (3). It is likely that stable ergodicity is actually equivalent to the condition that $\overline{HB} = G$. In the next two propositions we state some special cases of the theorem in which this is actually the case.

Proposition 1 [Brezin and Shub, 1995] Let G be a connected nilpotent Lie group and Γ a uniform discrete subgroup of G. Then the affine diffeomorphism aA of G/Γ is stably ergodic among left translations of G if and only if $\overline{H\Gamma} = G$.

Proposition 2 [Brezin and Shub, 1995] Let G be a connected semi-simple Lie group and Γ a lattice in G.

a) If G has no compact factors, then the affine diffeomorphism aA of G/Γ is stably ergodic among left translations of G if and only if $H = G$.

b) If G has compact factors then the affine diffeomorphism aA of G/Γ is stably ergodic among left translations of G if and only if $\overline{H\Gamma} = G$.

5. Dissipative Systems

Now we drop the condition that f be measure preserving and simply assume that f is C^r for some finite r. For $m \in M$, $W^s(m) = (p \in M | d(f^n(m), f^n(p)) \to 0$ as $n \to \infty)$ and $W^u(m)$ is defined similarly. We define the partial order $>_1$, by $m >_1 n$ if $W^u(m)$ and $W^s(n)$ have non-empty intersection. Let $>$ denote the total order obtained by transitivizing $>_1$ and extending to a total order.

Conjecture 2: For the generic f and all $x \in M$ the set of $y \in M$ such that $x \geq y$ is a closed subset of M.

We say x is equivalent to y if $x \geq y$ and $y \geq x$.

Conjecture 3: Let dimension M ≥ 2. For the generic finite dimensional submanifold V contained in $Diff^r(M)$ and almost every $f \in V$ the equivalence classes of points in the chain recurrent set of f are open in the chain recurrent set.

Conjecture 3 would give a finite spectral decomposition for f where each piece of the decomposition has something akin to the accessibility property.

REFERENCES

Bonatti, C. and Diaz, L. (1994), *Persistent Nonhyperbolic Transitive Diffeomorphisms*, preprint.

Brezin, J. and Shub, M. (1995), *Stably Ergodicity in Homogeneous Spaces*, preprint.

Brin, M.I. and Pesin, J.B. (1974), *Partially Hyperbolic Dynamical Systems (English Translation)*, Math. USSR Izvestia **8**, 177–218.

Grayson, M., Pugh, C. and Shub, M. (1994), *Stably Ergodic Diffeomorphisms*, Annals of Math **140**, 295-329.

Hopf, E. (1971), *Ergodic Theory and the Geodesic Flow on Surfaces of Constant Negative Curvature*, Bull. Amer. Mat. Soc. **77**, 863–877.

Pugh, C. and Shub, M. (1996), *Stably Ergodic Dynamical Systems and Partial Hyperbolicity*, preprint.

Pugh, C., Shub, M. and Wilkinson, A. (1996), *Hölder Foliations*, preprint.

Wilkinson, A. (1995), *Thesis - University of California, Berkeley.*

Charles Pugh
Department of Mathematics
University of California at Berkeley
Berkeley, CA 94720

Michael Shub
Mathematical Sciences Department
IBM Watson Research Center
Yorktown Heights, NY 10598

D RUELLE
Sharp zeta functions for smooth interval maps

1. Introduction.

The present note is a complement to the paper on *sharp determinants* [1] by V. Baladi and the author, and shows how to take into account smoothness[*]. The techniques developed here may be of use beyond the case of maps of the interval. To give an idea of the results obtained we present immediately a simple example.

Let finite families (X_α), (ψ_ω) be given such that each X_α is a compact interval $[a_\alpha, b_\alpha] \subset \mathbb{R}$ and each ψ_ω is a C^∞ piecewise monotone map of some $X_{\alpha(\omega)}$ into some $X_{\beta(\omega)}$ [i.e., we assume that $X_{\alpha(\omega)}$ is the union of finitely many subintervals where ψ_ω is monotone]. Also let $g_\omega : X_{\alpha(\omega)} \to \mathbb{C}$ be C^∞. [We say that a function is C^∞ on a closed interval if it extends to a C^∞ function on a neighborhood of the interval.]

1.1. Proposition. *Assume that the derivatives ψ'_ω satisfy*

$$|\psi'_\omega| \leq 1 \qquad on \qquad X_{\alpha(\omega)}$$

$$|\psi'_\omega(x)| = 1 \quad implies \quad g_\omega(x) = 0$$

and write

$$T_m = \sum_{\omega_1 \ldots \omega_m}^{*} \prod_{k=0}^{m-1} g_{\omega_k}(\psi_{\omega_{k-1}} \circ \cdots \circ \psi_{\omega_1} \, x(\omega_1, \ldots, \omega_m))$$

where $\sum^{*}$ is restricted by $\beta(\omega_1) = \alpha(\omega_2), \ldots, \beta(\omega_m) = \alpha(\omega_1)$ and the condition that $\psi_{\omega_m} \circ \cdots \circ \psi_{\omega_1}$ has a unique fixed point $x(\omega_1, \ldots, \omega_m) \in X_{\alpha(\omega_1)}$. Then the zeta function

$$\zeta(z) = \exp \sum_{m=1}^{\infty} \frac{z^m}{m} T_m$$

is meromorphic in $\mathbb{C}$.

This is close to known results on zeta functions of smooth expanding (or contracting) maps (see Tangerman [9], Ruelle [4], [5], Pollicott [3]). The new

---

[*] I am indebted to Viviane Baladi for critical reading of the manuscript; her suggestions have been incorporated in the text of this note.

188

feature here is that the ψ_ω need not be uniformly contracting. As in the uniformly contracting case, the zeros and poles of the zeta function will be determined by the eigenvalues of generalized transfer operators $\mathcal{M}_0$, $\mathcal{M}_1$ (acting respectively on smooth functions and 1-forms), but here it is not known that the determinants $\mathrm{Det}^b(1 - z\,\mathcal{M}_0)$, $\mathrm{Det}^b(1 - z\,\mathcal{M}_1)$ will make sense. Specifically *the order of the meromorphic function ζ at z is the order of z^{-1} as eigenvalue of $\mathcal{M}_0$ minus the order of z^{-1} as eigenvalue of $\mathcal{M}_1$.*

For proofs of the above statements, and specification of the operators $\mathcal{M}_0$, $\mathcal{M}_1$, see subsection 3.6 below. [Note that a case with $\psi'(x) = 1$ and $g(x) \neq 0$ has been analyzed by S. Isola: Dynamical zeta functions for non-uniformly hyperbolic transformations, preprint (1995).]

1.2. Question. *Is the n-dimensional generalization of Proposition 1.1 true?*

2. Definition and spectral properties of transfer operators.

Fixing $a > 0$, we let $\mathcal{B}_{(0)}$ be the Banach space of functions $\mathbb{R} \to \mathbb{C}$ (mod countable sets) which are of bounded variation and vanish outside $(-a, a)^{*)}$. As norm on $\mathcal{B}_{(0)}$ we use $\|\cdot\|_{(0)} = \mathrm{Var}$: the total variation (mod countable sets) on $\mathbb{R}$.

Denote by $\Phi^{(k)}$ the k-th derivative of $\Phi : \mathbb{R} \to \mathbb{C}$ in the sense of distributions. [If $\Phi = \Phi^{(0)} \in \mathcal{B}_{(0)}$, then $\Phi' = \Phi^{(1)}$ is a bounded measure and $\mathrm{Var}\,\Phi$ is the total mass of $|\Phi'|$.] For integer $k \geq 0$ we let $\mathcal{B}_{(k)}$ be the Banach space of functions Φ such that $\Phi^{(0)}, \ldots, \Phi^{(k)} \in \mathcal{B}_{(0)}$ with the norm $\Phi \mapsto \|\Phi\|_{(k)} = \mathrm{Var}\,\Phi^{(k)}$.

Let $\mathrm{sgn}\,\xi = 0$ or ± 1 if $\xi = 0$ or $\xi \gtrless 0$. We define $\sigma_k(\cdot)$ by $\sigma_1(\xi) = \frac{1}{2}\,\mathrm{sgn}\,\xi$ and

$$\sigma_k(\xi) = \int_0^\xi \sigma_{k-1}(\eta)\,d\eta \quad \text{for} \quad k > 1.$$

[Therefore $\sigma_k(\xi) = \frac{\mathrm{sgn}\,\xi}{2(k-1)!} \cdot |\xi|^{k-1}$.] If $\Phi \in \mathcal{B}_{(0)}$ we define

$$\Phi^{(-k)}(x) = \int \sigma_k(x - y)\,\Phi(y)\,dy$$

so that $(\Phi^{(-k)})^{(\ell)} = \Phi^{(\ell-k)}$. [Note that $\Phi^{(-k)}$ need not vanish outside $(-a, a)$, or be bounded.] We let $\mathcal{B}_{(-k)}$ be the Banach subspace of $\mathcal{B}_{(0)}$ consisting of functions Φ with $\Phi^{(0)}, \ldots, \Phi^{(-k)} \in \mathcal{B}_{(0)}$.

$^{*)}$ We use here the notation $\mathcal{B}_{(0)}$ instead of the earlier $\mathcal{B}^{\#}$ (see [6]). Note that $\mathcal{B}_{(0)}$ is isomorphic to the space of functions of bounded variation (mod countable sets) in $(-a, a)$ or in $\mathbb{R}$ (using a homeomorphism $(-a, a) \to \mathbb{R}$).

2.1. Lemma. *The map* $\Phi \mapsto \Phi^{(-k)}$ *is a Banach space isomorphism* $\mathcal{B}_{(-k)} \to \mathcal{B}_{(k)}$ *with inverse* $\Phi \to \Phi^{(k)}$.

If $\Phi \in \mathcal{B}_{(-k)}$ then $\Phi^{(-k)}, \Phi^{(-k+1)}, \ldots, \Phi^{(0)} \in \mathcal{B}_{(0)}$, i.e., $\Phi^{(-k)} \in \mathcal{B}_{(k)}$ and $(\Phi^{(-k)})^{(k)} = \Phi$. If $\Phi \in \mathcal{B}_{(k)}$, then $(\Phi^{(k)})^{(-\ell)}$ vanishes outside $(-a, a)$ for $\ell = 0, \ldots, k$ because $\sigma_\ell(\xi)$ is, for $\xi > 0$ (or $\xi < 0$), proportional to $\xi^{\ell-1}$. Thus $\Phi^{(k)} \in \mathcal{B}_{(-k)}$ and $(\Phi^{(k)})^{(-k)} = \Phi$. Finally the norm of $\Phi \in \mathcal{B}_{(k)}$ is equal to the norm of the corresponding element $\Phi^{(k)} \in \mathcal{B}_{(-k)}$. $\square$

2.2. Assumptions. Let $\mu(d\omega) \geq 0$ be a measure on Ω. [We allow $\mu(\Omega) = +\infty$; an interesting case is when $\mu(\{\omega\}) = 1$ for each Ω, Ω countable, so that $\int \mu(d\omega) = \sum_\omega$.] For each $\omega \in \Omega$ we are given $\Lambda_\omega, \psi_\omega, g_\omega$. Here $\Lambda_\omega \subset (-a, a)$ is a closed interval of $\mathbb{R}$, ψ_ω is a homeomorphism of Λ_ω to $\psi_\omega \Lambda_\omega \subset (-a, a)$, $g_\omega : \mathbb{R} \to \mathbb{C}$ is such that $\operatorname{supp} g_\omega \subset \Lambda_\omega$. We write $\varepsilon_\omega = +1$ (resp. -1) if ψ_ω preserves (resp. reverses) the order.

Choosing an integer $r \geq 0$ we assume that $\psi_\omega^{(r)}$ restricted to the interior of Λ_ω has bounded variation, that $g_\omega \in \mathcal{B}_{(r)}$, and that $\int \mu(d\omega) \|P_\omega\|_{(0)} < \infty$ whenever

$$P_\omega(x) = g_\omega^{(k_0)}(x) \, (\psi_\omega^{(k_1)}(x))^{n_1} \ldots (\psi_\omega^{(k_\ell)}(x))^{n_\ell} \tag{*}$$

and the integers k_i, n_i satisfy $0 \leq k_0, \, 1 \leq k_1 < \cdots < k_\ell, \, n_1, \ldots, n_\ell \geq 0$ and $k_0 + n_1 k_1 + \cdots + n_\ell k_\ell \leq r$.

2.3. Remarks. (1) Our constructions will involve only the product $\mu(d\omega) \cdot g_\omega$. Therefore we may replace μ by a probability measure. For definiteness let us assume that (Ω, μ) is a standard probability space, and that $\omega \to \Lambda_\omega, \psi_\omega, g_\omega$ are measurable in some reasonable sense.

(2) Notice that only the restriction of ψ_ω to $\operatorname{supp} g_\omega$ will be used. Let $r \geq 1$ and suppose that ψ_ω' is extended to a function $\in \mathcal{B}_{(r-1)}$. If we assume

$$\sup_\omega \|\psi_\omega'\|_{(r-1)} < \infty \quad , \quad \int \mu(d\omega) \|g_\omega\|_{(r)} < \infty$$

then the conditions

$$\int \mu(d\omega) \|P_\omega\|_{(0)} < \infty$$

imposed above are automatically satisfied.

190

2.4. Definitions. Write

$$(\mathcal{M}_k \Phi)(x) = \int \mu(d\omega)\, g_\omega(x)\, (\psi'_\omega(x))^k\, \Phi(\psi_\omega x)$$

$$(\widehat{\mathcal{M}}_k \Phi)(x) = \int \mu(d\omega)\, \varepsilon_\omega\, g_\omega(\psi_\omega^{-1} x)\, (\psi'_\omega(\psi_\omega^{-1} x))^k\, \Phi(\psi_\omega^{-1} x).$$

This defines operators $\mathcal{M}_k$, $\widehat{\mathcal{M}}_k$ acting on the space $\mathcal{C}_0$ of bounded functions $\mathbb{R} \to \mathbb{C}$ with the uniform norm (mod a countable set) $\| \ \|_0$. We define R_k, $\widehat{R}_k$ to be the spectral radii of $\mathcal{M}_k$, $\widehat{\mathcal{M}}_k$ acting on $\mathcal{C}_0$:

$$R_k = \lim_{m\to\infty} (\|(\mathcal{M}_k)^m\|_0)^{1/m}$$

$$\widehat{R}_k = \lim_{m\to\infty} (\|(\widehat{\mathcal{M}}_k)^m\|_0)^{1/m}.$$

We also write $\mathcal{M} = \mathcal{M}_0$, $\widehat{\mathcal{M}} = \widehat{\mathcal{M}}_0$, $R = R_0$, $\widehat{R} = \widehat{R}_0$. In particular the *transfer operator* $\mathcal{M}$ is defined by

$$(\mathcal{M} \Phi)(x) = \int \mu(d\omega)\, g_\omega(x)\, \Phi(\psi_\omega x).$$

If $\Phi \in \mathcal{B}_{(r)}$ we may write the j-th derivative (for $j = 0, \ldots, r$) of $(\mathcal{M} \Phi)(x)$ as

$$(\mathcal{M} \Phi)^{(j)}(x) = \sum_{k=0}^{j} \int \mu(d\omega)\, P_\omega^{jk}(x)\, \Phi^{(k)}(\psi_\omega x)$$

where P_ω^{jk} is a polynomial in $g_\omega^{(0)}, \ldots, g_\omega^{(j)}, \psi_\omega^{(1)}, \ldots, \psi_\omega^{(j)}$, linear in $g_\omega^{(0)}, \ldots, g_\omega^{(j)}$, and such that

$$P_\omega^{00} = g_\omega$$

$$P_\omega^{j+1,k} = P_\omega^{j,k-1} \cdot \psi'_\omega + (P_\omega^{jk})'.$$

In particular the total order of derivatives in P^{jk} is j, and

$$P_\omega^{jj} = g_\omega \cdot (\psi'_\omega)^j.$$

Define an operator $\mathcal{M}^{(r)}$ on $\mathcal{B}_{(0)}$ by

$$(\mathcal{M}^{(r)} \Psi)(x) = (\mathcal{M}_r \Psi)(x)$$
$$+ \sum_{k=0}^{r-1} \int \mu(d\omega)\, P_\omega^{rk}(x) \int \sigma_{r-k}(\psi_\omega x - y)\, \Psi(y)\, dy.$$

Then

$$\mathcal{M}^{(r)}\,\Psi = (\mathcal{M}\,\Psi^{(-r)})^{(r)}\,.$$

Note that $\Psi^{(-r)}$ need not vanish outside of $(-a,a)$ but $\mathcal{M}\,\Psi^{(-r)}$ does, and therefore $\mathcal{M}^{(r)}\,\Psi \in \mathcal{B}_{(0)}$.

2.5. Proposition. *The essential spectral radius of $\mathcal{M}$ acting on $\mathcal{B}_{(0)}$ is $\leq \widehat{R}$.*

This is close to a result proved in [6], but we give here a new (simpler) proof.

Let M be the Banach space of measures on $\mathbb{R}$ with support in $[-a,a]$, and M_0 the subspace of measures with total mass 0. The map $\alpha : \Phi \mapsto \Phi'$ is an isomorphism $\mathcal{B}_{(0)} \to M_0$. Define $\mathcal{M}^* : M \to M$ by

$$\mathcal{M}^* = \mathcal{M}' + \mathcal{M}''$$

$$(\mathcal{M}'\,\mathbf{m})(dx) = \int \mu(d\omega)\,g_\omega(x)\cdot\varepsilon_\omega(\psi_\omega^{-1}\,\mathbf{m})(dx)$$

$$(\mathcal{M}''\,\mathbf{m})(dx) = \int \mu(d\omega)\,dg_\omega(x)\cdot\int \sigma_1(\psi_\omega\,x - y)\,\mathbf{m}(dy)$$

(where $\psi\,\mathbf{m}$ is the direct image of $\mathbf{m}$ by ψ). We have

$$\mathcal{M} = \alpha^{-1}\,\mathcal{M}^*\,\alpha$$

and therefore the proposition follows from Nussbaum's theorem [2] and the following facts:

(a) the spectral radius of $\mathcal{M}'$ is $\leq \widehat{R}$,

(b) $\mathcal{M}''$ is a compact operator.

Proof of (a). For bounded continuous $\Phi : \mathbb{R} \to \mathbb{C}$, we have

$$\int \Phi(x)\,(\mathcal{M}'\,\mathbf{m})(dx) = \iint \mu(d\omega)\,\varepsilon_\omega\,[(g_\omega\cdot\Phi)\circ\psi_\omega^{-1}]\,\mathbf{m}(dx)$$

$$= \int (\widehat{\mathcal{M}}\,\Phi)(x)\,\mathbf{m}(dx)\,.$$

Therefore

$$\|\mathcal{M}'^m\| \leq \|\widehat{\mathcal{M}}^m\|$$

which establishes (a).

Proof of (b). Note that

$$(\mathcal{M}''\,\mathbf{m})(dx) = \int \mu(d\omega)\,d(g_\omega\circ\psi_\omega^{-1})(\xi)\int \sigma_1(\xi - y)\,\mathbf{m}(dy)\,.$$

We write

$$\nu(d\xi) = \int \mu(d\omega) \, |d(g_\omega \circ \psi_\omega^{-1})(\xi)| \, .$$

Let $J = (J_i)$ be a finite decomposition of $(-a, a)$ into intervals J_i, each open or reduced to a point, and choose $x_i \in J_i$ for each i. Given $\varepsilon > 0$ we may assume that whenever J_i is not reduced to a point

$$\nu(J_i) < \varepsilon \, .$$

The map

$$\mathbf{m} \mapsto \mathbf{m}_J = \sum_i \mathbf{m}(J_i) \, \delta_{x_i}$$

(where δ_x is the unit mass at x) is of finite rank, and

$$\sup_{\xi \in J_i} \left| \int \sigma_1(\xi - y) \, \mathbf{m}_J(dy) - \int \sigma_1(\xi - y) \, \mathbf{m}(dy) \right| \leq |\mathbf{m}(J_i)| \, .$$

Therefore, writing $\sum^*$ or $\max^*$ for the sum or max over i such that J_i is not reduced to one point

$$\|\mathcal{M}'' \, \mathbf{m}_J - \mathcal{M}'' \, \mathbf{m}\|$$

$$\leq \int \mu(d\omega) \sum_i{}^* \int_{J_i} |d(g_\omega \circ \psi_\omega^{-1})(x)| \cdot |\mathbf{m}(J_i)|$$

$$\leq \|\mathbf{m}\| \ \max_i{}^* \ \nu(J_i) \leq \|\mathbf{m}\| \, \varepsilon \, .$$

Therefore the finite rank operators $\mathcal{M}''_J : \mathbf{m} \mapsto \mathcal{M}'' \, \mathbf{m}_J$ tend in norm to $\mathcal{M}'' : \mathbf{m} \mapsto \mathcal{M}'' \, \mathbf{m}$ when $\varepsilon \to 0$, which establishes (b). $\square$

2.6. Proposition. *The essential spectral radius of $\mathcal{M}^{(r)}$ acting on $\mathcal{B}_{(0)}$ is $\leq \widehat{R}_r$.*

Application of Proposition 2.5 proves this for $r = 0$, and also shows that the essential spectral radius of $\mathcal{M}_r$ is $\leq \widehat{R}_r$ when $r > 0$. For $r > 0$ we may write

$$\mathcal{M}^{(r)} = \mathcal{M}_r + \sum_{k=0}^{r-1} \mathcal{M}_{rk}$$

where

$$(\mathcal{M}_{rk} \, \Psi)(x) = \int \mu(d\omega) \, P_\omega^{rk}(x) \int \sigma_{r-k}(\psi_\omega x - y) \, \Psi(y) \, dy \, .$$

The proposition follows from Nussbaum's theorem and the following facts

(a) the essential spectral radius of $\mathcal{M}_r$ is $\leq \widehat{R}_r$ (see above).

(b) $\mathcal{M}_{rk}$ is a compact operator for $k = 0, \ldots, r-1$.

To prove (b) we decompose $(-a, a)$ into intervals J_i of length $< \varepsilon$ and choose $x_i \in J_i$. We approximate $\dot{\Psi}$ by Ψ_J piecewise constant equal to $\Psi(x_i)$ on J_i, and write

$$\Psi_\ell(\xi) = \int \sigma_\ell(\xi - y)\, \Psi(y)\, dy - \int \sigma_\ell(\xi - y)\, \Psi_J(y)\, dy \,.$$

Note that by changing the x_i we change $\Psi_\ell(0)$ by $\delta\, \Psi_\ell(0)$ such that

$$|\delta\, \Psi_\ell(0)| \leq \sum_i \int_{J_i} |\sigma_\ell(y)|\, dy \ \mathrm{var}(\Phi \mid J_i) \leq \varepsilon\, \sigma_\ell(a)\ \mathrm{Var}\, \Psi \,.$$

If Ψ is real valued one can choose the x_i such that $\Psi_\ell(0)$ is either ≥ 0 or ≤ 0, and thus in general

$$|\Psi_\ell(0)| \leq \sqrt{2} \cdot \varepsilon\, \sigma_\ell(a)\ \mathrm{Var}\, \Psi \,.$$

Therefore

$$\begin{aligned}
\mathrm{Var}\, \Psi_\ell &= \int_{-a}^{a} \left| \frac{d}{d\xi}\, \Psi_\ell(\xi) \right| d\xi = \int_{-a}^{a} |\Psi_{\ell-1}(\xi)|\, d\xi \\
&\leq \int_{-a}^{a} |\Psi_{\ell-1}(0) + \mathrm{Var}\, \Psi_{\ell-1}|\, d\xi \\
&\leq 2a \left[\sqrt{2} \cdot \varepsilon\, \sigma_{\ell-1}(a)\ \mathrm{Var}\, \Psi + \mathrm{Var}\, \Psi_{\ell-1} \right] .
\end{aligned}$$

Since

$$\mathrm{Var}\, \Psi_1 \leq \sum_i |J_i|\ \mathrm{var}(\Psi \mid J_i) \leq \varepsilon\ \mathrm{Var}\, \Psi$$

we see by induction on ℓ that

$$\mathrm{Var}\, \Psi_\ell \leq C_\ell\, \varepsilon\, a^{\ell-1}\ \mathrm{Var}\, \Psi$$

where C_ℓ is a numerical constant. Therefore the finite rank map $\Psi \to \mathcal{M}_{rk}\, \Psi_J$ is such that

$$\begin{aligned}
\mathrm{Var}(\mathcal{M}_{rk}\, \Psi - \mathcal{M}_{rk}\, \Psi_J) &\leq \int \mu(d\omega)\ \mathrm{Var}\, P_\omega^{rk} \cdot \mathrm{Var}\, \Psi_{r-k} \\
&\leq C\, \varepsilon \cdot \mathrm{Var}\, \Psi
\end{aligned}$$

showing that $\mathcal{M}_{rk}$ is compact. $\square$

2.7. Proposition. $\mathcal{M}$ *acting on* $\mathcal{B}_{(r)}$ *and* $\mathcal{M}^{(r)}$ *acting on* $\mathcal{B}_{(0)}$ *have the same estimate* $\widehat{R}_r$ *for the essential spectral radius, and the same eigenvalues* λ *such that* $|\lambda| > \widehat{R}_r$, *with the same multiplicity.*

This is because $\mathcal{M}^{(r)}$ maps $\mathcal{B}_{(0)}$ to the subspace $(\mathcal{B}_{(r)})^{(r)}$ and $\mathcal{M}^{(r)}$ acting on $(\mathcal{B}_{(r)})^{(r)} = \mathcal{B}_{(-r)}$ is conjugate to $\mathcal{M}$ acting on $\mathcal{B}_{(r)}$ by the isomorphism of Lemma 2.1. $\square$

2.8. Proposition. $\mathcal{M}$ *acting on* $\mathcal{B}_{(r)}$ *and* $\mathcal{M}^{(r)}$ *acting on* $\mathcal{B}_{(0)}$ *have spectral radius* $\leq \max{(R, \widehat{R}_r)}.$

It suffices to consider $\mathcal{M}$ acting on $\mathcal{B}_{(r)}$. Let $\mathcal{R}$ be the corresponding spectral radius. It suffices to consider the case $\mathcal{R} > \widehat{R}_r$; this implies that $\mathcal{M}$ acting on $\mathcal{B}_{(r)}$ has an eigenfunction Φ corresponding to an eigenvalue λ with $|\lambda| = \mathcal{R}$; Φ is also an eigenvalue of $\mathcal{M}$ acting on $\mathcal{C}_0$, hence $|\lambda| \leq R$, and $\mathcal{R} \leq R$ as announced. $\square$

2.9. Remark. (Piecewise monotone ψ_ω). Let $\mathcal{B}^c_{(0)}$ be the subspace of $\mathcal{B}_{(0)}$ consisting of continuous functions and, for $k > 0$, write

$$\mathcal{B}^c_{(k)} = \{\Phi \in \mathcal{B}_{(k)} : \Phi^{(k)} \in \mathcal{B}^c_{(0)}\}.$$

We have up to now assumed that the ψ_ω are homeomorphisms (hence continuous and strictly monotone). In the present subsection we want to *weaken* this to the condition that the ψ_ω are continuous and piecewise monotone maps $\Lambda_\omega \to (-a, a)$, with a finite number $N(\omega)$ of laps (monotonicity intervals)*). We *strengthen* our requirements on ψ_ω, g_ω by imposing that $\psi_\omega^{(r)}$ restricted to the interior of Λ_ω is continuous and of bounded variation, that $g_\omega \in \mathcal{B}^c_{(r)}$, and that

$$\int \mu(d\omega)\, N(\omega)\, \|P_\omega\|_{(0)} < \infty$$

when P_ω is of the form $(*)$. (This can be replaced by the stronger conditions of Remark 2.3 (2).) We modify thus our setup by the following requirement:

(C) *The ψ_ω are piecewise monotone, with $N(\omega)$ laps, $\psi_\omega^{(r)}$ restricted to the interior of Λ_ω is continuous of bounded variation, $g_\omega \in \mathcal{B}^c_\omega$, and $\int \mu(d\omega)\, N(\omega)\, \|P_\omega\|_{(0)} < \infty$ for P_ω of the form $(*)$.*

In view of defining an operator $\widehat{\mathcal{M}}_k$ on $\mathcal{C}_0$ we may decompose, for each ω, $(-a, a)$ into $N(\omega)$ laps $\Lambda_{\omega,i}$ and define $\psi_{\omega,i}^{-1}$ to be a monotone right inverse of $\psi_\omega \mid \Lambda_{\omega,i}$. Also let $\varepsilon_{\omega,i} = \pm 1$ depending on whether $\psi_\omega \mid \Lambda_{\omega,i}$ is increasing or decreasing. Note that

---

*) We do not require strict monotonicity; certain situations where $N(\omega) = +\infty$ are tractable, but will not be discussed here.

$\psi_{\omega,i}^{-1}$ is strictly monotone, and that its definition may involve choices at discontinuity points. We let now

$$(\widehat{\mathcal{M}}_k \Phi)(x) = \int \mu(d\omega) \sum_{i=1}^{N(\omega)} \varepsilon_{\omega,i} \, g_\omega(\psi_{\omega,i}^{-1} x) \, (\psi_\omega'(\psi_{\omega,i}^{-1} x))^k \cdot \chi_{\omega,i}(x) \, \Phi(\psi_{\omega,i}^{-1} x)$$

where $\chi_{\omega,i}$ is the characteristic function of $\psi_\omega \Lambda_{\omega,i}$. Since $\mathcal{C}_0$ is defined up to a countable set, $\widehat{\mathcal{M}}_k$ does not depend on the choices made in the definition of $\psi_{\omega,i}^{-1}$. As before we let $\widehat{R}_k$ be the spectral radius of $\widehat{\mathcal{M}}_k$ on $\mathcal{C}_0$.

We may now state the variants of Propositions 2.5, 2.6, 2.7, 2.8 which are valid in the present setup.

2.10. Proposition. *Under condition* (C) *the following hold:*

(1) *The essential spectral radius of* $\mathcal{M}$ *acting on* $\mathcal{B}_{(0)}^c$ *is* $\leq \widehat{R}$.

(2) *The essential spectral radius of* $\mathcal{M}^{(r)}$ *acting on* $\mathcal{B}_{(0)}^c$ *is* $\leq \widehat{R}_r$.

(3) $\mathcal{M}$ *acting on* $\mathcal{B}_{(r)}^c$ *and* $\mathcal{M}^{(r)}$ *acting on* $\mathcal{B}_{(0)}^c$ *have the same estimate* $\widehat{R}_r$ *for the essential spectral radius, and the same eigenvalues* λ *such that* $|\lambda| > \widehat{R}_r$, *with the same multiplicity.*

(4) $\mathcal{M}$ *acting on* $\mathcal{B}_{(r)}^c$ *and* $\mathcal{M}^{(r)}$ *acting on* $\mathcal{B}_{(0)}^c$ *have spectral radius* $\leq \max(R, \widehat{R}_r)$.

Writing $\mathcal{B}_{(-k)}^c = \mathcal{B}_{(-k)} \cap \mathcal{B}_{(0)}^c$ we see that Lemma 2.1 remains true with $\mathcal{B}_{(k)}$, $\mathcal{B}_{(-k)}$ replaced by $\mathcal{B}_{(k)}^c$, $\mathcal{B}_{(-k)}^c$.

In the proof of Proposition 2.5 it suffices to replace the space M by the subspace M^c of nonatomic measures, noting that $\mathcal{M}^*$ maps M^c to M^c. For Proposition 2.6, simply replace $\mathcal{B}_{(k)}$, $\mathcal{B}_{(-k)}$ by $\mathcal{B}_{(k)}^c$, $\mathcal{B}_{(-k)}^c$, and insert a factor $N(\omega)$ in the last formula of the proof. Similarly for Propositions 2.7, 2.8. □

3. Sharp trace and zeta function.

In this section, besides the assumptions already made we let $r \geq 1$. Since the g_ω are continuous we may as in [1] define the *sharp trace*

$$\mathrm{Tr}^{\#} \, \mathcal{M} = \int \mu(d\omega) \int d(g_\omega(x)) \, \sigma_1(\psi_\omega(x) - x)$$

and let

$$\zeta^{\#}(z) = \exp \sum_{m=1}^{\infty} \frac{z^m}{m} \, \mathrm{Tr}^{\#} \, \mathcal{M}^m$$

$$= [\mathrm{Det}^{\#}(1 - z\,\mathcal{M})]^{-1}.$$

In the discrete case where $\int \mu(d\omega)$ can be replaced by $\sum\limits_{\omega}$, it is known (see [1]) that $\mathrm{Tr}^{\#}$ is uniquely defined, i.e. $\mathrm{Tr}^{\#} \mathcal{M}$ does not depend on the presentation of $\mathcal{M}$. Easy counterexamples show however that $\mathrm{Tr}^{\#} \mathcal{M}$, and thus $\zeta^{\#}(z)$, may depend on the presentation of $\mathcal{M}$ (i.e., the choice of $\mu(d\omega)$, g_ω, ψ_ω) when μ is a continuous measure[*]. This will however not affect our results.

3.1. Kneading operators.

Following [1] we define the *kneading operator* $\mathcal{D} = \mathcal{D}(z)$ as an operator on $L^2(\nu)$ where ν is Lebesgue measure on $(-a, a)$. [This choice of ν is possible because the g_ω are smooth.] We let

$$(\mathcal{D}\varphi)(y) = \int \mu(d\omega) \int_{-a}^{a} \varphi(x)\, z\, d(g_\omega(x)) \left[(1 - z\,\mathcal{M})^{-1} \sigma_1(\cdot - y)\right](\psi_\omega\, x)$$

where $\mathcal{D}$ has the kernel

$$\mathcal{D}_{xy} = \sum_{k=1}^{\infty} z^k \int \mu(d\omega_1)\ldots\mu(d\omega_k)\, g'_{\omega_1}(x)\, g_{\omega_2}(\psi_{\omega_1}\, x)\ldots$$
$$\ldots g_{\omega_k}(\psi_{\omega_{k-1}}\ldots\psi_{\omega_1}\, x)\, \sigma_1(\psi_{\omega_k}\ldots\psi_{\omega_1}\, x - y).$$

For suitable z this kernel is bounded, and defines $\mathcal{D}$ as a Hilbert-Schmidt operator on $L^2(\nu)$. As proved (for countable Ω) in [1] we have then, as an identity between power series in z,

$$[\mathrm{Det}^{\#}(1 - z\,\mathcal{M})]^{-1} = \mathrm{Det}_*(1 + \mathcal{D}(z)).$$

Here

$$\mathrm{Det}_*(1 + \mathcal{D}) = 1 + \sum_{m=1}^{\infty} \frac{1}{m!} \int \nu(dx_1)\ldots\int \nu(dx_m)\, \Delta_m(x_1, \ldots, x_m)$$
$$= \left(\exp \int \nu(dx)\, \mathcal{D}_{xx}\right) \cdot \mathrm{Det}_2(1 + \mathcal{D})$$

where Δ_n is the determinant of the $n \times n$ matrix with elements $\mathcal{D}_{x_i x_j}$ $(i, j = 1, \ldots, n)$; the integral $\int \nu(dx)\, \mathcal{D}_{xx}$ is well defined and plays the role of a trace even though $\mathcal{D}$ is not of trace class; Det_2 is a *regularized determinant* defined by the power series

$$\mathrm{Det}_2(1 + \mathcal{D}) = \exp \sum_{m=2}^{\infty} \frac{(-1)^{m-1}}{m} \mathrm{Tr}\, \mathcal{D}^m$$

---

[*] See for instance Ruelle [7].

(see Simon [8]). Regularizing the kernel of $\mathcal{D}$ (by convolution) one obtains operators $\mathcal{D}_{*n}$ such that $\mathrm{Det}_*(1 + \mathcal{D}_{*n})$ is a bona fide Fredholm determinant $\mathrm{Det}(1 + \mathcal{D}_{*n})$ tending to $\mathrm{Det}_*(1 + \mathcal{D})$ when $n \to \infty$. Therefore, as discussed in [1], $\mathrm{Det}_*(1 + \mathcal{D})$ has essentially the properties of a Fredholm determinant.

The identity

$$\zeta^\#(z) = \mathrm{Det}_*(1 + \mathcal{D}(z))$$

yields a meromorphic extension of $\zeta^\#(z)$ to $|z| < \widehat{R}^{-1}$ because $\widehat{R}$ is an upper bound to the essential spectral radius of $\mathcal{M}$ acting on $\mathcal{B}_{(0)}$.

We may replace $\mathcal{D}$ by a modified kneading operator $\mathcal{D}^{(r)}$, while respecting the power series for the determinant. Let $\mathcal{D}^{(r)}$ have the kernel

$$\mathcal{D}^{(r)}_{xy} = \sum_{k=1}^\infty z^k \int \mu(d\omega_1) \dots \mu(d\omega_k) \frac{d^r}{dx^r} \left[g'_{\omega_1}(x)\, g_{\omega_2}(\psi_{\omega_1} x) \dots \right.$$
$$\left. \dots g_{\omega_k}(\psi_{\omega_{k-1}} \dots \psi_{\omega_1} x)\, \sigma_{r+1}(\psi_{\omega_k} \dots \psi_{\omega_1} x - y) \right]$$
$$= \sum_{k=1}^\infty z^k \int \mu(d\omega_1) \dots \mu(d\omega_k) \sum_{s=0}^r \frac{r!}{s!(r-s)!}\, g^{(s+1)}_{\omega_1}(x) \cdot$$
$$\cdot \frac{d^{r-s}}{dx^{r-s}} \left[g_{\omega_2}(\psi_{\omega_1} x) \dots g_{\omega_k}(\psi_{\omega_{k-1}} \dots \psi_{\omega_1} x)\, \sigma_{r+1}(\psi_{\omega_k} \dots \psi_{\omega_1} x - y) \right].$$

Since

$$(-1)^r \frac{d^r}{dy^r} \sigma_{r+1}(\psi_{\omega_k} \dots \psi_{\omega_1} x - y) = \sigma_1(\psi_{\omega_k} \dots \psi_{\omega_1} x - y)$$

we see that the replacement $\mathcal{D} \to \mathcal{D}^{(r)}$ does not change the determinant. Note that $g^{(r)}_\omega(\cdot)$ is of bounded variation, with derivative a measure which need not be absolutely continuous with respect to Lebesgue. To take care of this fact let now ν be the measure on $(-a, a)$ defined by

$$\nu(dx) = dx + \sum_\omega |d\, g^{(r)}_\omega(x)|.$$

We interpret $g^{(s+1)}_\omega$ as the Radon-Nikodym derivative of $d\left(\frac{d^s\, g_\omega(x)}{dx^s} \right)$ with respect to $\nu(dx)$. In this way, $\mathcal{D}^{(r)}_{xy}$ is the (bounded) kernel of a Hilbert-Schmidt operator on $L^2(\nu)$, with a z-dependence to be discussed in a minute, and such that

$$\zeta^\#(z) = \mathrm{Det}_*(1 + \mathcal{D}^{(r)}(z)).$$

We have

$$\mathcal{D}_{xy}^{(r)} = \sum_{s=0}^{r} \frac{r!}{s!(r-s)!} \int \mu(d\omega)\, z\, g_\omega^{(s+1)}(x) \cdot \frac{d^{r-s}}{dx^{r-s}}$$

$$\left[(1 - z\,\mathcal{M})^{-1}\, \sigma_{r+1}(\cdot - y)\right](\psi_\omega\, x).$$

It is readily seen that $\{\sigma_{r+1}(\cdot - y) : y \in (-a, a)\}$ is a bounded set in $\mathcal{B}_{(r)}$. Therefore, if z^{-1} is outside of the spectrum of $\mathcal{M}$ acting on $\mathcal{B}_{(r)}$, then $\mathcal{D}_{xy}^{(r)}$ is bounded in x, y, and $\mathcal{D}^{(r)}$ is thus a Hilbert-Schmidt operator on L^2, depending holomorphically on z. As in [1], it then follows that $\mathrm{Det}_*(1 + \mathcal{D}^{(r)}(z))$ is holomorphic in z when z^{-1} is outside of the spectrum of $\mathcal{M}$. In fact, using spectral theory one sees as in [1] that $\mathrm{Det}_*(1 + \mathcal{D}^{(r)}(z))$ is meromorphic when $|z|^{-1} >$ essential spectral radius of $\mathcal{M}$, and has a pole of order $\le k$ at λ^{-1} if λ is an eigenvalue of order k [check this first for $k = 1$, then use a perturbation argument].

3.2. Theorem. *The zeta function*

$$\zeta^{\#}(z) = \exp \sum_{m=1}^{\infty} \frac{z^m}{m}\, \mathrm{Tr}^{\#}\, \mathcal{M}^m$$

is meromorphic when $|z| < \widehat{R}_r^{-1}$, *and if* λ^{-1} *is a pole of order* ℓ, *then* λ *is an eigenvalue of order* $\ge \ell$ *of* $\mathcal{M}$ *acting on* $\mathcal{B}_{(r)}$.

This follows from the considerations above, and the fact that the essential spectral radius of $\mathcal{M}$ is $\le \widehat{R}_r$ (Proposition 2.6). $\square$

3.3. Other formulae for $\zeta^{\#}$.

Using the change of variable $v = \psi_{\omega_k} \ldots \psi_{\omega_1} x$, it is readily seen that the power series expansion for $\mathrm{Det}_*(1 + \mathcal{D}(z))$ is unchanged if $\mathcal{D}$ is replaced by $\mathcal{D}^*$ where

$$\mathcal{D}_{uv}^* = \sum_{k=1}^{\infty} z^k \int \mu(d\omega_1) \ldots \mu(d\omega_k) \left| \frac{d}{dv}\, \psi_{\omega_1}^{-1} \ldots \psi_{\omega_k}^{-1}\, v \right|$$

$$\cdot\, \sigma_1\left(u - \psi_{\omega_1}^{-1} \ldots \psi_{\omega_k}^{-1}\, v\right) g_{\omega_1}'\left(\psi_{\omega_1}^{-1} \ldots \psi_{\omega_k}^{-1}\, v\right)$$

$$\cdot\, g_{\omega_2}\left(\psi_{\omega_2}^{-1} \ldots \psi_{\omega_k}^{-1}\, v\right) \ldots g_{\omega_k}\left(\psi_{\omega_k}^{-1}\, v\right).$$

[Remember that we integrate $x, v, \ldots$ with respect to $\nu =$ Lebesgue measure on $(-a, a)$.] The substitutions $\psi_\omega \to \psi_\omega^{-1}$, $g_\omega \to \varepsilon_\omega \cdot g_\omega \circ \psi_\omega^{-1}$ transform $\mathcal{D}(z)$ into an

199

operator $\widehat{\mathcal{D}}(z)$ and we have $\mathrm{Det}_*(1 + \widehat{\mathcal{D}}(z)) = \mathrm{Det}_*(1 + \widehat{\mathcal{D}}^*(z))$ where

$$\widehat{\mathcal{D}}^*_{uv} = \sum_{k=1}^{\infty} z^k \int \mu(d\omega_1) \ldots \mu(d\omega_k) \frac{d}{dv} (\psi_{\omega_1} \ldots \psi_{\omega_k} v)$$

$$\cdot \sigma_1 (u - \psi_{\omega_1} \ldots \psi_{\omega_k} v) \, g'_{\omega_1} (\psi_{\omega_2} \ldots \psi_{\omega_k} v) (\psi'_{\omega_1} (\psi_{\omega_2} \ldots \psi_{\omega_k} v))^{-1}$$

$$\cdot g_{\omega_2} (\psi_{\omega_3} \ldots \psi_{\omega_k} v) \ldots g_{\omega_k} (v).$$

$$= \sum_{k=1}^{\infty} z^k \int \mu(d\omega_1) \ldots \mu(d\omega_k) [(\psi'_{\omega_k} g_{\omega_k})(v)] \ldots [(\psi'_{\omega_2} g_{\omega_2})(\psi_{\omega_3} \ldots \psi_{\omega_k} v)]$$

$$g'_{\omega_1} (\psi_{\omega_2} \ldots \psi_{\omega_k} v) \, \sigma_1 (u - \psi_{\omega_1} \ldots \psi_{\omega_k} v)$$

$$= \int \mu(d\omega) [(1 - z \, \mathcal{M}_1)^{-1} (g'_\omega (\cdot) \, \sigma_1 (u - \psi_\omega \cdot))](v)$$

$$= -((1 - z \, \mathcal{M}_1)^{-1} \Psi_u)(v)$$

with

$$\Psi_u (w) = \int \mu(d\omega) \, g'_\omega(w) \, \sigma_1(\psi_\omega w - u).$$

More generally

$$\mathrm{Det}_*(1 + \widehat{\mathcal{D}}(z)) = \mathrm{Det}_*(1 + \mathcal{D}_{(r)}(z))$$

where

$$\mathcal{D}_{(r)xy} = (-1)^r \frac{d^{r-1}}{dx^{r-1}} ((1 - z \, \mathcal{M}_1)^{-1} \Psi_{(r)y})(x)$$

$$\Psi_{(r)y}(x) = \int \mu(d\omega) \, g'_\omega(x) \, \sigma_r(\psi_\omega x - y)$$

and we have (see [1])

$$\zeta^{\#}(z)^{-1} = \mathrm{Det}_*(1 + \widehat{\mathcal{D}}(z)) = \mathrm{Det}_*(1 + \mathcal{D}_{(r)}(z)).$$

Since $r \geq 1$, $\{\Psi_{(r)y} : y \in (-a, a)\}$ is bounded in $\mathcal{B}_{(r-1)}$. As in Section 3.1 we see that $\mathrm{Det}_*(1 + \mathcal{D}_{(r)}(z))$ is meromorphic when $|z| < \widehat{R}_r$ and has a pole of order $\leq k$ at λ^{-1} if λ is an eigenvalue of order k of $\mathcal{M}_1$ acting on $\mathcal{B}_{(r-1)}$. We have thus proved the following result:

3.4. Theorem. *The zeta function*

$$\zeta^{\#}(z) = \exp \sum_{m=1}^{\infty} \frac{z^m}{m} \, \mathrm{Tr}^{\#} \mathcal{M}^m$$

is meromorphic when $|z| < \widehat{R}_r^{-1}$ and if λ^{-1} is a zero of order ℓ, then λ is an eigenvalue of order $\geq \ell$ of $\mathcal{M}_1$ acting on $\mathcal{B}_{(r-1)}$. $\square$

Comparison of Theorem 3.2 and Theorem 3.4 suggest the following:

3.5. Conjecture. *The zeta function*

$$\zeta^\#(z) = \exp \sum_{m=1}^{\infty} \frac{z^m}{m} \, \mathrm{Tr}^\# \, \mathcal{M}^m$$

is meromorphic when $|z| < \widehat{R}_r^{-1}$ and its order at z is the order of z^{-1} as an eigenvalue of $\mathcal{M} = \mathcal{M}_0$ minus the order of z^{-1} as an eigenvalue of $\mathcal{M}_1$, where $\mathcal{M}$ acts on $\mathcal{B}_{(r)}$ and $\mathcal{M}_1$ on $\mathcal{B}_{(r-1)}$.

We shall prove a special case of this conjecture in Corollary 3.8.

Let $\mathcal{D}_k$, $\widehat{\mathcal{D}}_k$ be obtained by the replacement $g_\omega \to (\psi'_\omega)^k \, g_\omega$ in the definition of $\mathcal{D}$ and $\widehat{\mathcal{D}}$. We have then

$$\zeta^\# = \frac{\mathrm{Det}_*(1+\widehat{\mathcal{D}}_1)\ldots\mathrm{Det}_*(1+\widehat{\mathcal{D}}_r)}{\mathrm{Det}_*(1+\widehat{\mathcal{D}}_0)\ldots\mathrm{Det}_*(1+\widehat{\mathcal{D}}_r)}.$$

To prove the conjecture it would suffice to show that $\mathrm{Det}_*(1+\widehat{\mathcal{D}}_0)\ldots\mathrm{Det}_*(1+\widehat{\mathcal{D}}_r)$ is holomorphic for $|z| < \widehat{R}_r^{-1}$ and has zeros λ^{-1} corresponding to the eigenvalues λ of $\mathcal{M}$, with the same multiplicity. Note that $\mathrm{Det}_*(1+\widehat{\mathcal{D}}_0)\ldots\mathrm{Det}_*(1+\widehat{\mathcal{D}}_r)$ is equal to a modified sharp determinant $\mathrm{Det}_r^\#(1-z\,\mathcal{M})$ associated with the modified sharp trace $\mathrm{Tr}_r^\# \, \mathcal{M} = \int \mu(d\omega) \int d\,[g_\omega(x)(1+\psi'_\omega(x)+\cdots+(\psi'_\omega(x))^r)]\,\sigma_1(\psi_\omega\,x-x)$.

3.6. Corollary of Theorem 3.2. *Let the $\psi_\omega : \Lambda_\omega \to \psi_\omega \Lambda_\omega$ be C^∞ diffeomorphisms, and the g_ω be C^∞ functions with support in Λ_ω such that*

$$\int \mu(d\omega)\,\|P_\omega\|_0 < \infty$$

for all P_ω of the form $(*)$ and all r. Assume that $|\psi'_\omega(x)| \leq 1$ for all x, ω and that if $|\psi'_\omega(x)| = 1$ then $g_\omega(x) = 0$. Under these conditions the zeta function $\zeta^\#$ is meromorphic in $\mathbb{C}$, and if λ^{-1} is a pole of order ℓ, then λ is an eigenvalue of order $\geq \ell$ of $\mathcal{M}$ acting on $\mathcal{B}_{(k)}$ for sufficiently large k.

It suffices to show that $\widehat{R}_k \to 0$ when $k \to \infty$. Remember that $\widehat{R}_k$ is the spectral radius of $\widehat{\mathcal{M}}_k$ acting on $\mathcal{C}_0$, where

$$(\widehat{\mathcal{M}}_k \, \Phi)(x) = \int \mu(d\omega) \, [\varepsilon_\omega \, g_\omega(\psi'_\omega)^k \cdot \Phi](\psi_\omega^{-1}\,x).$$

In particular

$$\widehat{R}_k \leq \|\widehat{\mathcal{M}}_k\|_0 \leq \int \mu(d\omega) \sup_x |g_\omega(x)(\psi'_\omega(x))^k|.$$

The functions $\varphi_k \geq 0$ on Ω defined by

$$\varphi_k(\omega) = \sup_x |g_\omega(x)(\psi'_\omega(x))^k|$$

are uniformly bounded by φ_0, which is μ-integrable [because $\int \mu(d\omega)\varphi_0(\omega) \leq \int \mu(d\omega)\|g_\omega\|_0 < \infty$], and tend pointwise to 0 [fix ω; for each $\varepsilon > 0$ there is $\delta > 0$ such that $|g_\omega(x)| \geq \varepsilon \varphi_0(\omega)$ implies $|\psi'_\omega(x)| \leq 1 - \delta$; also for sufficiently large k we have $(1-\delta)^k < \varepsilon$, therefore $|g_\omega(x)(\psi'_\omega(x))^k| < \varphi_0(\omega)\varepsilon$ for all x, hence $\varphi_k(\omega) \to 0$ when $k \to \infty$]. We have thus $\widehat{R}_k \leq \int \mu(d\omega)\varphi_k(\omega) \to 0$ when $k \to \infty$. $\square$

We now give a stronger version of this Corollary corresponding to the situation of Remark 2.9 (piecewise monotone ψ_ω with condition (C)).

3.7. Corollary. *Let the $\psi_\omega : \Lambda_\omega \to (-a, a)$ be piecewise monotone C^∞ maps with $N(\omega)$ laps, and the g_ω be C^∞ functions with support in Λ_ω such that $\int \mu(d\omega)N(\omega)$ $\|P_\omega\|_0 < \infty$ for all P_ω of the form $(*)$ and all r. Assume that $|\psi'_\omega(x)| \leq 1$ for all ω, x, and that if $|\psi'_\omega(x)| = 1$ then $g_\omega(x) = 0$. Under these conditions the zeta function $\zeta^{\#}$ is meromorphic in $\mathbb{C}$, and if λ^{-1} is a pole of order ℓ, then λ is an eigenvalue of order $\geq \ell$ of $\mathcal{M}$ acting on $\mathcal{B}_{(k)}$ for sufficiently large k.*

Again we have to show that $\widehat{R}_k \to 0$ when $k \to \infty$; we have now

$$\widehat{R}_k \leq \|\widehat{\mathcal{M}}_k\|_0 \leq \int \mu(d\omega)N(\omega) \sup_x |g_\omega(x)(\psi'_\omega(x))^k|$$

and the proof of Lemma 3.6 readily adapts to the present case. $\square$

Let us keep the assumptions of Corollary 3.7, and introduce C^∞ approximations $\widetilde{\psi}_\omega : \Lambda_\omega \to (-a, a)$ to the ψ_ω. Given $\varepsilon > 0$ we assume, as we may, that

(a) *$\widetilde{\psi}_\omega$ and ψ_ω are ε-close, as well as a finite number of their derivatives,*

(b) *$\widetilde{\psi}_\omega(x) = \psi_\omega(x)$ unless $|\psi'_\omega(x)|$ is close to 1, and the set of x such that the interval between $\psi_\omega(x)$ and $\widetilde{\psi}_\omega(x)$ contains any given y has Lebesgue measure $\leq N(\omega)\varepsilon$,*

(c) *also $\widetilde{\psi}_\omega(x) = \psi_\omega(x)$ unless $|g_\omega(x)| < \varepsilon \|g_\omega\|_{(0)}$,*

(d) *finally, for some $\delta(\omega) > 0$,*

$$\sup_x |(\widetilde{\psi}_\omega)'(x)| \leq 1 - \delta(\omega).$$

For $\widetilde{\Omega} \subset \Omega$, we write

$$(\mathcal{M}^* \Phi)(x) = \int_{\widetilde{\Omega}} \mu(d\omega)\, g_\omega(x)\, \Phi(\psi_\omega\, x)$$

$$(\widetilde{\mathcal{M}} \Phi)(x) = \int_{\widetilde{\Omega}} \mu(d\omega)\, g_\omega(x)\, \Phi(\widetilde{\psi}_\omega\, x).$$

Choose an integer $k \geq 0$, and remember that we can assume μ to be a probability measure. If $\mu(\Omega \backslash \widetilde{\Omega})$ is sufficiently small, we have

$$\|\mathcal{M} - \mathcal{M}^*\|_{(k)} < \varepsilon.$$

We may take $\widetilde{\Omega}$ such that for some $\delta > 0$,

$$\omega \in \widetilde{\Omega} \Rightarrow \delta(\omega) > \delta$$

and therefore (d) gives

$$\sup_x |(\widetilde{\psi}_\omega)'(x)| \leq 1 - \delta \quad \text{if} \quad \omega \in \widetilde{\Omega}.$$

We want to estimate now the norm of $\mathcal{M}^* - \widetilde{\mathcal{M}}$ acting on $\mathcal{B}_{(k)}$, where

$$((\mathcal{M}^* - \widetilde{\mathcal{M}})\, \Phi)(x) = \int_{\widetilde{\Omega}} \mu(d\omega)\, g_\omega(x)\, (\Phi(\psi_\omega\, x) - \Phi(\widetilde{\psi}_\omega\, x)).$$

Starting with $k = 0$, we estimate

$$\mathrm{Var}\, [g_\omega \cdot ((\Phi \circ \psi_\omega) - (\Phi \circ \widetilde{\psi}_\omega))]$$

$$\leq \int dx\, |g_\omega'(x)| \cdot |\Phi(\psi_\omega\, x) - \Phi(\widetilde{\psi}_\omega\, x)|$$

$$+ \int dx\, |g_\omega(x)| \cdot \left| \frac{d}{dx}(\Phi(\psi_\omega\, x) - \Phi(\widetilde{\psi}_\omega\, x)) \right|$$

$$\leq \|g_\omega\|_{(1)} \int dx \int_{J(x)} dy\, |\Phi'(y)|$$

$$+ \varepsilon \|g_\omega\|_{(0)} \int dx\, \left| \frac{d}{dx}(\Phi(\psi_\omega\, x) - \Phi(\widetilde{\psi}_\omega\, x)) \right|$$

where $J(x)$ is the interval between $\psi_\omega(x)$ and $\widetilde{\psi}_\omega(x)$, and we have made use of assumption (c) above. Using assumption (b) we have

$$\int dx \int_{J(x)} dy\, |\Phi'(y)| \leq \varepsilon\, N(\omega)\, \mathrm{Var}\, \Phi.$$

203

Furthermore

$$\int dx \left| \frac{d}{dx} \left(\Phi(\psi_\omega x) - \Phi(\tilde{\psi}_\omega x) \right) \right| \leq 2\,N(\omega)\,\text{Var}\,\Phi\,.$$

Therefore

$$\|(\mathcal{M}^* - \widetilde{\mathcal{M}})\,\Phi\|_{(0)} \leq \varepsilon\,N(\omega)\,\text{Var}\,\Phi\,(\|g_\omega\|_{(1)} + 2\|g_\omega\|_{(0)})\,.$$

Similar results hold for larger k, so that

$$\|\mathcal{M}^* - \widetilde{\mathcal{M}}\|_{(k)} \leq C_k\,\varepsilon\,.$$

Finally

$$\|\mathcal{M} - \widetilde{\mathcal{M}}\|_{(k)} \leq (1 + C_k)\,\varepsilon\,.$$

From this it follows that the kernel of the kneading operator $\widetilde{\mathcal{D}}^{(k)}$ associated with $\widetilde{\mathcal{M}}$ tends to $\mathcal{D}^{(k)}$ uniformly for z in any compact K such that K^{-1} is disjoint from the spectrum of $\mathcal{M}$ acting on $\mathcal{B}_{(k)}$. Therefore the meromorphic function

$$\tilde{\zeta}(z) = \text{Det}(1 + \widetilde{\mathcal{D}}(z))$$

tends to $\zeta^\#(z) = \text{Det}_*(1 + \mathcal{D}(z))$ under the same conditions.

On the other hand, because of the uniform bound $|\tilde{\psi}'_\omega| \leq 1 - \delta$ we may write (see [5])

$$\tilde{\zeta}(z) = \frac{\text{Det}^\flat\,(1 - z\,\widetilde{\mathcal{M}}_1)}{\text{Det}^\flat\,(1 - z\,\widetilde{\mathcal{M}}_0)}$$

where the numerator and denominator are entire analytic functions; the zeros λ^{-1} of $\text{Det}^\flat(1 - z\,\widetilde{\mathcal{M}}_i)$ correspond to the eigenvalues λ of $\mathcal{M}_i$ acting on $\mathcal{B}_{(k)}$ for k sufficiently large, with the same multiplicity. Therefore the order of the meromorphic function $\tilde{\zeta}$ at z is the order of z^{-1} as eigenvalue of $\widetilde{\mathcal{M}}_0$ minus the order of z^{-1} as eigenvalue of $\widetilde{\mathcal{M}}_1$.

Letting $\widetilde{\mathcal{M}}$ tend to $\mathcal{M}$ we see that we have proved the following result.

3.8. Corollary. *Under the conditions of Corollary 3.7, the order of $\zeta^\#$ at z is the order of z^{-1} as eigenvalue of $\mathcal{M} = \mathcal{M}_0$ minus the order of z^{-1} as eigenvalue of $\mathcal{M}_1$ (both acting on $\mathcal{B}_{(k)}$ for sufficiently large k).* $\square$

3.9. Proof of Proposition 1.1, and complements. We show now how to put Proposition 1.1 in the framework of Corollaries 3.7 and 3.8.

It is no restriction of generality to assume that $X_\alpha \subset Y_\alpha$, where the Y_α are disjoint open subintervals of $(-a, a)$. For each ω let Λ_ω be a closed interval contained in $Y_{\alpha(\omega)}$ and containing $X_{\alpha(\omega)}$ in its interior. We may then choose an extension $\overline{\psi}_\omega$ of ψ_ω to Λ_ω such that

(a) $\overline{\psi}_\omega$ is C^∞

(b) $\overline{\psi}_\omega$ has a finite number $N(\omega)$ of laps

(c) $\overline{\psi}_\omega \Lambda_\omega \subset Y_{\beta(\omega)}$

(d) $|\overline{\psi}'_\omega(x)| < 1$ for $x \in \Lambda_\omega \setminus X_{\alpha(\omega)}$.

We also choose a C^∞ extension $\overline{g}_\omega$ of g_ω to $\mathbb{R}$ such that $\operatorname{supp} \overline{g}_\omega \subset \Lambda_\omega$.

The extensions $\overline{\psi}_\omega, \overline{g}_\omega$ now satisfy the conditions of Corollary 3.7. Let us show that the corresponding zeta function $\zeta^\#$ is in fact the zeta function ζ of Proposition 1.1. We have

$$\operatorname{Tr}^\# \mathcal{M}^m = \sum_{\omega_1 \ldots \omega_m} \int d\left(\overline{g}_{\omega_1 \ldots \omega_m}(x)\right) \sigma_1(\overline{\psi}_{\omega_m} \ldots \overline{\psi}_{\omega_1} x - x)$$

where

$$\overline{g}_{\omega_1 \ldots \omega_m}(x) = \prod_{k=0}^{m-1} \overline{g}_{\omega_k}(\overline{\psi}_{\omega_{k-1}} \ldots \overline{\psi}_{\omega_1} x)$$

and we may replace $\sum$ by $\sum^{**}$ restricted by $\beta(\omega_1) = \alpha(\omega_2), \ldots, \beta(\omega_m) = \alpha(\omega_1)$. With this restriction $\psi_{\omega_m} \circ \cdots \circ \psi_{\omega_1}$ has a fixed point in $X_{\alpha(\omega_1)}$. The set

$$I(\omega_1, \ldots, \omega_m) = \{x : \overline{\psi}_{\omega_m} \ldots \overline{\psi}_{\omega_1} x = x\}$$

contains thus a point of $X_{\alpha(\omega_1)}$ and since (d) gives $|(\overline{\psi}_{\omega_m} \circ \ldots \circ \overline{\psi}_{\omega_1})'(x)| < 1$ for $x \notin X_{\alpha(\omega_1)}$ we see that

$$I(\omega_1, \ldots, \omega_m) \subset X_{\alpha(\omega_1)}.$$

This set I is either an interval $[x', x'']$ or a point $\{x(\omega_1, \ldots, \omega_m)\}$. In the first case $\overline{\psi}'_{\omega_1}(x') = \overline{\psi}'_{\omega_1}(x'') = 1$, hence $\overline{g}_{\omega_1 \ldots \omega_n}(x') = \overline{g}_{\omega_1 \ldots \omega_m}(x'') = 0$ and the corresponding contribution to $\operatorname{Tr}^\# \mathcal{M}^m$ vanishes. Therefore

$$\operatorname{Tr}^\# \mathcal{M}^m = \sum^*_{\omega_1 \ldots \omega_m} \overline{g}_{\omega_1 \ldots \omega_m}(x(\omega_1, \ldots, \omega_m))$$

$$= T_m$$

which shows that $\zeta^{\#} = \zeta$.

From Corollary 3.7 we obtain thus the conclusion that $\zeta(z)$ is meromorphic in $\mathbb{C}$. From Corollary 3.8 we obtain the relation of the zeros and poles of $\zeta(z)$ and the eigenvalues of $\mathcal{M}_0$, $\mathcal{M}_1$; these are the transfer operators constructed from the $\overline{\psi}_\omega$, $\overline{g}_\omega$, and involve therefore a (noncanonical) extension outside of the sets X_α originally given. $\square$

Bibliography.

[1] Baladi, V and Ruelle, D *Sharp determinants*, to appear.

[2] Nussbaum, R D (1970) *The radius of the essential spectrum*, Duke Math. J. **37**, 473-478.

[3] Pollicott, M (1991) *On the Ruelle-Tangerman theorem for zeta functions* in European Conference on Iteration Theory, World Scientific, Singapore, 201-209.

[4] Ruelle, D (1989) *The thermodynamic formalism for expanding maps*, Commun. Math. Phys. **125**, 239-262.

[5] Ruelle, D (1990) *An extension of the theory of Fredholm determinants*, Publ. Math. IHES **72**, 175-193.

[6] Ruelle, D *Functional equation for dynamical zeta functions of Milnor-Thurston type*, Commun. Math. Phys., to appear.

[7] Ruelle, D (1995) *Functional determinants related to dynamical systems and the thermodynamic formalism*, unpublished Lecture Notes, Pisa.

[8] Simon, B (1979) *Trace ideals and their applications*, Cambridge University Press, Cambridge.

[9] Tangerman, F (1986) *Meromorphic continuation of Ruelle zeta functions*, Boston University thesis (unpublished).

SLOBODAN SIMIĆ
Lyapunov functions and Anosov flows

1 Introduction and background

Our main concern in this paper is existence of global cross sections to codimension one Anosov flows. However, instead of attacking the problem directly (which we did elsewhere: see [S1], [S2]), we focus our attention to the lift of the codimension one Anosov flow to the universal covering space of the underlying manifold. But first, let us briefly recall the main notions.

A nonsingular C^1 flow $\{f_t\}$ on a compact connected Riemannian manifold M is called **Anosov** if the tangent bundle of M splits continuously into the Tf_t-invariant Whitney sum $TM = E^c \oplus E^{ss} \oplus E^{uu}$ such that the **strong stable bundle** E^{ss} is exponentially contracted and the **strong unstable bundle** E^{uu} is exponentially expanded by Tf_t for $t > 0$; $E^c = \mathbf{R}X$ is the line bundle tangent to the flow, where X is the generating vector field. We call $E^{cs} := E^c \oplus E^{ss}$ and $E^{cu} := E^c \oplus E^{uu}$ the **center stable** and **center unstable bundle**, respectively. It was shown by Anosov [An] that all the bundles mentioned above are uniquely integrable, that is tangent to unique foliations denoted here by W^{ss}, W^{uu}, W^{cs} and W^{cu}, respectively. For basic facts about Anosov flows see also [Pl1]. An Anosov flow is of **codimension one** if $\dim E^{uu} = 1$ or $\dim E^{ss} = 1$; we will always assume the former. It is well known [Ve] that if $\dim M > 3$, then any codimension one Anosov flow on M is topologically transitive; moreover, its center stable bundle E^{cs} is of class C^1 (also if $\dim M = 3$; see [HP] and also [HK]).

Verjovsky conjectured that if $\{f_t\}$ is Anosov of codimension one and $\dim M > 3$, then $\{f_t\}$ admits a global cross section and is therefore topologically conjugate to the suspension of a linear Anosov diffeomorphism on a torus. This conjecture was verified in several special cases; see [Pl2], [S1] and [S2]. In this paper we consider the universal covering $p : \tilde{M} \to M$ of M and the lifts $\tilde{f}_t$, $\tilde{X}$, $\tilde{E}^{cs}$, $\tilde{W}^{cs}$, etc. by p of the corresponding objects on M. It was shown by Palmeira [Pa] (as a consequence of a more general result) that $\tilde{M} \approx \mathbf{R}^n$, and from now on we identify $\tilde{M}$ with $\mathbf{R}^n$ via this diffeomorphism.

Later on we will need the following notion of the generalized exterior differential of a differential form.

Definition A locally integrable form β of degree $r + 1$ is the **generalized differential** of a locally integrable r-form α if for every C^∞ form ϕ of degree $n - r - 1$ with compact support the following holds:

$$\int_{\mathbf{R}^n} \phi \wedge \beta = (-1)^{n-r} \int_{\mathbf{R}^n} d\phi \wedge \alpha.$$

It is easy to see by Stokes' theorem that if α is of class C^1, then its ordinary differential is also its generalized differential. For details on this matter we refer the reader to [Rs], II.4.

2 Results

For the rest of the paper $\{f_t\}$ will denote a codimension one Anosov flow on a compact manifold M. Consider the following:

Condition (L) *The lift $\tilde{W}^{cs}$ of the center stable foliation of $\{f_t\}$ is a trivial foliation, i.e. given by a C^1 submersion $\mathbf{R}^n \to \mathbf{R}$.*

This condition was studied in [Ve], and an incorrect proof of it was given. (More precisely, Proposition 3.4 of [Ve] is incorrect. This was pointed out to me by S. Fenley.) We do not know how to prove condition (L), but since it will lead to some interesting consequences, we will adopt it as an additional assumption. It is easy to see that (L) is satisfied if, for instance, $\tilde{W}^{cs}$ admits a transversal intersecting all of its leaves. By Solodov [So] this is the case when the center of the fundamental group of M is free cyclic.

From now on we take (L) as our standing hypothesis. Thus without loss of generality we may assume that the covering projection p is of class C^1 and $\tilde{W}^{cs}$ is the hyperplane foliation given by $x_n = $ constant. Define a 1-form ω on M by requiring

$$\mathrm{Ker}(\omega) = E^{cs}, \quad \omega(Y) = 1,$$

where Y is a continous vector field in E^{uu}. (We made such a choice for the value of ω in the transverse direction to E^{cs} because we want formula (1) below to hold.) Since E^{cs} is a C^1 bundle, by being a little careful in choosing Y (for instance, take Y of the form $Z + V$, where $V \in E^{cs}$, and Z is a nonsingular C^∞ vector field transverse to E^{cs}), we can make sure that ω is of class C^1. By Frobenius theorem for the integrable bundle E^{cs}, there exists a continuous 1-form η on M such that $d\omega = \omega \wedge \eta$. Define $u : M \to \mathbf{R}$ by $u = \eta(X)$. Also let $\lambda(x,t) = \det T_x f_t|_{E^{uu}}$, with respect to some previously chosen Riemann structure $\mathcal{R}$ on M. Note that the form η is not unique: if $\eta' = \eta + h\omega$, then $\eta' \wedge = \eta \wedge \omega$. However, thanks to the fact that $\omega(X) = 0$, u does not depend on the choice of η. It was shown in [S2] that for all $x \in M$ and $t \in \mathbf{R}$,

$$\lambda(x,t) = \exp\left\{ \int_0^t u(f_s x)\, ds \right\}, \tag{1}$$

and it is possible to choose ω and a continuous Riemann structure $\mathcal{R}$ on M such that the corresponding u is C^1 and strictly positive. Note that this implies that the flow is "immediately expanding" on E^{uu}.

208

Since the 1-forms $\tilde{\omega}$ and dx_n have the same nullspace, namely $\tilde{E}^{cs}$, there exists a continuous function $g : \mathbf{R}^n \to \mathbf{R}$ such that

$$\tilde{\omega} = g\, dx_n.$$

Without loss we will assume that g is everywhere positive; otherwise, replace ω by $-\omega$.

Now we can state the following result.

Theorem 1 (a) *g has partial derivatives with respect to $x_1, \ldots, x_{n-1}$ and they are continuous.*

(b) *There exists a continuous function $b : \mathbf{R}^n \to \mathbf{R}$ such that*

$$\tilde{\eta} = d^{cs}(\log g) + b(x)\, dx_n.$$

Here d^{cs} denotes the leafwise $\tilde{W}^{cs}$-differential: $d^{cs}(\log g) = \sum_{i=1}^{n-1}(\partial \log g/\partial x_i)\, dx_i$.

(c) *We have: $u(p(x)) = \tilde{X}_x(\log g)$ and*

$$\lambda(p(x), t) = \frac{g(\tilde{f}_t(x))}{g(x)},$$

for every $x \in \mathbf{R}^n$ and $t \in \mathbf{R}$.

(d) *For every deck transformation T and every $x \in \mathbf{R}^n$,*

$$g(Tx)\, T_n'(x_n) = g(x).$$

(Here T_n denotes the nth coordinate function of T. Since T preserves $\tilde{W}^{cs}$, T_n depends only on x_n, so $T_n'(x_n) = \frac{\partial T_n}{\partial x_n}(x_n)$.)

PROOF (a), (b) We will use the notion of generalized differentials defined above.

Consider the form $\tilde{\omega} = p^*\omega$. Since p is only of class C^1, we cannot claim that $\tilde{\omega}$ is a C^1 form. However, using p as a local change of variables, it is not difficult to see that $\tilde{\omega}$ is actually differentiable in the generalized sense. Indeed, let ϕ be a C^∞ form of degree $n-2$ with compact support in $\mathbf{R}^n$. Without loss of generality we can assume the support of ϕ is contained in a small open set U on which p is 1–1 (otherwise use a partition of unity). Thus there exists an $(n-2)$-form ψ on M such that $\phi = (p|_U)^*\psi$ and $d\psi$ exists in the generalized sense. It follows

$$
\begin{aligned}
\int_{\mathbf{R}^n} d\phi \wedge \tilde{\omega} &= \int_U p^*(d\psi \wedge \omega) \\
&= \int_M d\psi \wedge \omega \\
&= (-1)^{n-1}\int_M \psi \wedge d\omega \\
&= (-1)^{n-1}\int_{\mathbf{R}^n} \phi \wedge p^*(d\omega).
\end{aligned}
$$

Therefore, $p^*(d\omega)$ is the generalized differential of $\tilde{\omega}$. Clearly, $p^*(d\omega) = p^*(\eta \wedge \omega) = \tilde{\eta} \wedge \tilde{\omega}$. There exist continuous functions $a_i : \mathbf{R}^n \to \mathbf{R}$ $(1 \le i \le n)$ such that

$$\tilde{\eta} = \sum_{i=1}^{n} a_i \, dx_i.$$

Let $h : \mathbf{R}^n \to \mathbf{R}$ be an arbitrary smooth function with compact support. Define $\phi = h \, dx_2 \wedge \ldots \wedge dx_{n-1}$. Then we have:

$$
\begin{aligned}
\int_{\mathbf{R}^n} g \frac{\partial h}{\partial x_1} dx_1 \wedge \ldots \wedge dx_n &= \int_{\mathbf{R}^n} d\phi \wedge \tilde{\omega} \\
&= (-1)^{n-1} \int_{\mathbf{R}^n} \phi \wedge \tilde{\eta} \wedge \tilde{\omega} \\
&= -\int_{\mathbf{R}^n} g a_1 h \, dx_1 \wedge \ldots \wedge dx_n.
\end{aligned}
$$

This shows that the generalized (i.e. Sobolev) derivative of g with respect to x_1 is $g a_1$. Thus

$$a_1 = \frac{\partial (\log g)}{\partial x_1},$$

in the generalized sense. Similarly, we can prove analogous formulas for $a_2, \ldots, a_{n-1}$. Since a_i's are continuous hence locally bounded functions, g actually belongs to the Sobolev space $W_{\text{loc}}^{1,\infty}$ on every hyperplane $x_n = $ constant. By the theory of Sobolev spaces and Rademacher's theorem (see, for instance, [EG]), g is differentiable in the ordinary sense with respect to the first $n-1$ variables and its weak partials equal its ordinary partials. It follows that g is continuously differentiable with respect to the first $n-1$ variables in the ordinary sense. This completes the proof of both (a) and (b).

(c) We have:

$$
\begin{aligned}
u(p(x)) &= \eta(X_{p(x)}) \\
&= \tilde{\eta}(\tilde{X}_x) \\
&= d^{cs}(\log g)(\tilde{X}_x) \qquad (2) \\
&= \tilde{X}_x(\log g).
\end{aligned}
$$

Note that (2) follows from (b). The second equality of part (c) is an easy consequence of (1).

(d) Let T be a deck transformation of the covering $p : \mathbf{R}^n \to M$. Since T preserves the foliation $\tilde{W}^{cs} : x_n = $ constant, it follows that its nth component, T_n, depends only on x_n. Note that $T^* \tilde{\omega} = \tilde{\omega}$ and $T^*(dx_n) = T_n'(x_n) \, dx_n$. Part (d) now easily follows by applying the pullback T^* to the relation $\tilde{\omega} = g \, dx_n$.

This completes the proof of the theorem. $\square$

210

Since $u > 0$, we see that g strictly increases along the orbits of the lifted flow, so g is a global Lyapunov function for $\tilde{f}_t$. To strengthen this, we will need another result from [S2]. Namely, there we showed that every codimension one Anosov flow can be "synchronized", i.e. reparametrized by a C^1 function (actually, u) so that for every t, the determinant of the time t map of the synchronized flow f_t' on its strong unstable bundle is identically equal to e^t. Since the center stable bundles of f_t and f_t' are identical, Theorem 1 implies

$$g(\tilde{f}_t' x) \equiv e^t g(x),$$

where $\tilde{f}_t'$ is the lift of the synchronized flow. Now assume $y \in \tilde{W}^{ss}(x)$, where $\tilde{W}^{ss}$ denotes the lift by p of the strong stable foliation of the synchronized flow. Let γ be a path in $\tilde{W}^{ss}(x)$ connecting x to y. Without loss of generality we can assume that x and y are so close that the restriction of p to γ is one-to-one. Then we have:

$$
\begin{aligned}
|\log g(x) - \log g(y)| &= |\log g(\tilde{f}_t' x) - \log g(\tilde{f}_t' y)| \\
&= \left| \int_{\tilde{f}_t'(\gamma)} \tilde{\eta} \right| \\
&= \left| \int_{f_t'(p(\gamma))} \eta \right| \qquad (3) \\
&\leq \|\eta\|_\infty \, \ell(f_t'(p(\gamma))) \\
&\to 0,
\end{aligned}
$$

as $t \to \infty$. Here ℓ denotes arclength. Note that (3) is just a change of variables formula. Thus $g(x) = g(y)$. Therefore, the foliation by the level sets of g of each leaf of the lifted center stable foliation for f_t' coincides with the lift of the strong stable foliation of f_t'. Thus we have proved the following result:

Theorem 2 *If a codimension one Anosov flow satisfies condition (L), then its lift to the universal covering space (that is $\mathbf{R}^n$) admits a continuous global Lyapunov function which strictly increases along its orbits and is constant on the leaves of $\tilde{W}^{ss}$ foliation for the synchronized flow.*

Let $\mathcal{G}$ be the partition of $\mathbf{R}^n$ by the level sets of g. (Note that since g is not necessarily C^1, we cannot claim that the leaves of $\mathcal{G}$ are smooth.) It is natural to ask whether $\mathcal{G}$ is invariant with respect to the deck transformations T. If so, by projecting $\mathcal{G}$ to M, we would obtain a partition $p(\mathcal{G})$ of M which is invariant with respect to the synchronized flow f_t. That would in turn imply that $p(\mathcal{G})$ is a continuous *foliation* tangent to $E^{ss} \oplus E^{uu}$, hence (by [Pl1]) the flow would be a suspension. Unfortunately, considering Theorem 1 (d), it does not seem easy to prove that $\mathcal{G}$ is projectable to M.

3   Remarks

1) Plante [Pl1] showed that an Anosov flow has a global cross section if its strong stable and strong unstable foliations are jointly integrable. This means that, locally speaking, the projection from a center stable leaf to a center stable leaf along the strong unstable manifolds maps the strong stable leaves to strong stable leaves. In view of that result, it is easy to see that if the function g is constant on the leaves of the $\tilde{W}^{uu}$ foliation for the synchronized flow, then the flow $\{f_t\}$ that we started with admits a global cross section.

2) T. Barbot [Ba] showed that a sufficient condition for a codimension one Anosov flow (in any dimension) to have a cross section is that every leaf of $\tilde{W}^{uu}$ intersects every leaf of $\tilde{W}^{cs}$ exactly once. We state this condition in terms of the function g. As above, let Y be a continuous vector field in E^{uu} such that $\omega(Y) = 1$, let $\tilde{Y}$ be its lift by p, and denote by $\{\tilde{\phi}_t\}$ the lift of of flow of Y (i.e. the flow of $\tilde{Y}$). Then we have:

Lemma *A codimension one Anosov flow which satisfies condition (L) has a global cross section if for all $x \in \mathbf{R}^n$,*

$$\int_0^{\pm\infty} \frac{1}{g(\tilde{\phi}_s x)} \, ds = \pm\infty.$$

PROOF Let x_n denote as before the projection to the nth coordinate in $\mathbf{R}^n$. Then we have:

$$
\begin{aligned}
\frac{d}{dt} x_n(\tilde{\phi}_t x) &= dx_n(\tilde{Y}_{\tilde{\phi}_t x}) \\
&= \frac{1}{g(\tilde{\phi}_t x)} \tilde{\omega}(\tilde{Y}) \\
&= \frac{1}{g(\tilde{\phi}_t x)}.
\end{aligned}
$$

Thus

$$x_n(\tilde{\phi}_t x) = x_n(x) + \int_0^t \frac{1}{g(\tilde{\phi}_s x)} \, ds. \tag{4}$$

Since the condition of Barbot is satisfied if the integral on the right hand side of (4) diverges to $+\infty$ as $t \to +\infty$, and to $-\infty$ as $t \to -\infty$, the proof is complete. $\square$

It is not difficult to see that the divergence of this integral is equivalent to the complete integrability of the vector field $Z = g\,\tilde{Y}$. In other words, that its flow lines through all points of $\mathbf{R}^n$ are defined for all time. One could hope to prove this by finding the "true meaning" of g in terms of objects in the base manifold M.

Acknowledgement

This paper constitutes Chapter 4 of author's doctoral dissertation [S2] completed at the University of California at Berkeley in 1995. The author thanks his advisor, Professor Charles C. Pugh, for his constant support, inspiring criticism and encouragement.

References

[An] D.V. Anosov: Geodesic flows on closed Riemannian manifolds with negative curvature, *Proceedings of the Steklov Math. Inst.*, no. **90** (1967), AMS Translations (1969)

[Ba] T. Barbot: Caractérisation des flots d'Anosov en dimension 3 par leurs feuilletages faibles, preprint, 1994

[EG] L.C. Evans and R.F. Gariepy: *Measure theory and fine properties of functions*, Studies in Advanced Math., CRC Press, Boca Raton, 1992

[HP] M.W. Hirsch, C.C. Pugh: Stable manifolds and hyperbolic sets, *Proc. Symp. in Pure Math.*, vol. **14** (1970), pp.133-163, AMS, Providence

[HK] S. Hurder, A. Katok: Differentiability, rigidity and Godbillon-Vey classes for Anosov flows, *Publicationes Math. IHES*, **72** (1990), pp.5-61

[Pa] C.F.B. Palmeira: Open manifolds foliated by planes, *Annals of Math.*, **107** (1978), pp.109-131

[Pl1] J.F. Plante: Anosov flows, *American J. of Math.*, vol. **94** (1972), pp.729-754

[Pl2] J.F. Plante: Anosov flows, transversely affine foliations, and a conjecture of Verjovsky, *J. London Math. Soc.* (**2**), (1981), pp.359-362

[Rs] Yu.G. Reshetnyak: *Space mappings with bounded distortion*, Translations of math. monographs, vol. **73** (1989), AMS, Providence, RI

[S1] S. Simić: Lipschitz distributions and Anosov flows, to appear in *Proc. of the AMS*

[S2] S. Simić: Anosov flows of codimension one, Ph.D. Thesis, University of California at Berkeley, 1995

[So] V.V. Solodov: Topological topics in dynamical systems theory, *Russian Math. Surveys* (translated from *Uspekhi Mat. Nauk*), **46**:4 (1991), pp.107-134

[Ve] A. Verjovsky: Codimension one Anosov flows, *Boletin de la Sociedad Matematica Mexicana* (2) **19** (1974), no.2, pp.49-77

Address: Department of Mathematics, University of Illinois at Chicago, Chicago, IL 60607-7045

Current address: Department of Mathematics, University of Southern California, Los Angeles, CA 90089-1113

RAÚL URES[*]

Hénon attractors: SBR measures and Dirac measures for sinks

Introduction In this paper we study the following family of diffeomorphisms

$$T_{(a,b)} = (1 - ax^2 + y, by) \text{ with } 0 < a < 2 \text{ and } 0 < |b| < 1$$

This family is known as the Hénon family.

In [1] (see also [6]) and [3] it was proved that for a positive Lebesgue measure set E of parameters (a, b), $T_{(a,b)}$ presents a transitive nonhyperbolic attractor that supports a unique ergodic SBR-measure $\lambda_{a,b}$.

In [8] Thunberg proved that the absolutely continuous measures obtained in [2] (see also [11]) for the quadratic family on the real line can be approximated by Dirac measures supported on superstable periodic orbits. Then, it is a natural question if there is a similar result for the Hénon-like attractors with the Bowen-Ruelle-Sinai (SBR) measures constructed in [3].

The aim of this paper is to prove the following theorem.

Theorem. *There exist a set $E \subset \mathbf{R}^2$ with $Leb(E) > 0$ such that for all $(a, b) \in E$, $T_{(a,b)}$ presents a transitive nonhyperbolic attractor that supports an ergodic SBR-measure $\lambda_{a,b}$. and there exists a sequence $\{\tilde{a}_n\}$ such that $\tilde{a}_n \to a$, $T_{\tilde{a}_n,b}$ exhibits a sink s_n for all n and the Dirac measures δ_{s_n} supported on s_n weakly converges to the measure $\lambda_{a,b}$.*

We think that the same result is valid in the setting of [6] (one-parameter families unfolding homoclinic tangencies) at least for generic families.

Finally we want to mention that when Hénon [4], by a numerical study, detected what seemed to be a non-trivial attractor in the Hénon family one could not exclude the possibility that this attractor were just a sink of a very high period. Benedicks and Carleson [1] showed that there exists a non-trivial attractor for a positive Lebesgue

---

[*]Partially supported by CSIC and CONICYT, Uruguay.

214

measure set of parameter values but the theorem above seems to show that it is very difficult to distinguish, by numerical studies, between this two cases.

Acknowledgements. I am in debt with M. Viana that suggest me to study this problem . I am also grateful to IMPA this fine institution where part of this paper was written.

Preliminaries

In our setting an ergodic SBR-measure will mean a measure that verifies the Pesin's formula (i.e. its future Liapunov exponent is equal to its entropy).

We will need some properties about the attractors obtained in the theorems of Benedicks and Carleson [1] and Benedicks and Young [3]. This properties are:

a) For all $(a, b) \in E$, the transitive nonhyperbolic attractor is the closure of $W^u(P_{a,b})$ where $P_{a,b}$ is a fixed point for $T_{a,b}$.(see [1])

b) For any $(a, b) \in E$ there exists a sequence $\{a_n\}$ such that $a_n \to a$ and $T_{a_n,b}$ generically unfolds a quadratic homoclinic tangency associated to $P_{a_n,b}$. (see [10])

The other property we need, that we call the "rectangle property", is the existence of a "rectangle" fenced by two compact segments of $W^s(P_{a,b})$ and two compact segments of $W^u(P_{a,b})$ such that the vertices are transversal intersections between them and there is some point of the attractor in its interior. It is not difficult to verify that the Hénon family has this property for the parameter values we are interested in (see [1] and [6]).

Then, the theorem is a consequence of the following proposition:

Proposition. Let $(f_\mu)_\mu$ be a real analytic one parameter family of diffeomorphisms of a surface and μ_0 a parameter value such that:

i) f_{μ_0} presents a hyperbolic fixed point P_{μ_0} such that the closure of $W^u(P_{\mu_0})$ is a compact set that supports an ergodic SBR measure λ_{μ_0} with a positive Liapunov exponent.

ii) f_{μ_0} is dissipative on the attractor (i.e. Jacobian less that one).

iii) There exists a sequence $\{\mu_n\}$ such that $\mu_n \to \mu_0$ and f_{μ_n} generically unfolds a quadratic homoclinic tangency associated to P_{μ_n} (P_{μ_n} is the analytic continuation of P_{μ_0}).

iv) P_{μ_0} verifies the "rectangle property".

then there exists a sequence $\{\mu'_n\}$ such that $\mu'_n \to \mu_0$, $f_{\mu'_n}$ exhibits a sink s_n for all n and the Dirac measures δ_{s_n} supported on s_n weakly converges to the measure λ_{μ_0}.

Remark. In [9] it was proved that for any ergodic SBR-measure μ there exists a hyperbolic periodic point p such that

$$\text{supp}(\mu) = \text{closure}(W^u(o(p)) \cap W^u(o(p)))$$

Then, what seems to be crucial to obtain the approximation by Dirac measures of sinks is property iii).

Proof of the theorem As we showed in the latter section to prove the theorem it is enough to prove the proposition.

An important tool for the proof of this proposition is the following theorem of Mendoza.

Theorem [5, Th.4.1.] *Suppose that μ is an ergodic f-invariant measure for a C^2 diffeomorphism $f : M \to M$ of a surface M. If its entropy $h_\mu(f)$ is equal to its future Liapunov exponent $\chi_\mu(f)$(i.e. it verifies Pesin's formula) and they are positive, then there exists a collection of ergodic measures μ_n supported on hyperbolic basic sets Ω_n satisfying $\mu_n \to \mu$.*

Before we prove the proposition we need some lemmas.

From now on, let us denote by δ_p the Dirac measure supported on the orbit of a periodic point p.

Lemma 1. *Let Ω be a hyperbolic basic set for f and ν be an ergodic measure supported on it. Then, there exists a sequence of periodic points $\{p_i\}_{i \in \mathbf{N}} \subset \Omega$ such that the Dirac measures δ_{p_i} supported on the orbits of p_i converge to ν.*

Proof: Take a ν-generic point $x \in \Omega$ (i.e. $\frac{1}{n}\sum_{j=0}^{n-1} \alpha(f^j(x)) \to n \to \infty \int \alpha d\nu$ for any continuous function $\alpha : \Omega \to \mathbf{R}$). As $x \in \omega(x)$ we have that for any $m \in \mathbf{N}^+$ there exists $n(m) \in \mathbf{N}^+$ such that $\mathrm{dist}(x, f^{n(m)}(x)) < \frac{1}{m}$. Then, by the shadowing lemma, given $i \in \mathbf{N}^+$ there exists $m = m(i) \in \mathbf{N}^+$ such that the orbit of x up to $n(m)$ is $\frac{1}{i}$-shadowed by a periodic orbit p_i. It is not difficult to see that the sequence $\{p_i\}$ verifies the conclusion of the lemma.

Lemma 2. *Let $(\varphi_\mu)_{\mu \in \mathbf{R}}$ be a family of two-dimensional C^2 diffeomorphisms that generically unfolds a quadratic homoclinic tangency associated to a dissipative hyperbolic periodic point p_0 of φ_0 ($|\lambda_1 \lambda_2| < 1$ where λ_i are the eigenvalues of $D_{p_0}(\varphi_0^k)$ and k the period of p_0). Then, there exist $\mu_n \to 0$ and sinks $\{s_n(\mu_n)\}$ of φ_{μ_n} such that $\delta_{s_n(\mu_n)} \to \delta_{p_0}$.*

Proof: By [7, Sec.3.4] there exists a sequence of sinks $\{s_n(\mu_n)\}$ with periods going to infinity and $\mu_n \to 0$ such that given any neighbourhood U of p, s_n has a uniformly bounded number of iterates out of U. Then all limit measures of δ_{s_n} are supported in U. As for sufficiently small U the unique invariant measure with support in it is δ_{p_0} this implies the lemma (recall that if f_n uniformly converge to f and m_n are f_n-invariant probability measures that weakly converge to m then m is f-invariant).

A consequence of the two lemmas above is the following lemma.

Lemma 3. *Let $(\varphi_\mu)_{\mu \in \mathbf{R}}$ be a family of two dimensional C^2 diffeomorphisms and Ω_0 a hyperbolic basic set for φ_0 that unfolds generically homoclinic tangencies at $\{\bar{\mu}_i\}$, $\bar{\mu}_i \to 0$ and ν an ergodic measure supported on it. Then, there exists a sequence of sinks $\{s_n(\mu_n)\}$, $\mu_n \to 0$, such that $\delta_{s_n(\mu_n)} \to \nu$.*

216

Proof: By lemma 1 there exist p_i such that $\delta_{p_i} \to \nu$ and as Ω_0 unfolds generically homoclinic tangencies the same occurs with p_i for all $i \in \mathbf{N}$ at parameter values $\mu_n(p_i) \to 0$. Now lemma 2 implies the existence of sequences of parameter values converging to $\mu_n(p_i)$ and exhibiting sinks such that their Dirac mesures converge to $\delta_{p_i(\mu_n(p_i))}$ for all $i \in \mathbf{N}$ and for all $n \in \mathbf{N}$ ($p_i(\mu_n(p_i))$ means the analytic continuation of p_i at $\mu_n(p_i)$). As $\delta_{p_i(\mu_n(p_i))} \to \delta_{p_i}$ the lemma is obtained by taking an appropiate subsequence.

Proof of the Proposition: First of all, let us recall that SBR-measures verify Pesin's formula. By lemma 3 and Mendoza's theorem we only have to prove that for any Ω_n, with n big enough, there exists a sequence of parameter values $\mu_{n_k} \to \mu_0$ such that Ω_n generically unfolds a quadratic homoclinic tangency at each μ_{n_k}. To obtain this it is enough to prove that $W^s(P_{\mu_0})$ and $W^u(P_{\mu_0})$ have transversal intersections with $W^u(z)$ and $W^s(z)$, being z any periodic point of Ω_n. Then, taking a subsequence of all the sinks obtained applying lemma 3, we prove the proposition.

Observe that, for n large enough, any $W^u(z)$ and $W^s(z)$ with $z \in \Omega_n$ a periodic point has at least topologically transversal intersections with $W^s(P_{\mu_0})$ and $W^u(P_{\mu_0})$ respectively. In fact, by Mendoza's theorem, Ω_n for n large enough have points inside and outside the "rectangle" given by the "rectangle property". Then, as Ω_n is a hyperbolic basic set, the stable and unstable manifold of any point in it are dense in Ω_n, thus, they have points inside and outside the mentioned "rectangle" (see figure). As all these manifolds are real analytic we obtain points of intersection of $W^u(z)$ and $W^s(z)$ with $W^s(P_{\mu_0})$ and $W^u(P_{\mu_0})$ having contact of odd type.

As $W^s(P_{\mu_0})$ and $W^u(P_{\mu_0})$ have transversal intersections, λ-lemma type estimates imply that $W^s(z)$ and $W^u(z)$ have transversal intersections with $W^u(P_{\mu_0})$ and $W^s(P_{\mu_0})$. More precisely, taking appropiate coordinates in a neighbourhood U of P_{μ_0} we obtain that the tangent space of $W^u(z)$ at the point having second coordinate y has slope less than $\frac{C}{|y|}$ with respect to $W^u(P_{\mu_0})$ and that this point stays in U, at least, $\frac{\log|y|^{-1}}{\log|\mu|}$ iterates where μ is the expansive eighenvalue of P_{μ_0}. Then, after $\frac{\log|y|^{-1}}{\log|\mu|}$ iterates and neglecting higher order terms, we obtain that the slope of the iterates of the tangent space of $W^u(z)$ at the mentioned point is less than

$$C|y|^{-1}(|\mu|^{-1}|\lambda|)^{\frac{\log|y|^{-1}}{\log|\mu|}} = C|y|^{-1}|y|^{1+\frac{\log|\lambda|^{-1}}{\log|\mu|}} = C|y|^{\frac{\log|\lambda|^{-1}}{\log|\mu|}}$$

where λ is the contractive eighenvalue of P_{μ_0}. It is not difficult to see that this implies that if $W^u(P_{\mu_0})$ transversally intersects $W^s(P_{\mu_0})$ the same is true for $W^u(z)$.

This implies, as mentioned above, the proposition.

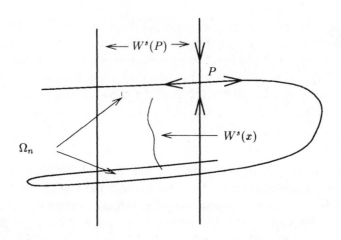

References

[1] M. Benedicks, L. Carleson *The dynamic of the Hénon map*, Ann. of Math. **133** (1991), 73-169.

[2] M.Benedicks, L.S.Young *Absolutely continuous invariant measures and random perturbations for certain one-dimensional maps*, Erg. Th. and Dyn. Syst. **12** (1992), 13-37.

[3] M.Benedicks, L.S.Young *SBR-measures for certain Hénon maps*, Inventiones Math. **112** (1993), 541–576.

[4] M.Hénon *A two-dimensional mapping with a strange attractor*, Comm. Math. Phys. **50** (1976), 69-77.

[5] L.Mendoza *Ergodic attractors for diffeomorphisms of surfaces*, J. London Math. Soc.(2) **37** (1988), 362–374.

[6] L. Mora, M. Viana *The abundance of strange attractors*, Acta Math. **171** (1993), 1-72.

[7] J. Palis, F. Takens Hyperbolicity and sensitive-chaotic dynamics at homoclinic bifurcations, fractal dimensions and infinitely many attractors, Cambridge University Press, 1993.

[8] H.Thunberg *Absolutely continuous invariant measures and superstable periodic orbits: weak*-convergence of natural measures*, preprint Royal Inst. Tech.,(1993).

[9] M. Tsujii *Regular points for ergodic Sinai measures*, Trans. Amer. Math. Soc. **328** (1991), 747-766.

[10] R.Ures *On the approximation of Hénon-like attractors by homoclinic tangencies* Erg. Th. Dyn. Sys. **15** (1995), 1223-1229.

[11] M.Yakobson *Absolutely continuous invariant measures for one-dimensional maps,* Comm. Math. Phys. **81** (1981), 39-88.

IMERL,Facultad de Ingeniería
C.C.30, Montevideo
Uruguay
email: ures fing.edu.uy

J J P VEERMAN
Two dimensional generalizations of Haar bases

1 Introduction

Let $M : \mathbb{R}^n \to \mathbb{R}^n$ be an linear isomorphism with eigenvalues strictly outside the unit circle and preserving $\mathbb{Z}^n$ (that is: has integer entries). Define further $R \subset \mathbb{Z}^n$ a complete set of residues modulo M (that is: R contains precisely one representative in $\mathbb{Z}^n$ of each of the classes $\mathbb{Z}^n/M\mathbb{Z}^n$). By performing a translation we may assume that R contains the origin. Note that R contains $|\det M|$ elements.

We consider the set Λ of expansions on the base M using the set of digits R, or

$$\Lambda(M, R) = \{x \in \mathbb{R}^n \,|\, x = \sum_{i=1}^{\infty} M^{-i} r_i \text{ with } r_i \in R\} \quad .$$

Definition 1.1 *Let $N : \mathbb{R}^n \to \mathbb{R}^n$ be a linear map preserving $\mathbb{Z}^n$ and $\pi_N : A \to \mathbb{R}^n/N\mathbb{Z}^n$ the canonical projection. A compact set A in $\mathbb{R}^n$ of positive measure is called a tile by $N\mathbb{Z}^n$ if $\pi_N : A \to \mathbb{R}^n/N\mathbb{Z}^n$ is a bijection for Lebesgue almost every point of A.*

When the matrix N is not specified (as in most of this paper), we assume it to be the identity. In this case, we see that a tile is a compact set such that the union of its translates by $\mathbb{Z}^n$ covers $\mathbb{R}^n$, but two translates by distinct elements of $\mathbb{Z}^n$ may intersect in sets of measure zero only.

One of the main results was proved by Gröchenig and Haas [4]. It states that in one dimension with the conditions given here, $\pi : \Lambda(M, R) \to \mathbb{R}^n/\mathbb{Z}^n$ covers exactly q times (almost everywhere) where q is the greatest common divisor of R. The proof of this result was substantially simplified in [6]. In this work, we simplify the proof further (by introducing intersection numbers) and generalize it to include many two-dimensional cases. We also give some counter-examples in the last section.

In their paper Gröchenig and Haas [4] also give a weak extension of their results to two dimensions. For any 2 by 2 matrix M satisfying our standing assumptions plus another condition, they construct a digit set R such that (M, R) has the tiling property. The extra condition on the matrix M is that if it has two distinct real eigenvalues, these have to be rational.

In higher dimension, there is the result in [6] which states that when $|\det M| = 2$ then, in many cases, Λ is a tile In particular this includes all two- and three-dimensional cases. The method of proof of this latter result is very different from the reasonings we shall use in the present work.

The main results in the present work can be summarized as follows:

Theorem 1.2 *Let M and R as before and suppose they also satisfy*

- *R generates $\mathbb{Z}^n$ under addition and multiplication by $\mathbb{Z}$.*

- *$\phi(x) = \sum_{r \in R} e^{2\pi i r \cdot x}$ has finitely many zeroes.*

Then $\Lambda(M, R)$ is a tile.

The first hypothesis is a clear characterization. Assuming it holds, we will prove that the second hypothesis holds if

- The dimension is one.

- The dimension is two and one of the following statements holds:

  - $|\det M|$ is three.

  - $|\det M|$ is four and in addition the elements of R satisfy a certain "non-resonance" condition (see proposition 3.2).

  (In section four these condition are relaxed somewhat by allowing for 'common divisors'.)

After finishing the first draft of this paper, we became aware of several related preprints by Lagarias and Wang. In of these [10], they independently proved results that are essentially equivalent to our theorem 1.2 and proposition 3.2. In another paper [12], they proved the following remarkable result. Let M and R satisfy the standing hypotheses and denote by $\mathbb{Z}[M, R]$ the smallest M-invariant sublattice of $\mathbb{Z}^n$ that contains $D = R - R$ (smallest in the sense that it contains no subset satisfying the same requirements). Then $\Lambda(M, R)$ is a tile by $\mathbb{Z}[M, R]$ unless the following is true. There exists an integer matrix $P \in GL(n, \mathbb{Z})$ such that

- PMP^{-1} is a 'block-triangular' matrix $\begin{pmatrix} A & B \\ \emptyset & C \end{pmatrix}$.

- $P(R)$ is of so-called quasi-product form (for the definition see [12]).

In fact, they continue to prove that these exceptional cases also tile $\mathbb{R}^n$ by a lattice, albeit possibly a different one. Curiously, this still leaves one problem unresolved. It is a problem of algebraic nature: Is it true that for any matrix M there is a digit set R of complete residues such that $\mathbb{Z}[M, R] = \mathbb{Z}^n$? Lagarias and Wang seem to have resolved this problem for dimension 2 and 3 (personal communication by Wang, see also [11] for partial results).

The outline of this article is as follows. In the next section we start by deriving an equation very similar to the Perron-Frobenius equation for densities of measures. To prove that Λ is a tile we will have to prove that this equation has only one solution (namely the constant). Most of the discussion of this section can be found in [6], the exception being the notion and subsequent use of intersection numbers to simplify the reasoning. In the third section, we analyze the Perron-Frobenius(-like) equation and prove that in certain cases the only solution is indeed the constant. These cases include the ones mentioned above. In the last section, we sharpen a criterion given in [6] to decide whether $\Lambda(M, R)$

is a tile, and then use that criterion to look at some two-dimensional examples of $\Lambda(M,R)$ that are not tiles.

We end this introduction by indicating a few applications. Let M, R be given such that $\Lambda(M,R)$ is a tile by $\mathbb{Z}^n$ and denote $|\det M|$ by m. Wavelets are widely used in image- and sound-processing as an alternative for Fourier decomposition. They are functions with compact support forming a basis of the square integrable functions on $\mathbb{R}^n$. As additional properties one requires that if $f(x)$ is a basis-function then so is $f(Mx)$ (scaling property). For a detailed account of these notions we refer to [2]. With the help of the above self-similar tiles we can now easily construct such a basis. This construction gives what is called a generalized Haar basis by analogy with a certain one-dimensional construction. Let $\chi_\Lambda(x)$ be the characteristic function on Λ and U an $m \times m$ unitary matrix whose first column consists of the vector with constant entries (namely $m^{-1/2}$). Then the functions

$$\psi_i \overset{\text{def}}{=} \sum_{r \in R} U_{ij} \chi_\Lambda(Mx - r)$$

clearly form an orthonormal basis of the function with support on $\cup_{r \in R} M^{-1}(\Lambda + r)$ and for each $r \in R$ constant on $M^{-1}(\Lambda + r)$. Now define

$$f_{ijk} \overset{\text{def}}{=} |\det M|^{j/2} \psi_i(M^j x - k)$$

where $j \in \mathbb{Z}^+ i$, $i \in \{0, 1, \cdots |\det M| - 1\}$, and $k \in \mathbb{Z}^n$. Then the set $\{f_{ijk}\}$ forms an orthonormal basis of the square integrable functions on $\mathbb{R}^n$ with the required properties. This was first proved by Gröchenig and Madych [5] (see also [3] and [2] for additional information).

The second application concerns the measure of the compact set obtained in the following way. For every compact set $A \subset \mathbb{R}^n$ define the following affine iterated function system

$$\tau(A) = \cup_{r \in R_t} M^{-1}(A + r) \quad ,$$

where R_t depends on the parameter t. The simplest non-trivial example is when M multiplication by 3 in $\mathbb{R}$ and

$$R_t = \{0, t, 2\}, \text{ with } t \in [0, 1] \quad .$$

As we will see in the following section, there is a unique compact set invariant under τ. Denoting this set by $\Lambda(t)$, one obtains that its Lebesgue measure $\mu(\Lambda(t))$ has the following properties:

$$\mu(\Lambda(t)) = \frac{2}{q} \text{ if } t = \frac{2p}{q} \text{ and } pq \bmod 3 = 2$$
$$\mu(\Lambda(t)) = 0 \text{ else} \quad .$$

The first case is an easy consequence of the theorem by [4] already mentioned, although multiplication by integers greater than 3 introduce some extra problems. The second property follows from a result by [9].

Acknowledgements: I am grateful to Wiesław Szlenk for bringing proposition 2.7 to my attention and to Antonio Falcó for reviewing this manuscript. I am also indebted to the department of mathematics and the Centre de Recerca Matemàtica of the Universidad Autónoma of Barcelona for their kind hospitality. I thank the referee for pointing out the papers [10] and [11] to me. Finally, I'd like to thank Yang Wang for several helpful comments.

2   The Perron Frobenius Equation

In this section we define the intersection numbers and show that these satisfy an elementary self-similarity relation. This relation can in turn be translated to an equation of the Perron-Frobenius type. If this equation admits only a constant solution, then $\Lambda(M, R)$ is a tile. Most of the reasoning here has appeared in [6], but the use of intersection numbers simplify that proof.

Let X be a a closed ball in $\mathbb{R}^n$, or, more generally a complete compact metric space. Define the space $H(X)$ of closed subsets of X. The following construction now defines the so-called distance on $H(X)$. For $A, B \in H(X)$, let $N_\epsilon(A)$ denote the open ϵ n eighborhood of a set A. The Hausdorff distance between A and B is $\mathrm{Hd}(A, B)$:

$$\mathrm{Hd}(A, B) = \inf\{\epsilon > 0 | A \subset N_\epsilon(B) \text{ and } B \subset N_\epsilon(A)\} \ .$$

This distance induces a topology on $H(X)$ so that $H(X)$ is a complete compact metric space [7]. Limits in this topology will be denoted by Hlim. In $H(\mathbb{R}^n)$ we define:

$$\tau : H(\mathbb{R}^n) \to H(\mathbb{R}^n)$$

by

$$\tau(A) = \cup_{r \in R} M^{-1}(A + r) \ .$$

It is easy to prove that τ is a contraction (see [7]) and its unique fixed point is precisely the set Λ as defined before. Hence we obtain that Λ is 'self similar' or:

$$\Lambda = \cup_{r \in R} M^{-1}(\Lambda + r)$$

or, equivalently:

$$M\Lambda = \cup_{r \in R}(\Lambda + r) \ . \tag{2.1}$$

The (Lebesgue) measure of the set of this last equation is, of course, $|\det M|$ times the measure of Λ. From the righthand side of the equation one then concludes easily that translates of Λ by distinct elements of $\mathbb{R}$ intersect in sets of measure zero.

Denote $\mathbb{T}^n = \mathbb{R}^n/\mathbb{Z}^n$ and let $\pi : H(\mathbb{R}^n) \to H(\mathbb{T}^n)$ be induced by the usual canonical projection. It is easy to verify that π is continuous. Let

$$W : H(\mathbb{T}^n) \to H(\mathbb{T}^n)$$

be induced by the usual complete inverse of M on the torus. Clearly, W has $|\det M|$ branches.

Lemma 2.1 $\pi\Lambda = \mathbb{T}^n$.

Proof: (see also [6].) The following diagram commutes:

$$H(\mathbb{R}^n) \overset{\tau}{\to} H(\mathbb{R}^n)$$

$$\downarrow \pi \qquad \downarrow \pi$$

$$H(\mathbb{T}^n) \overset{W}{\to} H(\mathbb{T}^n)$$

Noting that for any compact K we have

$$\pi\Lambda = \pi \, \mathrm{Hlim}_{k\to\infty}\tau^k K = \mathrm{Hlim}_{k\to\infty}W^k K = \mathbb{T}^n \quad,$$

the lemma is easily implied.   ∎

Thus $\cup_{v\in\mathbb{Z}^n}(\Lambda+v) = \mathbb{R}^n$ and it follows from Baire's theorem that Λ has non-empty interior.

Self-similarity (2.1) implies that for each $y \in \Lambda$ there is at least one $r \in \mathbb{R}$ such that $My - r \in \Lambda$. So define $t : \Lambda \to \Lambda$ as

$$t(y) = [\cup_{r\in R}\{My - r\}]\cap\Lambda \quad.$$

On the other hand each point $x \in \Lambda$ has $|\det M|$ preimages $M^{-1}(x+r)$. Thus t preserves Leabesgue measure:

$$\sum_{ty=x}|dy| = |dx| \quad. \tag{2.2}$$

In fact, since t is expanding, we have the following result.

Proposition 2.2 t *is ergodic with respect to the Lebesgue measure.*

Proof: For a detailed proof see [6].   ∎

This last fact has important consequences. Define for $k \geq 1$:

$$\Lambda^{(k)} = \{x \in \Lambda | \{x + \mathbb{Z}^n\}\cap\Lambda \text{ has at least } k \text{ points }\} \quad.$$

Then $\Lambda^{(k)}$ is t-invariant. So, by the ergodic theorem $\Lambda^{(k)}$ has either full measure or measure zero. Denoting Lebesgue measure by μ, we have

$$\mu(\Lambda) = \max\{k \in \mathbb{N}|\mu(\Lambda^{(k)}) > 0\} \quad.$$

Call this number ℓ. Thus the canonical projection of Λ to the torus is ℓ to 1 in almost every point of the torus.

224

Definition 2.3 (Intersection Numbers) *The intersection numbers are the values of the function $\nu : \mathbb{Z}^n \to \mathbb{R}^+$, defined as follows:*

$$\nu(k) = \mu((\Lambda + k) \cap \Lambda) \quad .$$

This function ν satisfies a simple property due to self-similarity (2.1). First define the difference set <u>with multiplicity</u>:

$$D = R - R \stackrel{\text{def}}{=} \{d \in \mathbb{Z}^n | \exists r_1, r_2 \in R \text{ such that } d = r_1 - r_2\} \quad . \tag{2.3}$$

(For instance, the element 0 occurs at least $|\det M|$ times in D, namely $0 = r - r$ for all elements r in R.)

Proposition 2.4 *The function ν satisfies the following equation:*

$$\frac{1}{|\det M|} \sum_{d \in D} \nu(Mj - d) = \nu(j) \quad .$$

Proof: We have:

$$\sum_{r_1, r_2 \in R} \mu((\Lambda + r_1 - r_2 + Mj) \cap \Lambda) \quad = \quad \sum_{r_1, r_2 \in R} \mu((\Lambda + r_1 + Mj) \cap (\Lambda + r_2)) =$$

$$\mu(M(\Lambda + j) \cap M\Lambda) \quad = \quad |\det M| \, \mu((\Lambda + j) \cap \Lambda) \quad .$$

$\blacksquare$

Let Ω be the space

$$\{\sum a(k) z^k | k \in \mathbb{Z}^n, \, a(k) \in \mathbb{C}\} \quad .$$

Define the 'transition operator' $T : \Omega \to \Omega$:

$$T(\sum a(k) z^k) = \frac{1}{|\det M|} \sum_{j \in \mathbb{Z}^n} \sum_{d \in D} a(Mj - d) z^j \quad . \tag{2.4}$$

Note that T operates on functions defined on the torus $\mathbb{T}^n$. By definition, the functions $f = z^0$ (the constant function) and $f = \sum \nu(k) z^k$ are eigenfunctions of the transition operator both with eigenvalues 1. It is not clear whether there are other eigenfunctions associated with the eigenvalue 1.

Let us now return to equation (2.2) for the existence of an invariant measure. Let $\lambda(x)$ be a probability measure with a continuous density $h(x)$:

$$|d\lambda(x)| = h(x)|dx|$$

The lefthand side of (2.2) now becomes the Perron-Frobenius operator:

$$\sum_{ty=x} |dy| = \sum_{r\in R} h(M^{-1}(x+r)) \cdot \frac{dy}{|\det M|} \quad .$$

The following theorem reinterprets the operator T just defined as a (generalized) Perron-Frobenius operator. Before stating the theorem, we need some notation. Define

$$\phi(x) = \sum_{r\in R} e^{2\pi i r \cdot x} \quad , \tag{2.5}$$

Define the real, non-negative weight function

$$w(x) = \frac{|\phi(x)|^2}{|\det M|^2} \quad . \tag{2.6}$$

Theorem 2.5 *The transition operator (2.4) can be extended to continuous functions of the torus and is given by:*

$$Tf(x) = \sum_{j\in J} f((M^\dagger)^{-1}(x+j)) \, w((M^\dagger)^{-1}(x+j)) \quad ,$$

where $(M^\dagger)$ denotes the transpose of M and J is any complete set of residues modulo $(M^\dagger)$. Moreover,

$$\sum_{j\in J} w((M^\dagger)^{-1}(x+j)) = 1 \quad ,$$

and

$$T(\sum_{k\in \mathbb{Z}} \nu(k)e^{2\pi i k \cdot x}) = \sum_{k\in \mathbb{Z}} \nu(k)e^{2\pi i k \cdot x} \quad .$$

Proof: Define

$$z^k \;=\; e^{2\pi i k \cdot x}$$

where the $\cdot$ denotes the usual innerproduct of $\mathbb{R}^n$. Now it is an unpleasant, although straightforward, exercise to write 2.4 in the correct form, but the details are in [4]. The extension to the continuous functions is immediately clear by the Stone-Weierstrass theorem.

For the second part, note that by construction $T1 = 1$. Substituting 1 for f yields the relation. The operator has the specified eigenfunction by construction. ∎

In passing we remark that since $\nu(k) = \nu(-k)$, the eigenfunction can be written as:

$$f(x) = \sum_{k\in \mathbb{Z}^n} \nu(k)\cos(2\pi k \cdot x) \quad ,$$

and is a function from the torus ($\mathbb{T}^n$) to the reals. We wish to establish conditions under which $Tf = f$ only admits the constant solution. If those conditions are satisfied then all intersection numbers except $\nu(0)$ must be zero and $\Lambda(M, R)$ is a tile.

The following seems to be a folklore result and was brought to our attention by Wiesław Szlenk.

Proposition 2.6 *If $w(x) > 0$, then the only continuous solution of $Tf = f$ is the constant solution.*

Proof: A non-constant continuous function on the torus has at least a maximum and minimum. Let x be an absolute extremum. From theorem 2.5 we conclude that the value of f at x is the weighted average average of the values of f at the inverse images of x under $M^\dagger$. Then these inverse images must also be absolute extrema. Going backward indefinitely, it is well-known that these preimages form a dense set. ∎

Unfortunately, this proposition doesn't get us very far. Since $w(0) = 1$, we must have that
$w(M^{\dagger -1}(j)) = 0$ when $j \in J - \{0\}$. In [1] a more general result is stated in which zeroes of $w(x)$ are allowed. However, they do require other conditions on $w(x)$ which are not satisfied here. Indeed, by only allowing $w(x)$ to have zeroes the otherwise simple problem of proposition 2.6 becomes highly complicated. In fact, in general one does not have uniqueness as we shall see.

3 Solving the Perron-Frobenius Equation

In this section we prove our main result in three steps. We will use the notation established in the previous sections. In the first of these, we establish that in a number of interesting cases the zeroes of the weight $w(x)$ of the Perron-Frobenius operator are isolated. We then derive the condition on M and R for which $Tf = f$ has a unique solution in the class of trigonometric polynomials. Finally, we verify these conditions in a number of cases in dimension one and two.

It is conceivable that our methods extend to a more general two-dimensional context, but certainly they cannot be generalized to higher dimensions without substantial change.

We will limit ourselves to a special case by requiring that the zeroes of $w(x)$ be isolated. Clearly, since $w(x)$ is a trigonometric function, this requirement is satisfied in dimension 1. In general, we have that (see equations (2.5) and (2.6))

$$w(x) = 0 \Leftrightarrow \phi(x) = 0$$

and $\phi : \mathbb{T}^n \to \mathbb{C}$ is a smooth periodic function. Clearly, we expect ϕ to have isolated zeroes only in non-degenerate one- or two-dimensional cases. So these will be the only cases we deal with.

Proposition 3.1 *Suppose that the dimension* $n = 2$ *and* $|\det M| = 3$. $w(x)$ *has only isolated zeroes if and only if* R *spans* $\mathbb{R}^2$.

Proof: Let $R = \{(0,0), r_1, r_2\}$ and suppose that r_1 and r_2 are independent. Interpreting these two vectors as column vectors, define the matrix

$$N = (r_1, r_2) \quad ,$$

and the corresponding coordinate change $y = N^\dagger x$, where $N^\dagger$ is the transposed of N. Then

$$\phi(y) = 1 + e^{2\pi i y_1} + e^{2\pi i y_2} \quad .$$

(Note that the $y_i = r_i \cdot x$ are independent.) The image of ϕ in $\mathbb{C}$ can easily be visualized as a 'flattened' torus $\mathcal{T}$ (see figure 3.1). Notice that $y \in N^\dagger[0,1]^2$. Thus ϕ covers the flattened torus $|\det N^\dagger|$ times. Therefore, the point $0 \in \mathcal{T}$ has exactly $2|\det N \dagger|$ pre-images. $\blacksquare$

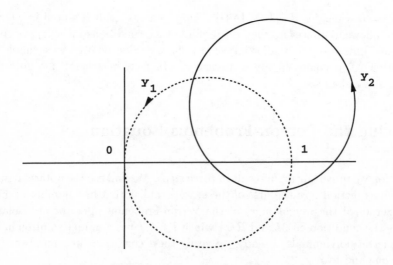

Figure 3.1: A flattened torus

Proposition 3.2 *Suppose that the dimension* n *is two and* m *is four. Suppose that* $\{r_1, r_2\} \subset R$ *spans* $\mathbb{R}^2$ *and write* $R = \{(0,0), r_1, r_2, a_1 r_1 + a_2 r_2\}$. *Then* w *has finitely many zeroes if and only if for all* $j, k \in \mathbb{Z}$:

$$(a_1, a_2) \notin \left\{ (1, \frac{2k+1}{2j+1}), (\frac{2k+1}{2j+1}, 1), (\frac{2k+1}{2j+1}, -\frac{2k+1}{2j+1}) \right\} \quad .$$

Proof: Using the conventions and notation of the previous proof and $r_3 = a_1 r_1 + a_2 r_2$, we get

$$w(y) = 1 + e^{2\pi i y_1} + e^{2\pi i y_1} + e^{2\pi i (a_1 y_1 + a_2 y_2)} \quad .$$

Rewrite this as:

$$\phi(y) = (e^{-2\pi i \frac{y_1}{2}} + e^{2\pi i \frac{y_1}{2}})e^{2\pi i \frac{y_1}{2}}$$

$$+(e^{-2\pi i \frac{1}{2}(a_1 y_1 + (a_2-1)y_2)} + e^{2\pi i \frac{1}{2}(a_1 y_1 + (a_2-1)y_2)})e^{2\pi i \frac{1}{2}(a_1 y_1 + (a_2+1)y_2)} .$$

$$= 2e^{2\pi i \frac{y_1}{2}}\left\{\cos(2\pi \frac{y_1}{2}) + \cos(2\pi \frac{a_1 y_1 + (a_2-1)y_2}{2})e^{2\pi i \frac{1}{2}((a_1-1)y_1 + (a_2+1)y_2)}\right\} .$$

This gives zero if one of the two following holds

- A. Both cosines yield zero.

- B. The exponential is real and the cosines cancel.

Case A:

$$\begin{cases} \dfrac{y_1}{2} = \dfrac{n + \frac{1}{2}}{2} \\[2mm] \dfrac{a_1 y_1 + (a_2-1)y_2}{2} = \dfrac{m + \frac{1}{2}}{2} \end{cases} .$$

This is equivalent to

$$\begin{pmatrix} 1 & 0 \\ a_1 & a_2 - 1 \end{pmatrix}\begin{pmatrix} y_1 \\ y_2 \end{pmatrix} = \begin{pmatrix} n + \frac{1}{2} \\ m + \frac{1}{2} \end{pmatrix} .$$

These equations are dependent if and only if $a_2 = 1$ and

$$a_1(2n+1) = 2m+1 \Leftrightarrow a_1 = \frac{2m+1}{2n+1} .$$

Case B:

$$\begin{cases} \dfrac{(a_1-1)y_1 + (a_2+1)y_2}{2} = n + \dfrac{\delta}{2} \\[2mm] \epsilon\dfrac{y_1}{2} = \dfrac{a_1 y_1 + (a_2-1)y_2}{2} + m + \dfrac{1-\delta}{2} \end{cases} ,$$

where $\delta \in \{0,1\}$ and $\epsilon \in \{-1,1\}$. Or, equivalently:

$$\begin{pmatrix} a_1 - 1 & a_2 + 1 \\ -(a_1 - \epsilon) & -(a_2 - 1) \end{pmatrix}\begin{pmatrix} y_1 \\ y_2 \end{pmatrix} = \begin{pmatrix} 2n + \delta \\ 2m + 1 - \delta \end{pmatrix} .$$

where $\epsilon \in \{-1,1\}$ and $\delta \in \{0,1\}$. These equations are dependent if and only if

$$\text{determinant} = (1 - \epsilon)a_2 + 2a_1 - 1 - \epsilon = 0 .$$

Thus when $\epsilon = 1$ we have dependency when

$$a_1 = 1 \quad\text{and}\quad a_2 = \frac{2n - 2m + 2\delta - 1}{2n + 2m + 1} = \frac{2k+1}{2j+1} .$$

When $\epsilon = -1$

$$a_1 = -a_2 \quad \text{and} \quad a_2 = \frac{2m - 2n - 2\delta + 1}{2m + 2n + 1} = \frac{2k + 1}{2j + 1} \quad .$$

∎

Let f be a non-constant eigenfunction of the transition operator. The set of absolute extrema of f will be denoted by $\mathcal{E}_f$. The set of zeroes of the weight function w will be denoted by $\mathcal{Z}_w$.

In the following result we use the fact that the eigenfunction $f(x) = \sum \nu(k)e^{2\pi i k \cdot x}$ we are looking for, has finitely many Fourier components (that is: f is a trigonometric polynomial).

Proposition 3.3 *Let f be a non-constant trigonometric eigenfunction of the transition operator. If $\mathcal{Z}_w$ is finite then $\mathcal{E}_f$ is a finite union of cycles under $M^\dagger$.*

Proof: Suppose that $\mathcal{E}_f$ is not finite, then it is a finite collection of arcs (because f is a trigonometric polynomial). From the proof of proposition 2.6 we conclude that

$$M^{\dagger - 1}(\mathcal{E}_f) \subseteq \mathcal{E}_f \cup \mathcal{Z}_w \quad .$$

Define

$$K = \mathcal{E}_f - \cup_{i=0}^{\infty} M^{\dagger i}(\mathcal{Z}_w) \quad .$$

Since $\cup_{i=0}^{\infty} M^{\dagger i}(\mathcal{Z}_w)$ is countable, we see that K is not empty and we have:

$$M^{\dagger - 1}(K) \subseteq K \quad .$$

However, since $M^{\dagger - 1} : H(\mathbb{T}^2) \to H(\mathbb{T}^2)$ is a contraction K contains a compact set, we have that

$$\text{Hlim}_{n \to \infty} M^{\dagger - n}(K) = \mathbb{T}^2 \quad .$$

This gives a contradiction, proving that $\mathcal{E}_f$ is finite.

Notice that the pre-image of each extremum contains at least one extremum and the complete pre-images of distinct points are distinct. It then follows that the pre-image of each extremum contains exactly one extremum. This implies the result. ∎

Theorem 3.4 *Let $\mathcal{Z}_w$ is finite. Then there exists $j \in \mathbb{Z}$ such that $x_0 \in \mathcal{E}_f$ implies the following statements:*
i) There are at least two $k \in \mathbb{Z}^n$ such that

$$x_0 = (M^j - I)^{\dagger - 1} k \in [0, 1]^n \quad .$$

ii) For all $r \in R$, we have

$$r \cdot x_0 \in \mathbb{Z}^n \quad .$$

Remark: Note that the hypothesis of this theorem can only hold if the dimension n takes the value 1 or 2.

Proof: Let j be the product of the periods of the finitely many cycles referred to in the previous proposition. Then the first part follows from the periodicity of f and the fact that for both the minimum and the maximum we have:

$$M^{\dagger j} x_0 = x_0 + k \quad.$$

The second equation comes because $M^{\dagger - 1} x_0$ contains exactly one point x_1 of $\mathcal{E}_f$ by the previous proposition and so $w(x_1) = 1$. The statement for x_0 follows by periodicity. ∎

We consider some applications of this theorem. The first is due to [4], but with a different proof. The second is an extension of this result.

Proposition 3.5 *In one dimension $\pi : \Lambda(M, R) \to I\!R/\mathbb{Z}$ covers exactly q times (almost everywhere) where q is the greatest common divisor of R.*

Proof: First divide R by q. Now assume that $\Lambda(M, R)$ is not a tile. Then the transition operator for the new system still has a non constant eigenfunction f associated with the eigenvalue 1. Note that by the previous theorem, f has one extremum in $(0, 1)$. Call this extremum x_0. Thus

$$x_0 = \frac{k}{M^n - 1} = \frac{p}{q}$$

in smallest terms. By the second part of the theorem:

$$r \cdot \frac{p}{q} \in \mathbb{Z} \quad.$$

Thus all r are divisible by q. This is a contradiction. ∎

Notice that the theorem implies that $\Lambda(M, R)$ is a tile if and only if the greatest common divisor of R is 1. In fact, $\Lambda(M, qR)$ is a tile by $q\mathbb{Z}$ (see definition 1.1). A more general version of this last statement is proved in lemma 4.4.

Proposition 3.6 *Suppose M and R are such that w has only isolated zeroes. If two elements of R satisfy $|r_1 \times r_2| = 1$, then $\Lambda(M, R)$ is a tile.*

Proof: As before, if $\Lambda(M, R)$ is not a tile, then the transition operator for the system has a non constant eigenfunction f associated with the eigenvalue 1. It must have an extremum $x_0 \in [0, 1]^2 - \mathbb{Z}^2$. Interpreting the vectors r_1 and r_2 of the proposition as column vectors, define the matrix

$$N = (r_1, r_2) \quad.$$

By the second part of the theorem, we have that $N^{\dagger} x_0 \in \mathbb{Z}^2$. But by definition $N^{\dagger}$ restricted to $\mathbb{Z}^2$ is a bijection. So this gives a contradiction. ∎

In special cases one can derive stronger results then the previous corollary. For example, when $M = 2I$. For j as in the theorem, define $\Gamma_j = [\frac{1}{2^j-1} \cdot \mathbb{Z}]^2 \cap [0,1]^2$. Now we have that

$$\mathcal{E}_f \subseteq \Gamma_j \quad .$$

It is easy to check whether, for a given R, $r \cdot \Gamma_j$ is in $\mathbb{Z}^2$.

4   Examples

In this section we state a sharp criterion for $\Lambda(M, R)$ to be a tile by $\mathbb{Z}^n$. Using that criterion we give some two dimensional examples of sets $\Lambda(M, R)$ that are not tilings by $\mathbb{Z}^2$.

Recall the definition of the difference D of R (equation (2.3)). Define $G \subseteq \mathbb{Z}^n$ as

$$G(M, R) = \{x \in \mathbb{Z}^n \,|\, x = \sum_{i=1}^{\infty} M^i d_i \text{ with } d_i \in D\} \quad .$$

The first proposition is an easy consequence of results of [6].

Proposition 4.1 $\Lambda(M, R)$ *is a tile by* $\mathbb{Z}^n$ *if and only if* $G = \mathbb{Z}^n$.

Proof: $\Leftarrow$: See [6].
$\Rightarrow$: Suppose that Λ is a tile and there is a $k \in \mathbb{Z}^n$ such that $k \notin G$. Then by the tiling property, we have that for all $v \in G$, $\mu((\Lambda + k) \cap (\Lambda + v)) = 0$. Thus

$$\cup_{v \in G}\{\Lambda + v\} \subseteq \mathbb{R}^n - \{\Lambda + k\}$$

This contradicts property 1.13 of [6]. ∎

Definition 4.2 *Let* M *and* R *be as usual and* A *a linear isomorphism whose matrix has integer entries. We call* A *a common divisor of* (M, R) *if* $A^{-1}MA \in GL(n, \mathbb{Z})$ *and* $A^{-1}R \in \mathbb{Z}^n$.

In one dimension, the definition reduces to the usual one. Furthermore, the definition is consistent in that the following lemma holds.

Lemma 4.3 *If* R *is a complete set of residues modulo* $M\mathbb{Z}^n$, *then* $A^{-1}R$ *is a complete set of residues modulo* $A^{-1}MA\mathbb{Z}^n$.

Proof: Recall that D is the difference set of R. A set R of cardinality $|\det M|$ is a complete set of residues if and only if

$$D \cap M\mathbb{Z}^n = 0 \quad .$$

Thus the elements $A^{-1}r_i$ of $A^{-1}R$ satisfy:

$$A^{-1}D \cap A^{-1}M\mathbb{Z}^n = 0 \quad .$$

Since $\mathbb{Z}^n \subset A\mathbb{Z}^n$, the result follows. ∎

Lemma 4.4 *Let (M, R) have common divisor A and suppose that $\Lambda(A^{-1}MA, A^{-1}R)$ is a tile by $\mathbb{Z}^n$. Then $\Lambda(M, R)$ is a tile by $A\mathbb{Z}^n$.*

Proof: One easily verifies that

$$\Lambda(M, R) \ = \ A \cdot \Lambda(A^{-1}MA, A^{-1}R) \quad .$$

$$G(M, R) \ = \ A \cdot G(A^{-1}MA, A^{-1}R) \quad .$$

By proposition 4.1, $G(A^{-1}MA, A^{-1}R) = \mathbb{Z}^n$, thus $G(M, R) = A\mathbb{Z}^n$. Now our result follows from [6], theorem 1.22. ∎

As examples define

$$M_n \ = \ \begin{pmatrix} 2 & n \\ 0 & 2 \end{pmatrix} \quad ,$$

$$R_s \ = \ \left\{ \begin{pmatrix} 0 \\ 0 \end{pmatrix}, \begin{pmatrix} s \\ 0 \end{pmatrix}, \begin{pmatrix} 0 \\ 1 \end{pmatrix}, \begin{pmatrix} s \\ 1 \end{pmatrix} \right\} \quad ,$$

$$A_m \ = \ \begin{pmatrix} m & 0 \\ 0 & 1 \end{pmatrix} \quad .$$

Notice that we cannot use the results of section 3 to prove that $\Lambda(M_n, R_1)$ is a tile, because $w(x) = (1 + e^{2\pi i x})(+e^{2\pi i y})$, whose zeroes are not isolated. However, we will employ the criterion of proposition 4.1 to prove this.

Proposition 4.5 $\Lambda(M_n, R_1)$ *is a tile.*

Proof: The difference set D_1 of R_1 consists of the points $(\delta_1, \delta_2) \in \mathbb{Z}^2$ with $\delta_i \in \{-1, 0, 1\}$. (The multiplicities are of no consequence here.) Denote by G_i all those points in $\mathbb{Z}^2$ whose second coordinate lies in the interval $[-2^i + 1, 2^i - 1]$. To prove the proposition we will prove that for all i, $G_i \subset G$.

We first show that $G_0 \subset G$. G_0 is the set $\mathbb{Z} \times \{0\}$. The matrix M_n acts on this sets as multiplication by 2 (in $\mathbb{Z}$). So representing G_0 on the basis M_n with digits $(\delta_1, 0)$ where $\delta_1 \in \{-1, 0, 1\}$ is thus equivalent to representing all numbers in $\mathbb{Z}$ as

$$\sum_{i=0}^{\infty} 2^i \mathcal{E} \ \text{ where } \ \mathcal{E} = \{-1, 0, 1\} \quad .$$

Now we show that if $G_i \subset G$ then $G_{i+1} \subset G$. We are done if for every point $(c, d) \in G_{i+1}$ there is a point $(a, b) \in G_i$ such that

$$M_n \begin{pmatrix} a \\ b \end{pmatrix} - \begin{pmatrix} \delta_1 \\ \delta_2 \end{pmatrix} = \begin{pmatrix} 2a + nb - \delta_1 \\ 2b - \delta_1 \end{pmatrix} = \begin{pmatrix} c \\ d \end{pmatrix} .$$

These equations can be solved by determining b and δ_2 from the second equation and then a and δ_1 from the first equation.

By induction on i, we obtain that for all $i \in \mathbb{Z}$, $G_i \subset G$. ∎

Corollary 4.6 *For* $n, s \in \mathbb{Z}$, $\Lambda(M_{ns}, R_s)$ *is a tile by* $A_s \mathbb{Z}^2$.

Proof: Apply lemma 4.4 to obtain that

$$\Lambda(M_{ns}, R_s) = A_s \cdot \Lambda(M_n, R_1) .$$

Then use proposition 4.5. ∎

We remark that only for s odd is R_s a complete set of residues modulo $M_{ns}\mathbb{Z}^2$.

In the above examples G is a group. From proposition 3.5 it easily follows that this is always the case in one dimension. In higher dimensions, the situation is more complicated. For instance, in dimension two, consider the system (M_1, R_3) (see [8]). One easily verifies that this system does not admit a common divisor with determinant of absolute value greater than one. The reason being, of course, that matrix multiplication is not commutative. In this case one also easily verifies that $G(M_1, R_3)$ is not a group: This set contains $(0, 1)$ and $M(0, 1) = (1, 2)$. The cross-product of these two vectors is 1, so the only additive subgroup of $\mathbb{Z}^2$ containing both of them is $\mathbb{Z}^2$. It is easy to verify that $(1, 0)$ and $(2, 0)$ are not in $G(M_1, R_3)$. What happens in this case is that $G + \Lambda$ covers $\mathbb{R}^2$ more than once. One checks easily that $3 + 4\cos(2\pi x) + 2\cos(4\pi x)$ is a non-trivial eigenfunction, with eigenvalue 1, of the operator T associated with this problem as discussed in section 2.

References

[1] M. F. Barnsley, S. Demko, J. Elton, J. S. Geronimo, *Invariant Measures for Markov Processes arising from Iterated Functions Systems* ..., Ann. Inst. H. Poincaré 24, 3-31, 1988.

[2] I. Daubechies, *Ten Lectures on Wavelets*, SIAM, 1993.

[3] H. Gröchenig, *Orthogonality Criteria for Compactly Supported Scaling Functions*, Appl. and Comp. Harm. An. 1, 242-245, 1994.

[4] K. Gröchenig, A. Haas, *Self-similar Lattice Tilings*, J. Fourier An. Appl. 1, 131-170, 1994.

[5] K. Gröchenig, W. R. Madych, *Multiresolution Analysis, Haar Bases, and Self-Similar Tilings*, IEEE Trans. Information Th. 38, No 2, part 2, 558-568, 1992.

[6] D. Hacon, N. C. Saldanha, J. J. P. Veerman, *Some Remarks on Self-Affine Tilings*, J. Exp. Math. 3, 317-327, 1994.

[7] J. E. Hutchinson, *Fractals and Self-Similarity*, Indiana Univ. Math. J., 30, No 5, 713-747, 1981.

[8] J. C. Lagarias, Y. Wang, *Self-Affine Tiles in IR^n*, Adv. in Math., to appear.

[9] J. C. Lagarias, Y. Wang, *Tiling the Line with One Tile*, Preprint Georgia Tech, 1994.

[10] J. C. Lagarias, Y. Wang, *Haar Type Orthonormal Wavelet Bases in IR^2*, J. Fourier An. Appl., to appear.

[11] J. C. Lagarias, Y. Wang, *Haar Bases for $L^2(IR^n)$ and Algebraic number Theory*, J. Number Th., to appear.

[12] J. C. Lagarias, Y. Wang, *Integral Self-Affine Tilings in IR^n II, Lattice Tilings*, Preprint Georgia Tech, 1994.

J. J. P. Veerman
Departamento de matemática, UFPE, Recife, Brazil*

*e-mail: veerman@dmat.ufpe.br

R F WILLIAMS
Spaces that won't say no

Dedicated to Ricardo Mañé

Introduction. A class of spaces K and their natural maps $f : K \to K'$ were introduced in [W4] which satisfies the defining property that $f_* : H_n(K) \to H_n(K')$ is always a "positive matrix." Here n is the dimension of both K and K'; we follow the usage in symbolic dynamics and say that a matrix is *positive* provided all its entries are ≥ 0. The purpose of this paper is to show how a natural construction, related to the "DA" maps of Smale [W2] often leads to SOB's, and in particular shows that they exist in all dimensions. Here we present the easier conceptual parts of this theory and elicit a geometric property, which when it holds, leads to quite satisfactory proofs.

1. Definitions.
This class is determined by the following properties. First of all, K is a branched manifolds of dimension n, say, and is oriented. Secondly, $H_n(K)$ (with integral coefficients) is free abelian of rank k say, and generators $g_1, g_2, ..., g_k$ are chosen for $H_n(K)$. Lastly, the homomorphism $f_* : H_n(K) \to H_n(K')$ induced by any orientation preserving immersion $f : K \to K'$ is a positive $k' \times k$ matrix, relative to the generators chosen for K and K'. Thus, in addition to being an oriented, branched, n-manifold, K is *sensed* by a choice of generators for its top dimensional homology group. Sensed, oriented, branched manifolds are called SOB's for short.

Complete definitions and basic properties of branched manifolds can be found in [W4]; for completeness, we include enough about branched manifolds for most readers. Local charts in dimensions 1 and 2 are (open subsets of) the spaces indicated in figure 1.

Local projections are given as an integral part of the structure of a branched manifold. In detail, each point has a neighborhood U which is the union of finitely many, say m, (smooth) disks D_i, provided with a projection map $\pi : U \to V, V$ open in $\mathbb{R}^n$, where $\pi|D_i$ is a diffeomorphism for each i. This neighborhood is *normally branched* (see [W4] and compare [C]) provided $k \leq n + 1$, the disks $D_i = (i, \mathbb{R}^n)$,

1991 *Mathematics Subject Classification.* 57

Supported in part by a grant from the National Science Foundation.

We wish to thank the Aspen Center for Physics and the Mathematics Department of Montana State University for their hospitality.

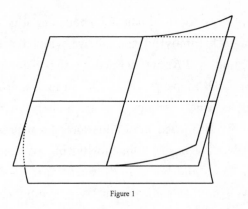

Figure 1

Figure 1

say $i = 1,..k$, and $(i, x) \equiv (i + 1, x)$, for all $x \in \mathbb{R}^n$ with $x_i \leq 0, i = 1, 2, ..n$. This has further consequences: for example, $(1, x) \equiv (2, x) \equiv (3, x)$, if both $x_1, x_2 \leq 0$. Branch manifolds arize as quotients of foliations, as in figure 2.

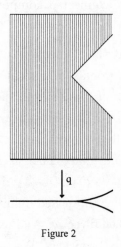

Figure 2

Figure 2

These quotients are as smooth as the foliation one starts with. In this paper they will be quite smooth (analytic) though we will have use only for their C^1 structure. Tangent bundles TK are defined in the usual way and may or may not be orientable. Smooth maps $f : K \to K'$ induce maps $df : TK \to TK'$ of tangent bundles, and as usual, we say f is an *immersion* provided df is an isomorphism at each point. Branched manifolds have been used to study hyperbolic attractors, [W1,W3,W4,C], Anosov diffeomorphisms [W2], pseudo-Anosov diffeomorphisms [T,F-L-P], flows and knots in 3-dimensions, [B-W1,B-W2,F,F-W,G,G-H,H,H-W,S1,S2,W8,W9], and to some extent [W5], symbolic dynamics. The reader is warned that immersions need not be locally 1-to-1; in particular, the projection maps $\pi : U \to \mathbb{R}^n$ are immersions. They *are* locally 1-to-1 on the smooth disks D_i which make up a chart.

2. Basic Examples.

Let n be an integer ≥ 3 and let P and N be positive vectors in $\mathbb{R}^n$. That is each coordinate of P, N is positive. Let $\Pi = N^\perp$, the plane through the origin, O, perpendictular to N, and let Q be an open, round disk in $\mathbb{R}^n$, of radius r centered at O. Let $\mathcal{L}$ be the family of all lines parallel to the vector P. In the quotient space, $\mathbb{T}^n = \mathbb{R}^n / \mathbb{Z}^n$, let Q_t and $\mathcal{L}_t$ be the images of Q and $\mathcal{L}$ respectively. Then $\mathbb{T}^n - Q_t$ is foliated by the set $\mathcal{C}_t$ of all the connected components C of $L \cap (\mathbb{T}^n - Q_t)$, all $L \in \mathcal{L}_t$.

Assumption. We assume that the elements of $\mathcal{C}_t$ are all line intervals. This is always true if the vector P is irrational; the exceptional case we need to rule out occurs when some of the elements of $\mathcal{C}$ are simple closed curves, which happens if the disk Q is small and the vector P is rational. Finally, let $K = (\mathbb{T}^n - Q_t)/\mathcal{C}_t$ be the quotient space in which all the intervals of $\mathcal{C}_t$ are collapsed to points, and let $q : \mathbb{T}^n - Q_t \to K$ be the quotient map. Then K has the homotopy type of the $(n-1)$-skeleton of $\mathbb{T}^n$, (which consists of n, $(n-1)$-dimensional tori.) Furthermore, K is an orientible, $(n-1)$-dimensional branched manifold. This follows from general results in [W4], or can easily be seen directly. And except for isolated values of the radius of the cell Q, K is normally branched. We prove this below as part of our analysis of the geometric, algebraic, and homological structure of K.

Conjecture. For many choices of the disk Q, $K = (\mathbb{T}^n - Q_t)/\mathcal{C}_t$ is an SOB.

Meanwhile this is known to be true in some special cases:

Proposition 1. *If N is close enough to P, then for Q a round ball of radius $\frac{1}{2}$, the quotient space K is an SOB.*

Proof. For then, the projection $\rho : \mathbb{R}^n \to \Pi$ along P sends each of the standard basis vectors e_i to points within 1 unit of the origin. The projection q is a homotopy equivalence, and the $(n-1)$-dimensional tori $T_1, ..., T_n$, dual to the standard basis $e_1, ..., e_n$, map down to generators $g_1, ..., g_n$ of $H_{n-1}(K)$. Furthermore, the lines L parallel to the vector P are all outwardly transverse to each of the tori T_i, so that the immersion $q|T_i : T_i \to K$ sends T_i in an orientation preserving manner. Thus the cycle g_i is a chain in the $(n-1)$-dimensional cells E_j of K, with all positive coefficients; that is, with orientations agreeing with the chosen orientation of K. As in [W6], one computes that

$$g_i = \sum_i I(z_i, T_i) E(z_i).$$

(We discuss this more fully below.) Thus

$$g_i = E(z_i) + \sum_{j>n} p_{ij} E(z_i),$$

and the theory of [W6] finishes the argument.

Proposition 2. *For $n = 3$, any two positive vectors N, P, and the disk $Q = \Pi \cap [-\frac{1}{2}, \frac{1}{2}]^3$, K is an SOB.*

This is proved in a lengthy preprint [W9]; the paper [W6] explains this argument and contains all details of [W9] except for the rather tedious estimates.

3. Geometry of $(\mathbb{T}^n - Q_t)/\mathcal{C}_t$.

We begin with some notation largely in the universal cover $\tau : \mathbb{R}^n \to \mathbb{T}^n$. Recall $Q \subset \Pi \subset \mathbb{R}^n$ is an open $(n-1)$-dimensional disk containing the origin. For each $z \in \mathbb{Z}^n$, let $Q(z) = Q + z$, be the translation of Q by z. Let

$$\mathcal{C} = \tau^{-1}(\mathcal{C}_t) = \{C : C \text{ is a component of } L \cap \mathbb{R}^n - \cup_z Q(z)\} .$$

Similarly, define $\mathcal{L} = \tau^{-1}(\mathcal{L}_t)$. Next let $\mathcal{C}_0 = \{C \in \mathcal{C} : C \text{ has its lower end point in } Q(O)$. We can now define the $(n-1)$-cells in our complex K :

$$E(z) = \{x \in Q(O) : C = (x, y) \in \mathcal{C} , \ x \in Q(O) , \ y \in Q(z), z \in \mathbb{Z}^n\}.$$

We introduce also the weight $w(E(z)) = \text{length}(C)$. This is independent of the choice of the point $x \in E(z)$.

Note that under our assumption, there are only finitely many cells, $E(z)$, because each of the lines L is topologically dense: let $F = \{z \in \mathbb{Z}^n : E(z) \text{ is nonempty}\}$. We define the $(n-2)$ cells as $\alpha = E(z) \cap E(z')$, for all z, z' such that this intersection is $(n-2)$ dimensional. We may and do assume that the weights satisfy $w(E(z)) < w(E(z'))$. This implies, (see figure 3) that $\alpha \subset \rho(\partial Q(z))$, as the line intervals C with lower end point on $E(z)$ near α are shorter than those with lower end point on $E(z')$. We orient the outer boundary of $Q(O)$ as usual by the outward normal, and note, below, that this induces an orientation on the $(n-2)$ cells.

We introduce a relation among our cells $E(z)$ where $E(z_1) < E(z_2)$ provided

a) $\partial E(z_1) \cap \partial E(z_2)$ is $(n-2)$ dimensional (and thus an $(n-2)$ cell common to their boundaries);

b) their weights satisfy $w(E(z_1)) < w(E(z_2))$.

Proposition 3: Stucture Proposition. *The quotient space K is realized as the closure of $Q(O)$ with identifications along the $(n-2)$ cells as follows: for each $z \in F$ and each $x \in \partial E(z)$, x is identified with $x - \rho(z)$.*

Proof. First of all, C_0 covers a fundamental region of $\mathbb{T}^n - Q_t$, since each $C \in \mathcal{C}$ has its lower endpoint on some $Q(z)$. Secondly, $\partial E(z)$ is $(n-2)$ dimensional since P is transverse to the cells $Q(z)$, and thus $\rho|Q(z) : Q(z) \to \Pi$ is a diffeomorphism. For a point $x \notin \cup_F \partial E(z)$, the interval C_x with x as lower end point is the only interval of C_0 containing an integral translate of x. Finally the intervals C_x for $x \in \alpha = E(z) \cap E(z')$, are duplicated in the closure $\bar{Q}(0)$ by the point $T_z(x)$, where $T_z : E(z) \to \mathbb{R}^n$ is defined by $x \to x - \rho(z)$. Note that, as T_z preserves orientation, this orients our $(n-2)$ cells. Also that $T_z : \alpha \to \partial Q(O)$; this follows as for $x \in \partial E(z)$, there exists $y \in \partial Q(z)$ with $\rho(y) = x$. Thus $x - \rho(z) = \rho(y) - \rho(z) = \rho(y - z)$, which is a vector in Π of norm r. This completes the proof of our structure proposition.

We single out some special $(n-1)$ cells $E(z)$ as follows: a cell $E(z)$ is *lenticular*, provided $E(z) = \rho(Q(z)) \cap Q(0)$, or equivalently, provided its boundary $\partial E(z)$ is a subset of $\rho(\partial Q(z) \cup \partial Q(0))$. See figure 3.

In addition, for each lenticular cell $E(z)$, let $E^*(z)$ be the cochain dual to $E(z)$, that is the cochain which has the value 1 on $E(z)$ and zero on all other cells.

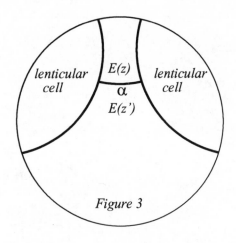

Figure 3

Figure 3

Main theorem.

a. *There are at least n lenticular cells;*

b. $\{q^*(E^*(z)) : E(z) \text{ is lenticular}\}$ *generates* $H^{n-1}(\mathbb{T}^n)$;

c. *if there are n and only n lenticular cells, then K is an SOB;*

b′ *if (c) holds then the map $z \mapsto E(z)$ is an isomorphism $H_1(\mathbb{T}^n) \to H^{n-1}(K)$ and is in fact, $(\tau^*)^{-1} \circ \mathcal{P}$, where $\mathcal{P}$ is Poincaré duality.*

Our proof begins with the following, central Lemma.

Lemma. *For $E(z_1) < E(z_2)$, there is a $z' \in F$ with $E(z') < E(z_2)$ and $z_2 = z_1 + z'$. In addition, there is an $(n-2)$ cell α with $\delta\alpha = E(z_1) + E(z') - E(z_2)$, where δ is the coboundary operator.*

Proof. This is very geometric: see figure 4, drawn for $n = 3$.

Note that directly above an interval C_x, with $x \in E(z_1)$ near $y \in \alpha = \partial E(z_1) \cap \partial E(z_2)$, there is an interval $C' \in \mathcal{C}$ (though not in $\mathcal{C}_0$) with C' cut off by $Q(z_1)$ and $Q(z_2)$. Thus length(C_x) + length(C') = length(C_y). Also, the translation by $-z_1$ moves C' to C'', say, and $C'' \in \mathcal{C}_0$, as its base is in $Q(O)$. As C'' was chosen to be near α but in the interior of $Q(z_1)$, it follows that the bottom point of C'' is in a unique cell, say $E(z')$; this shows that $z_2 - z_1 = z'$. Finally, the translation $\alpha - z_1 \subset \partial Q(O)$, so that in K, it is identified with α. Thus the coboundary $\delta\alpha = E(z_1) + E(z') - E(z_2)$. This completes the proof of the Lemma.

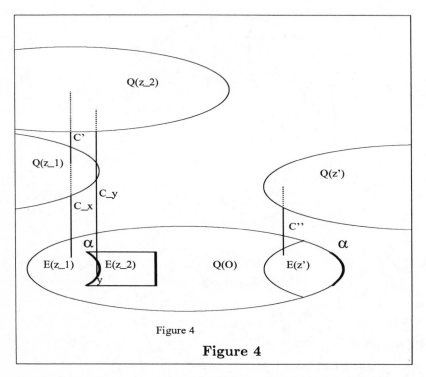

Figure 4

Figure 4

Proof of the Main Theorem. First, by repeated application of our Lemma we may write, for any $z \in F$,

$$E(z) = \sum_{i=1}^{k} p_i E(z_i) \quad , \quad z = \sum_i p_i z_i, p_i \geq 0$$

where the $E(z_i)$ are of minimal weight and thus lenticular. It follows that the cohomology classes of the lenticular cells, $E^*(z)$, generate $H^{n-1}(K)$. As $H^{n-1}(K) \approx H^{n-1}(\mathbb{T}^n)$, is n-dimensional, we see that there are at least n lenticular cells. If there are *only* n of these, then the set $\{E^*(z)\}, E(z)$ lenticular, forms a basis of $H^{n-1}(K)$.

We now specialize to this case, and number the $z \in F$ so that $E(z_1), ..., E(z_n)$ are lenticular, and $E(z_j), j > n$ are the remaining cells. Then for $j > n$,

$$E^*(z_j) = \sum_{i=1}^{n} a_{ij} E^*(z_i) \quad , \quad z_j = \sum_{i=1}^{n} a_{ij} z_j \ , \quad a_{ij} \geq 0 \ .$$

These coefficients are unique, of course. Let $Z_1, ... Z_n$ be $n-1$ chains "vector space

242

dual" to the $E^*(Z_i)$. That is

$$Z_i = E(z_i) + \sum_{j>n} b_{ij} E(z_j) \, ,$$

so that $E^*(z_i)$ evaluated on Z_j, $E^*(z_i) \cdot Z_j = \delta_{ij}$. We can compute b_{ij} by evaluation,

$$b_{ij} = E^*(z_i) \cdot E(z_j) = E^*(z_i) \cdot a_{ij} E(z_j) = a_{ij} \, ,$$

which we know is positive. Then $Z_1, ... Z_n$ forms a basis for $H_{n-1}(K)$, as there is no torsion.

One can describe specific $n-1$ chains $Y_i = q_*^{-1}(Z_i)$, whose classes form a basis of $H_{n-1}(K)$, as the situation is so simple; however, we know they exist, as q is a homotopy equivalence. To compute the coefficients a_{ij}, consider an interval $C = (x, y)$ which has its end points on $Q(O)$ and $Q(z_j)$ respectively. Then C together with short arcs joining x to O and y to z_j defines a 1-cycle, w_i, homologous to z_j, considered as a 1-cycle on $\mathbb{T}^n$. Then the intersections of Y_i and w_j occur only along the arc C and thus $I(z_i, Y_j)$ is the multiplicity of the map q at x. It follows that $I(z_i, Y_j)$ is the coefficient in the sum,

$$q_* Y_i = Z_i = \sum_i I(z_i, Y_j) E(z_j) \, .$$

In particular, $I(z_i, Y_j) = \delta_{ij}$, so that $z_1, ..., z_n$ are the Poincaré duals of $Y_1, ..., Y_n$. Therefore $F_l = \{z_1, ..., z_n\}$ forms a basis of $H_1(\mathbb{T}^n)$, when the z_i are considered as 1 dimensional cycles, and in turn, a basis of $\mathbb{Z}^n$, when they are considered as elements of $\mathbb{R}^n$.

It remains only to verify that K is an SOB. This follows quickly from the equation

$$Z_i = E(z_i) + \sum_{j>n} a_{ij} E(z_j) \, , \qquad a_{ij} > 0 \, ,$$

as shown in [W6]. The crucial fact is that $E(z_i), i = 1, ... n$, occurs in the cycle Z_j, only for $i = j$. (these cells are *unshared*. We repeat this short argument: now let K' be another member of this class with notation as above, decorated with "primes," and let $f : K \to K'$ be an orientation preserving immersion. Then the induced map on the chain level sends each $n-1$-cell into K', preserving orientation, so that

$$f_*(Z_i) = \sum_{j=1}^{m'} p'_{i,j} E'(z_j) , \qquad i = 1, \dots k, p'_{i,j} \geq 0 ,$$

$$= \sum_{j=1}^{k'} A_{i,j} Z'_j ,$$

$$= \sum_{j=1}^{k'} A_{i,j} E'(z_j) + \sum_{j=k'+1}^{m'} B_{i,j} E'(z_j) ,$$

for some $B_{i,j}$, where the last equation uses the fact that the first k', $n-1$-cells of K' are "unshared." The second equality uses the fact that there are no boundaries, as we are in the top dimension. Comparing the first and third equation, we see that $A_{i,j} = p'_{i,j}$ and thus $A_{i,j} \geq 0$, $i = 1, \dots, k$; $j = 1, \dots, k'$.

Remark. There *are* examples, with $n = 3$, for which there are more than 3 lenticular cells. In every case known to the author, one can, by varying the radius of Q, obtain the appropriate number, that is 3. Finally, in these "excess" cases, we believe that the set $F_l = \{z : E(z) \text{is lenticular}\}$ generates $\mathbb{Z}^n$.

Proposition. *The branched manifold is normally branched except for bifurcation values of the radius r. For a range of r of the form $[a, b]$, $a > 0$, there are only finitely many bifurcation points.*

Proof. First of all, note that it suffices to prove this for an open interval J. Note that the set $F = F_r \subset \mathbb{Z}^n$ is finite for each $r > 0$. By a covering argument, $G = \cup_{r \in J} F_r$ is likewise finite. Now let $G' = \{\rho(z) : r \in J, \text{ and } z \in F_r\}$. Each subset $g \subset G'$ of n distinct elements, determines a unique $n-1$ dimensional sphere, of radius, say, $r(g)$. The set of all these, $B = \{r(g) : g \subset G' \text{ a subset of } n \text{ elements}\}$ is finite. But then B contains all the bifurcation points in J. To see this, note that at most $n-1$ of the $n-2$ dimensional spheres $\partial Q(z), z \in F_r$, could contain a point in common. For otherwise, their centers would be such a set $g \subset G'$ considered above, and their common point, the center of the $n-2$ sphere determined by g. This would contradict the definition of B. That is, for $r \in J - B$, the cells $Q(z), z \in F_r$, are in general position. It follows that K is normally branched, as this property is stable and generic, as proved in [W4].

244

We also include a direct proof of this last. Let y be a point of K which we assume to be a branch point, and say there are k values of $z \in F$ with $y \in \partial E(z)$. By the above, $k \leq n-1$. We order the k z's by weight, $z_1, ..., z_k$ and add $z_0 = O$. Enlarge $Q(z_i)$ to $Q^+(z_i)$, with a slightly larger radius. Let $y_0 = y$ and set

$$y_i = (\rho | Q^+(z_i))^{-1}(y_0) , \qquad i = 1, ...k .$$

Choose a neighborhood U_0 of y_0 in $Q(O)$, and let $U_i = (\rho | Q^+(z_i))^{-1}(y_0)$. If our enlargement is sufficiently small, the U_i are diffeomorphic to U_0 via ρ, and thus copies of $\mathbb{R}^{n-1}$. Then the map $\rho : \cup U_i \to U_0$ serves as the map π in our definition of normally branched, after the appropriate identifications are made. These identifications are as follows. Let v_i be the inward normal to $\partial Q(z_i)$ near y_i. Set

$$f_{ji} = (\rho | Q^+(z_i))^{-1} \circ \rho | Q^+(z_j) , \qquad df_{ji}(v_i) = v_{ij} .$$

Then $\{v_{ji}\}_{j=1}^n$ leads to a coordinate system $\{x_{ji}\}$ for U_j. Then the part of U_{j+1} which is not in $Q(z_{j+1})$ satisfies the equation $x_{j+1,j} \leq 0$. This is exactly the portion of U_{j+1} that needs to be identified with U_j. Thus with these identifications, we see that U is normally branched. This completes the proof of our proposition.

REFERENCES

[B-W1] Birman, J. and Williams, R.F., *Knotted periodic orbits in dynamical systems I: Lorenz's equations*, Topology **22** (1983), no. (1), 47–82.

[B-W2] Birman, J. and Williams, R.F., *Knotted periodic orbits in dynamical systems, II: Knot holders for fibered knots*, Contemporary Mathematics **20** (1983), 1–60.

[C] ‾‾‾‾‾Christy, J., *Branched Surfaces and Attractors I: Dynamic Branched Surfaces*, Transactions of the American Mathematical Society **336** (1993), 759–784.

[F] Franks, J., *Knots, links, and symbolic dynamics*, Annals of Math. **113** (1981), 529–552.

[F-W] Franks, J. and Williams, R.F., *Entropy and knots*, Transactions American Mathematical Society **291** (1985), no. (1), 241–253.

[F-L-P] Fathi, A., Laudenbach, F. and Poenaru, V., *Travaux de Thurston sur les surfaces*, Asterisque (1979).

[G] Ghrist, R., *Branched manifolds supporting all links*, Preprint, Princeton.

[G-H] Ghrist, R. and Holmes, P., *Bifurcations and periodic orbits of vector Fields*, NATO ASI series C, vol. 408, Kluwer Academic Press, 1993, pp. 185–239.

[H] Holmes, P., *Bifurcation sequences in the horseshoe map: infinitely many routes to chaos*, Physics Letters A **104** (1984), 299–302.

[H-W] Holmes, P. and Williams, R.F., *Knotted periodic orbits in suspensions of Smale's horseshoe: torus knots and bifurcation sequences*, Archive for Rational Mechanics and Analysis **90** (1985), no. (2), 115–193.

[S1] Sullivan, M.C., *Prime decomposition of knots in Lorenz-like templates*, J. Knot Theory and Its Ramifications **2** (1993), no. (4), 453–462.

[S2]   Sullivan, M.C., *The prime decomposition of knotted periodic orbits in dynamical systems*, J. Knot Theory and Its Ramifications **3** (1994), no. (1), 83–120.

[W1]   Williams, R., *One dimensional non wandering sets*, Topology **6** (1967), 473–487.

[W2]   _____, *The **DA** maps of Smale and structural stability*, Global Analysis, Proceedings of Symposia on Pure and Applied Mathematics, vol. 14, 1970, pp. 329–334.

[W3]   _____, *Expanding attractors*, Colloque de Topologie Differentielle, Mont-Aiguall 1969, Universite de Montpellier, 1969, pp. 79–89.

[W4]   _____, *Expanding attractors*, Publicationes des Institute des Hautes Etudes Scientifique **43** (1974), 169–203.

[W5]   _____, *Classification of subshifts of finite type*, Annals of Mathematics **68** (1973), 12–153; Errata **99** (1974), 380–381.

[W6]   _____, *Pisot-Vijayarghavan numbers and positive matrices*, Proceedings of the 1994 Tokyo Conference on Dynamical Systems and Chaos, Aioki-Shiraiwa-Takahashi, editors, vol. 1, 1995, pp. 268–277.

[W7]   _____, *The structure of Lorenz attractors*, Publications Mathematique I.H.E.S. **50** (1979), 321–342.

[W8]   _____, *Lorenz knots are prime*, Ergodic theory and dynamical systems **4** (1983), 147–163.

[W9]   _____, *The Lorenz attractor*, Turbulence Seminar, Springer Lecture Notes, vol. 615, 1977, pp. 94–112.

Address:

Department of Mathematics
The University of Texas at Austin
Austin, TX 78712-1082
`email:` bob@math.utexas.edu